Factorization Algebras in Quantum Field Theory
Volume 1

Factorization algebras are local-to-global objects that play a role in classical and quantum field theory that is similar to the role of sheaves in geometry: they conveniently organize complicated information. Their local structure encompasses examples such as associative and vertex algebras; in these examples, their global structure encompasses Hochschild homology and conformal blocks.

In this first volume, the authors develop the theory of factorization algebras in depth, but with a focus upon examples exhibiting their use in field theory, such as the recovery of a vertex algebra from a chiral conformal field theory and a quantum group from Abelian Chern–Simons theory. Expositions of the relevant background in homological algebra, sheaves, and functional analysis are also included, thus making this book ideal for researchers and graduates working at the interface between mathematics and physics.

KEVIN COSTELLO is the Krembil Foundation William Rowan Hamilton Chair in Theoretical Physics at the Perimeter Institute in Waterloo, Ontario.

OWEN GWILLIAM is a postdoctoral fellow at the Max Planck Institute for Mathematics in Bonn.

NEW MATHEMATICAL MONOGRAPHS

All the titles listed below can be obtained from good booksellers or from Cambridge University Press. For a complete series listing visit www.cambridge.org/mathematics.

Factorization Algebras in Quantum Field Theory
Volume 1

KEVIN COSTELLO
Perimeter Institute for Theoretical Physics, Waterloo, Ontario

OWEN GWILLIAM
Max Planck Institute for Mathematics, Bonn

CAMBRIDGE
UNIVERSITY PRESS

CAMBRIDGE
UNIVERSITY PRESS

University Printing House, Cambridge CB2 8BS, United Kingdom
One Liberty Plaza, 20th Floor, New York, NY 10006, USA
477 Williamstown Road, Port Melbourne, VIC 3207, Australia
4843/24, 2nd Floor, Ansari Road, Daryaganj, Delhi - 110002, India
79 Anson Road, #06-04/06, Singapore 079906

Cambridge University Press is part of the University of Cambridge.

It furthers the University's mission by disseminating knowledge in the pursuit of
education, learning and research at the highest international levels of excellence.

www.cambridge.org
Information on this title: www.cambridge.org/9781107163102
10.1017/9781316678626

First published 2017

A catalogue record for this publication is available from the British Library

ISBN 978-1-107-16310-2 Hardback

TO LAUREN
AND
TO SOPHIE

Contents

1

Introduction

This two-volume book provides the analog, in quantum field theory, of the deformation quantization approach to quantum mechanics. In this introduction, we start by recalling how deformation quantization works in quantum mechanics.

The collection of observables in a quantum mechanical system forms an associative algebra. The observables of a classical mechanical system form a Poisson algebra. In the deformation quantization approach to quantum mechanics, one starts with a Poisson algebra A^{cl} and attempts to construct an associative algebra A^q, which is an algebra flat over the ring $\mathbb{C}[[\hbar]]$, together with an isomorphism of associative algebras $A^q/\hbar \cong A^{cl}$. In addition, if $a, b \in A^{cl}$, and $\widetilde{a}, \widetilde{b}$ are any lifts of a, b to A^q, then

$$\lim_{\hbar \to 0} \frac{1}{\hbar}[\widetilde{a}, \widetilde{b}] = \{a, b\} \in A^{cl}.$$

Thus, A^{cl} is recovered in the $\hbar \to 0$ limit, i.e., the classical limit.

We will describe an analogous approach to studying perturbative quantum field theory. To do this, we need to explain the following.

- *The structure present on the collection of observables of a* classical *field theory.* This structure is the analog, in the world of field theory, of the commutative algebra that appears in classical mechanics. We call this structure a commutative factorization algebra.

- *The structure present on the collection of observables of a* quantum *field theory.* This structure is that of a factorization algebra. We view our definition of factorization algebra as a differential geometric analog of a definition introduced by Beilinson and Drinfeld. However, the definition we use is very closely related to other definitions in the literature, in particular to the Segal axioms.

1

- *The additional structure on the commutative factorization algebra associated to a classical field theory that makes it "want" to quantize.* This structure is the analog, in the world of field theory, of the Poisson bracket on the commutative algebra of observables.
- *The deformation quantization theorem we prove.* This states that, provided certain obstruction groups vanish, the classical factorization algebra associated to a classical field theory admits a quantization. Further, the set of quantizations is parametrized, order by order in $\hbar$, by the space of deformations of the Lagrangian describing the classical theory.

This quantization theorem is proved using the physicists' techniques of perturbative renormalization, as developed mathematically in Costello (2011b). We claim that this theorem is a mathematical encoding of the perturbative methods developed by physicists.

This quantization theorem applies to many examples of physical interest, including pure Yang–Mills theory and σ-models. For pure Yang–Mills theory, it is shown in Costello (2011b) that the relevant obstruction groups vanish and that the deformation group is one-dimensional; thus there exists a one-parameter family of quantizations. In Li and Li (2016), the topological B-model with target a complex manifold X is constructed; the obstruction to quantization is that X be Calabi–Yau. Li and Li show that the observables and correlations functions recovered by their quantization agree with well-known formulas. Grady, Li, and Li (2015) describe a one-dimensional σ-model with a smooth symplectic manifold as target and show how it recovers Fedosov quantization. Other examples are considered in Costello (2010, 2011a), Costello and Li (2011), and Gwilliam and Grady (2014).

We will explain how (under certain additional hypotheses) the factorization algebra associated to a perturbative quantum field theory encodes the correlation functions of the theory. This fact justifies the assertion that factorization algebras encode a large part of quantum field theory.

This work is split into two volumes. Volume 1 develops the theory of factorization algebras and explains how the simplest quantum field theories – free theories – fit into this language. We also show in this volume how factorization algebras provide a convenient unifying language for many concepts in topological and quantum field theory. Volume 2, which is more technical, derives the link between the concept of perturbative quantum field theory as developed in Costello (2011b) and the theory of factorization algebras.

1.1 The Motivating Example of Quantum Mechanics

The model problems of classical and quantum mechanics involve a particle moving in some Euclidean space $\mathbb{R}^n$ under the influence of some fixed field. Our main goal in this section is to describe these model problems in a way that makes the idea of a factorization algebra (Section 6.1 in Chapter 6) emerge naturally, but we also hope to give mathematicians some sense of the physical meaning of terms such as "field" and "observable." We will not worry about making precise definitions, since that's what this book aims to do. As a narrative strategy, we describe a kind of cartoon of a physical experiment, and we ask that physicists accept this cartoon as a friendly caricature elucidating the features of physics we most want to emphasize.

1.1.1 A Particle in a Box

For the general framework we want to present, the details of the physical system under study are not so important. However, for concreteness, we focus attention on a very simple system: that of a single particle confined to some region of space. We confine our particle inside some box and occasionally take measurements of this system. The set of possible trajectories of the particle around the box constitutes all the imaginable behaviors of this particle; we might describe this space of behaviors mathematically as Maps(I, Box), where $I \subset \mathbb{R}$ denotes the time interval over which we conduct the experiment. We say the set of possible behaviors forms a space of *fields* on the timeline of the particle.

The behavior of our theory is governed by an action functional, which is a function on Maps(I, Box). The simplest case typically studied is the massless free field theory, whose value on a trajectory $f : I \to$ Box is

$$ S(f) = \int_{t \in I} (f(t), \ddot{f}(t)) \, \mathrm{d}t. $$

Here we use $(-, -)$ to denote the usual inner product on $\mathbb{R}^n$, where we view the box as a subspace of $\mathbb{R}^n$, and $\ddot{f}$ to denote the second derivative of f in the time variable t.

The aim of this section is to outline the structure one would expect the observables – that is, the possible measurements one can make of this system – should satisfy.

1.1.2 Classical Mechanics

Let us start by considering the simpler case where our particle is treated as a classical system. In that case, the trajectory of the particle is constrained to be in a solution to the Euler–Lagrange equations of our theory, which is a differential equation determined by the action functional. For example, if the action functional governing our theory is that of the massless free theory, then a map $f : I \to$ Box satisfies the Euler–Lagrange equation if it is a straight line. (Since we are just trying to provide a conceptual narrative here, we will assume that Box becomes all of $\mathbb{R}^n$ so that we do not need to worry about what happens at the boundary of the box.)

We are interested in the observables for this classical field theory. Since the trajectory of our particle is constrained to be a solution to the Euler–Lagrange equation, the only measurements one can make are functions on the space of solutions to the Euler–Lagrange equation.

If $U \subset \mathbb{R}$ is an open subset, we will let Fields(U) denote the space of fields on U, that is, the space of maps $f : U \to$ Box. We will let

$$EL(U) \subset \text{Fields}(U)$$

denote the subspace consisting of those maps $f : U \to$ Box that are solutions to the Euler–Lagrange equation. As U varies, $EL(U)$ forms a sheaf of spaces on $\mathbb{R}$.

We will let $\text{Obs}^{cl}(U)$ denote the commutative algebra of functions on $EL(U)$ (the precise class of functions we consider discussed later). We will think of $\text{Obs}^{cl}(U)$ as the collection of observables for our classical system that depend only upon the behavior of the particle during the time period U. As U varies, the algebras $\text{Obs}^{cl}(U)$ vary and together constitute a cosheaf of commutative algebras on $\mathbb{R}$.

1.1.3 Measurements in Quantum Mechanics

The notion of measurement is fraught in quantum theory, but we will take a very concrete view. Taking a measurement means that we have a physical measurement device (e.g., a camera) that we allow to interact with our system for a period of time. The measurement is then how our measurement device has changed due to the interaction. In other words, we *couple* the two physical systems so that they interact, then decouple them and record how the measurement device has modified from its initial condition. (Of course, there is a symmetry in this situation: both systems are affected by their interaction, so a measurement inherently disturbs the system under study.)

The *observables* for a physical system are all the imaginable measurements we could take of the system. Instead of considering all possible observables, we might also consider those observables that occur within a specified time period. This period can be specified by an open interval $U \subset \mathbb{R}$.

Thus, we arrive at the following principle.

Principle 1. For every open subset $U \subset \mathbb{R}$, we have a set $\mathrm{Obs}(U)$ of observables one can make during U.

Our second principle is a minimal version of the linearity implied by, e.g., the superposition principle.

Principle 2. The set $\mathrm{Obs}(U)$ is a complex vector space.

We think of $\mathrm{Obs}(U)$ as being the collection of ways of coupling a measurement device to our system during the time period U. Thus, there is a natural map $\mathrm{Obs}(U) \to \mathrm{Obs}(V)$ if $U \subset V$ is a shorter time interval. This means that the space $\mathrm{Obs}(U)$ forms a precosheaf.

1.1.4 Combining Observables

Measurements (and so observables) differ qualitatively in the classical and quantum settings. If we study a classical particle, the system is not noticeably disturbed by measurements, and so we can do multiple measurements at the same time. (To be a little less sloppy, we suppose that by refining our measuring devices, we can make the impact on the particle as small as we would like.) Hence, on each interval J we have a commutative multiplication map $\mathrm{Obs}(J) \otimes \mathrm{Obs}(J) \to \mathrm{Obs}(J)$. We also have maps $\mathrm{Obs}(I) \otimes \mathrm{Obs}(J) \to \mathrm{Obs}(K)$ for every pair of disjoint intervals I, J contained in an interval K, as well as the maps that let us combine observables on disjoint intervals.

For a quantum particle, however, a measurement typically disturbs the system significantly. Taking two measurements simultaneously is incoherent, as the measurement devices are coupled to each other and thus also affect each other, so that we are no longer measuring just the particle. Quantum observables thus do not form a cosheaf of commutative algebras on the interval. However, there are no such problems with combining measurements occurring at different times. Thus, we find the following.

Principle 3. If U, U' are disjoint open subsets of $\mathbb{R}$, and $U, U' \subset V$ where V is also open, then there is a map

$$\star : \mathrm{Obs}(U) \otimes \mathrm{Obs}(U') \to \mathrm{Obs}(V).$$

If $O \in \text{Obs}(U)$ and $O' \in \text{Obs}(U')$, then $O \star O'$ is defined by coupling our system to measuring device O during the period U and to device O' during the period U'.

Further, there are maps for a finite collection of disjoint time intervals contained in a long time interval, and these maps are compatible under composition of such maps. (The precise meaning of these terms is detailed in Section 3.1 in Chapter 3.)

1.1.5 Perturbative Theory and the Correspondence Principle

In the bulk of this two-volume book, we will be considering perturbative quantum theory. For us, this adjective "perturbative" means that we work over the base ring $\mathbb{C}[[\hbar]]$, where at $\hbar = 0$ we find the classical theory. In perturbative theory, therefore, the space $\text{Obs}(U)$ of observables on an open subset U is a $\mathbb{C}[[\hbar]]$-module, and the product maps are $\mathbb{C}[[\hbar]]$-linear.

The correspondence principle states that the quantum theory, in the $\hbar \to 0$ limit, must reproduce the classical theory. Applied to observables, this leads to the following principle.

Principle 4. The vector space $\text{Obs}^q(U)$ of quantum observables is a flat $\mathbb{C}[[\hbar]]$-module such that modulo $\hbar$, it is equal to the space $\text{Obs}^{cl}(U)$ of classical observables.

These four principles are at the heart of our approach to quantum field theory. They say, roughly, that the observables of a quantum field theory form a factorization algebra, which is a quantization of the factorization algebra associated to a classical field theory. The main theorem presented in this two-volume book is that one can use the techniques of perturbative renormalization to construct factorization algebras perturbatively quantizing a certain class of classical field theories (including many classical field theories of physical and mathematical interest). As we have mentioned, this first volume focuses on the general theory of factorization algebras and on simple examples of field theories; this result is derived in Volume 2.

1.1.6 Associative Algebras in Quantum Mechanics

The principles we have described so far indicate that the observables of a quantum mechanical system should assign, to every open subset $U \subset \mathbb{R}$, a vector space $\text{Obs}(U)$, together with a product map

$$\text{Obs}(U) \otimes \text{Obs}(U') \to \text{Obs}(V)$$

if U, U' are disjoint open subsets of an open subset V. This is the basic data of a factorization algebra (see Section 3.1 in Chapter 3).

It turns out that in the case of quantum mechanics, the factorization algebra produced by our quantization procedure has a special property: it is *locally constant* (see Section 6.4 in Chapter 6). This means that the map $\text{Obs}((a, b)) \to \text{Obs}(\mathbb{R})$ is an isomorphism for any interval (a, b). Let A denote the vector space $\text{Obs}(\mathbb{R})$; note that A is canonically isomorphic to $\text{Obs}((a, b))$ for any interval (a, b).

The product map

$$\text{Obs}((a, b)) \otimes \text{Obs}((c, d)) \to \text{Obs}((a, d))$$

when $a < b < c < d$, becomes, via this isomorphism, a product map

$$m : A \otimes A \to A.$$

The axioms of a factorization algebra imply that this multiplication turns A into an associative algebra. As we will see in Section 4.2 in Chapter 4, this associative algebra is the Weyl algebra, which one expects to find as the algebra of observables for quantum mechanics of a particle moving in $\mathbb{R}^n$.

This kind of geometric interpretation of algebra should be familiar to topologists: associative algebras are algebras over the operad of little intervals in $\mathbb{R}$, and this is precisely what we have described. As we explain in Section 6.4 in Chapter 6, this relationship continues and so our quantization theorem produces many new examples of algebras over the operad E_n of little n-discs.

An important point to take away from this discussion is that *associative algebras appear in quantum mechanics because associative algebras are connected with the geometry of* $\mathbb{R}$. There is no fundamental connection between associative algebras and any concept of "quantization": associative algebras appear only when one considers one-dimensional quantum field theories. As we will see later, when one considers topological quantum field theories on n-dimensional space–times, one finds a structure reminiscent of an E_n-algebra instead of an E_1-algebra.

Remark: As a caveat to the strong assertion in the preceding (and jumping ahead of our story), note that for a manifold of the form $X \to \mathbb{R}$, one can push forward a factorization algebra Obs on $X \times \mathbb{R}$ to a factorization algebra $\pi_* \text{Obs}$ on $\mathbb{R}$ along the projection map $\pi : X \times \mathbb{R} \to \mathbb{R}$. In this case, $\pi_* \text{Obs}((a, b)) = \text{Obs}(X \times (a, b))$. Hence, a quantization of a higher dimensional theory will produce, via such pushforwards to $\mathbb{R}$, deformations of associative algebras, but knowing only the pushforward is typically insufficient to reconstruct the factorization algebra on the higher dimensional manifold. $\Diamond$

1.2 A Preliminary Definition of Prefactorization Algebras

Below (see Section 3.1 in Chapter 3) we give a more formal definition, but here we provide the basic idea. Let M be a topological space (which, in practice, will be a smooth manifold).

Definition 1.2.1 A prefactorization algebra $\mathcal{F}$ on M, taking values in cochain complexes, is a rule that assigns a cochain complex $\mathcal{F}(U)$ to each open set $U \subset M$ along with

  (i) A cochain map $\mathcal{F}(U) \to \mathcal{F}(V)$ for each inclusion $U \subset V$.
 (ii) A cochain map $\mathcal{F}(U_1) \otimes \cdots \otimes \mathcal{F}(U_n) \to \mathcal{F}(V)$ for every finite collection of open sets where each $U_i \subset V$ and the U_i are disjoint.
(iii) The maps are compatible in a certain natural way. The simplest case of this compatibility is that if $U \subset V \subset W$ is a sequence of open sets, the map $\mathcal{F}(U) \to \mathcal{F}(W)$ agrees with the composition through $\mathcal{F}(V)$).

Remark: A prefactorization algebra resembles a precosheaf, except that we tensor the cochain complexes rather than taking their direct sum. ◇

The observables of a field theory, whether classical or quantum, form a prefactorization algebra on the space–time manifold M. In fact, they satisfy a kind of local-to-global principle in the sense that the observables on a large open set are determined by the observables on small open sets. The notion of a factorization algebra (Section 6.1 in Chapter 6) makes this local-to-global condition precise.

1.3 Prefactorization Algebras in Quantum Field Theory

The (pre)factorization algebras of interest in this book arise from perturbative quantum field theories. We have already discussed in Section 1.1 how factorization algebras appear in quantum mechanics. In this section we will see how this picture extends in a natural way to quantum field theory.

The manifold M on which the prefactorization algebra is defined is the space–time manifold of the quantum field theory. If $U \subset M$ is an open subset, we will interpret $\mathcal{F}(U)$ as the collection of observables (or measurements) that we can make that depend only on the behavior of the fields on U. Performing a measurement involves coupling a measuring device to the quantum system in the region U.

One can bear in mind the example of a particle accelerator. In that situation, one can imagine the space–time M as being of the form $M = A \times (0, t)$, where A is the interior of the accelerator and t is the duration of our experiment.

In this situation, performing a measurement on an open subset $U \subset M$ is something concrete. Let us take $U = V \times (\varepsilon, \delta)$, where $V \subset A$ is some small region in the accelerator and where (ε, δ) is a short time interval. Performing a measurement on U amounts to coupling a measuring device to our accelerator in the region V, starting at time ε and ending at time δ. For example, we could imagine that there is some piece of equipment in the region V of the accelerator, which is switched on at time ε and switched off at time δ.

1.3.1 Interpretation of the Prefactorization Algebra Axioms

Suppose that we have two different measuring devices, O_1 and O_2. We would like to set up our accelerator so that we measure both O_1 and O_2.

There are two ways we can do this. We can insert O_1 and O_2 into disjoint regions V_1, V_2 of our accelerator. Then we can turn O_1 and O_2 on at any times we like, including for overlapping time intervals.

If the regions V_1, V_2 overlap, then we cannot do this. After all, it doesn't make sense to have two different measuring devices at the same point in space at the same time.

However, we could imagine inserting O_1 into region V_1 during the time interval (a, b); and then removing O_1, and inserting O_2 into the overlapping region V_2 for the disjoint time interval (c, d).

These simple considerations immediately suggest that the possible measurements we can make of our physical system form a prefactorization algebra. Let $\mathrm{Obs}(U)$ denote the space of measurements we can make on an open subset $U \subset M$. Then, by combining measurements in the way outlined in the preceding text, we would expect to have maps

$$\mathrm{Obs}(U) \otimes \mathrm{Obs}(U') \to \mathrm{Obs}(V)$$

whenever U, U' are disjoint open subsets of an open subset V. The associativity and commutativity properties of a prefactorization algebra are evident.

1.3.2 The Cochain Complex of Observables

In the approach to quantum field theory considered in this book, the factorization algebra of observables will be a factorization algebra of cochain complexes. That is, Obs assigns a cochain complex $\mathrm{Obs}(U)$ to each open U. One can ask for the physical meaning of the cochain complex.

We will repeatedly mention observables in a gauge theory, as these kinds of cohomological aspects are well known for such theores.

It turns out that the "physical" observables will be $H^0(\text{Obs}(U))$. If $O \in \text{Obs}^0(U)$ is an observable of cohomological degree 0, then the equation $O = 0$ can often be interpreted as saying that O is compatible with the gauge symmetries of the theory. Thus, only those observables $O \in \text{Obs}^0(U)$ that are closed are physically meaningful.

The equivalence relation identifying $O \in \text{Obs}^0(U)$ with $O + O'$, where $O' \in \text{Obs}^{-1}(U)$, also has a physical interpretation, which will take a little more work to describe. Often, two observables on U are physically indistinguishable (that is, they cannot be distinguished by any measurement one can perform). In the example of an accelerator outlined earlier, two measuring devices are equivalent if they always produce the same expectation values, no matter how we prepare our system, or no matter what boundary conditions we impose.

As another example, in the quantum mechanics of a free particle, the observable measuring the momentum of a particle at time t is equivalent to that measuring the momentum of a particle at another time t'. This is because, even at the quantum level, momentum is preserved (as the momentum operator commutes with the Hamiltonian).

From the cohomological point of view, if $O, O' \in \text{Obs}^0(U)$ are both in the kernel of the differential (and thus "physically meaningful"), then they are equivalent in the sense described previously if they differ by an exact observable.

It is a little more difficult to provide a physical interpretation for the other cohomology groups $H^n(\text{Obs}(U))$. The first cohomology group $H^1(\text{Obs}(U))$ contains anomalies (or obstructions) to lifting classical observables to the quantum level. For example, in a gauge theory, one might have a classical observable that respects gauge symmetry. However, it may not lift to a quantum observable respecting gauge symmetry; this happens if there is a nontrivial anomaly in $H^1(\text{Obs}(U))$.

The cohomology groups $H^n(\text{Obs}(U))$, when $n < 0$, are best interpreted as symmetries, and higher symmetries, of observables. Indeed, we have seen that the physically meaningful observables are the closed degree 0 elements of $\text{Obs}(U)$. One can construct a simplicial set, whose n-simplices are closed and degree 0 elements of $\text{Obs}(U) \otimes \Omega^*(\Delta^n)$. The vertices of this simplicial set are observables, the edges are equivalences between observables, the faces are equivalences between equivalences, and so on.

The Dold–Kan correspondence (see Theorem A.2.7 in Appendix A) tells us that the nth homotopy group of this simplicial set is $H^{-n}(\text{Obs}(U))$. This allows us to interpret $H^{-1}(\text{Obs}(U))$ as being the group of symmetries of the trivial observable $0 \in H^0(\text{Obs}(U))$, and $H^{-2}(\text{Obs}(U))$ as the symmetries of the identity symmetry of $0 \in H^0(\text{Obs}(U))$, and so on.

Although the cohomology groups $H^n(\mathrm{Obs}(U))$ where $n \geq 1$ do not have a clear physical interpretation as clear as that for H^0, they are mathematically very natural objects and it is important not to discount them. For example, let us consider a gauge theory on a manifold M, and let D be a disc in M. Then it is often the case that elements of $H^1(\mathrm{Obs}(D))$ can be integrated over a circle in M to yield cohomological degree 0 observables (such as Wilson operators).

1.4 Comparisons with Other Formalizations of Quantum Field Theory

Now that we have explained carefully what we mean by a prefactorization algebra, let us say a little about the history of this concept, and how it compares to other mathematical approaches to quantum field theory. We make no attempt to state formal theorems relating our approach to other axiom systems. Instead we sketch heuristic relationships between the various axiom systems.

1.4.1 Factorization Algebras in the Sense of Beilinson and Drinfeld

For us, one source of inspiration is the work of Beilinson and Drinfeld on chiral conformal field theory. These authors gave a geometric reformulation of the theory of vertex algebras in terms of an algebro-geometric version of the concept of factorization algebra. For Beilinson and Drinfeld, a factorization algebra on an algebraic curve X is, in particular, a collection of sheaves $\mathcal{F}_n$ on the Cartesian powers X^n of X. If $(x_1, \ldots, x_n) \in X^n$ is an n-tuple of distinct points in X, let $\mathcal{F}_{x_1,\ldots,x_n}$ denote the stalk of $\mathcal{F}_n$ at this point in X^n. The axioms of Beilinson and Drinfeld imply that there is a canonical isomorphism

$$\mathcal{F}_{x_1,\ldots,x_n} \cong \mathcal{F}_{x_1} \otimes \cdots \otimes \mathcal{F}_{x_n}.$$

In fact, their axioms tell us that the restriction of the sheaf $\mathcal{F}_n$ to any stratum of X^n (in the stratification by number of points) is determined by the sheaf $\mathcal{F}_1$ on X. The fundamental object in their approach is the sheaf $\mathcal{F}_1$. All the other sheaves $\mathcal{F}_n$ are built from copies of $\mathcal{F}_1$ by certain gluing data, which we can think of as structures put on the sheaf $\mathcal{F}_1$.

One should think of the stalk $\mathcal{F}_x$ of $\mathcal{F}_1$ at x as the space of local operators in a field theory at the point x. Thus, $\mathcal{F}_1$ is the sheaf on X whose stalks are the spaces of local operators. The other structures on $\mathcal{F}_1$ reflect the operator product expansions of local operators.

Let us now sketch, heuristically, how we expect their approach to be related to ours. Suppose we have a factorization algebra $\widetilde{\mathcal{F}}$ on X in our sense. Then, for every open $V \subset X$, we have a vector space $\widetilde{\mathcal{F}}(V)$ of observables on V. The space of local operators associated to a point $x \in X$ should be thought of as those observables that live on every open neighbourhood of x. In other words, we can define

$$\widetilde{\mathcal{F}}_x = \lim_{x \in V} \widetilde{\mathcal{F}}(V)$$

to be the limit over open neighbourhoods of x of the observables on that neighbourhood. This limit is the costalk of the pre-cosheaf $\widetilde{\mathcal{F}}$.

Thus, the heuristic translation between their axioms and ours is that the sheaf $\mathcal{F}_1$ on X that they construct should have stalks coinciding with the costalks of our factorization algebra. We do not know how to turn this idea into a precise theorem in general. We do, however, have a precise theorem in one special case.

Beilinson and Drinfeld show that a factorization algebra in their sense on the affine line $\mathbb{A}^1$, which is also translation and rotation equivariant, is the same as a vertex algebra. We have a similar theorem. We show in Chapter 5 that a factorization algebra on $\mathbb{C}$ that is translation and rotation invariant, and also has a certain "holomorphic" property, gives rise to a vertex algebra. Therefore, in this special case, we can show how a factorization algebra in our sense gives rise to one in the sense used by Beilinson and Drinfeld.

1.4.2 Segal's Axioms for Quantum Field Theory

Segal has developed and studied some very natural axioms for quantum field theory (see, e.g., Segal 2010). These axioms were first studied in the world of topological field theory by Atiyah, Segal, and Witten, and in conformal field theory by Kontsevich and Segal (2004).

According to Segal's philosophy, a d-dimensional quantum field theory (in Euclidean signature) is a symmetric functor from the category $\mathrm{Cob}_d^{\mathrm{Riem}}$ of d-dimensional Riemannian cobordisms. An object of the category $\mathrm{Cob}_d^{\mathrm{Riem}}$ is a compact $d - 1$-manifold together with a germ of a d-dimensional Riemannian structure. A morphism is a d-dimensional Riemannian cobordism. The symmetric monoidal structure arises from disjoint union. As defined, this category does not have identity morphisms, but they can be added in formally.

Definition 1.4.1 A Segal field theory is a symmetric monoidal functor from $\mathrm{Cob}_d^{\mathrm{Riem}}$ to the category of (topological) vector spaces.

We won't get into details about what kind of topological vector spaces one should consider, because our aim is just to sketch a heuristic relationship between Segal's picture and our picture.

In our approach to studying quantum field theory, the fundamental objects are not the Hilbert spaces associated to codimension 1 manifolds, but rather the spaces of observables. Any reasonable axiom system for quantum field theory should be able to capture the notion of observable. In particular, we should be able to understand observables in terms of Segal's axioms.

Segal, in lectures and conversations, has explained how to do this. Suppose we have a Riemannian manifold M and a point $x \in M$. Consider a ball $B(x, r)$ of radius r around x, whose boundary is a sphere $S(x, r)$. Segal explains that the Hilbert space $Z(S(x, r))$ should be thought of as the space of operators on the ball $B(x, r)$.

If $r < r'$, there is a cobordism $S(x, r) \to S(x, r')$ given by the complement of $B(x, r)$ in the closed ball $\overline{B}(x, r')$. This gives rise to maps $Z(S(x, r)) \to Z(S(x, r'))$. Segal defines the space of local operators at x to be the limit

$$\lim_{r \to 0} Z(S(x, r))$$

of this inverse system.

One can understand from this idea of Segal's how one should construct something like a prefactorization algebra on any Riemannian manifold M of dimension n from a Segal field theory. Given an open subset $U \subset M$ whose boundary is a codimension 1 submanifold ∂U, we define the space $\mathcal{F}(U)$ of observables on U to be $Z(\partial U)$. If U, V, W are three such opens in M, such that the closures of U and V are disjoint in W, then there is a cobordism

$$\overline{W} \setminus (U \amalg V) : \partial U \amalg \partial V \to \partial W.$$

This cobordism induces a map

$$\mathcal{F}(U) \otimes \mathcal{F}(V) = Z(\partial U) \otimes Z(\partial V) \to Z(\partial W) = \mathcal{F}(W).$$

There are similar maps defined when $U_1, \ldots, U_n, W$ are opens with smooth boundary such that the closures of the U_i are disjoint and contained in W. In this way, we can construct from a Segal field theory something that is very like a prefactorization algebra; the only difference is that we restrict our attention to those opens with smooth boundary, and the prefactorization algebra structure maps are defined only for collections of opens whose closures are disjoint.

Remark: In the text that follows we discuss how certain universal factorization algebras relate to *topological* field theories in the style of Atiyah–Segal–Lurie. Dwyer, Stolz, and Teichner have also proposed an approach to constructing

Segal-style non-topological field theories, such as Riemannian field theories, using factorization algebras.   ◊

1.4.3 Topological Field Theory

One class of field theories for which there exists an extensive mathematical literature is topological field theories (see, e.g., Lurie 2009c). One can ask how our axiom system relates to those for topological field theories.

There is a subclass of factorization algebras that appear in topological field theories, called *locally constant* factorization algebras. A factorization algebra $\mathcal{F}$ on a manifold M, valued in cochain complexes, is locally constant if, for any two discs $D_1 \subset D_2$ in M, the map $\mathcal{F}(D_1) \to \mathcal{F}(D_2)$ is a quasi-isomorphism. A theorem of Lurie (2016) shows that, given a locally constant factorization algebra $\mathcal{F}$ on $\mathbb{R}^n$, the complex $\mathcal{F}(D)$ has the structure of an E_n algebra.

This relationship matches with what one expects from the standard approach to the axiomatics of topological field theory. According to the standard axioms for topological field theories, a topological field theory (TFT) of dimension n is a symmetric monoidal functor

$$Z : \mathrm{Cob}_n \to \mathrm{Vect}$$

from the n-dimensional cobordism category to the category of vector spaces. Here, Cob_n is the category whose objects are closed smooth $n-1$-manifolds and the morphisms are cobordisms between them. (It is standard, following Freed (1994), to consider also higher-categorical objects associated to manifolds of higher codimension.)

For an $n-1$-manifold N, we should interpret $Z(N)$ as the Hilbert space of the TFT on N. Then, following standard physics yoga, we should interpret $Z(S^{n-1})$ as the space of local operators of the theory. There are natural cobordisms between disjoint unions of the $n-1$-sphere that make $Z(S^{n-1})$ into an E_n algebra. For example, in dimension 2, the pair of pants with k legs provides the k-ary operations for the E_2 algebra structure on $Z(S^1)$.

This story fits nicely with our approach. If we have a locally constant factorization algebra $\mathcal{F}$ on $\mathbb{R}^n$, then $\mathcal{F}(D^n)$ is an E_n algebra. Further, we interpret $\mathcal{F}(D^n)$ as being the space of observables supported on an n-disc. Since $\mathcal{F}$ is locally constant, this may as well be the observables supported on a point, because it is independent of the radius of the n-disc.

1.4.4 Correlation Functions

In some classic approaches to the axiomatics of quantum field theory – such as the Wightman axioms or the Osterwalder–Schrader axioms, their Euclidean counterpart – the fundamental objects are correlation functions. While we make no attempt to verify that a factorization algebra gives rise to a solution of any of these axiom systems, we do show that the factorization algebra has enough data to define the correlation functions. Let us briefly explain how this works for two different classes of examples.

Suppose that we have a factorization algebra $\mathcal{F}$ on a manifold M over the ring $\mathbb{R}[[\hbar]]$ of formal power series in $\hbar$. Suppose that

$$H^0(\mathcal{F}(M)) = \mathbb{R}[[\hbar]].$$

This condition holds in some natural examples: for instance, for Chern–Simons theory on $\mathbb{R}^3$ or for a massive scalar field theory on a compact manifold M.

Let $U_1, \ldots, U_n \subset M$ be disjoint open sets. The factorization product gives us a $\mathbb{R}[[\hbar]]$-multilinear map

$$\langle - \rangle : H^0(\mathcal{F}(U_1)) \times \cdots \times H^0(\mathcal{F}(U_n)) \to H^0(\mathcal{F}(M)) = \mathbb{R}[[\hbar]].$$

In this way, given observables $O_i \in H^0(\mathcal{F}(U_i))$, we can produce a formal power series in $\hbar$

$$\langle O_1, \ldots, O_n \rangle \in \mathbb{R}[[\hbar]].$$

In the field theories we just mentioned, this map does compute the expectation value of observables. (See Section 4.6 in Chapter 4, where we describe this map in terms of the Green's functions. We also recover the Gauss linking number there from Abelian Chern–Simons theory.)

This construction doesn't give us expectation values in every situation where we might hope to construct them. For example, if we work with a field theory on $\mathbb{R}^n$, it is rarely the case that $H^0(\mathcal{F}(\mathbb{R}^n))$ is isomorphic to $\mathbb{R}[[\hbar]]$. We would, however, expect to be able to define correlation functions in this situation. To achieve this, we define a variant of this construction that works well on $\mathbb{R}^n$. Given a factorization algebra on $\mathbb{R}^n$ with ground ring $\mathbb{R}[[\hbar]]$ as before, we define in Section 4.9 in Chapter 4 a *vacuum* to be an $\mathbb{R}[[\hbar]]$-linear map

$$\langle - \rangle : H^0(\mathcal{F}(\mathbb{R}^n)) \to \mathbb{R}[[\hbar]]$$

that is translation-invariant and satisfies a certain cluster decomposition principle. After choosing a vacuum, one can define correlation functions in the same way as in the preceding text.

1.5 Overview of This Volume

This two-volume work concerns, as the titles suggests, factorization algebras and quantum field theories. In this introduction so far, we have sketched what a prefactorization algebra is and why it might help organize and analyze the behavior of the observables of a quantum field theory. These two volumes develop these ideas further in a number of ways.

The first volume focuses on factorization algebras: their definition, some formal properties, and some simple constructions of factorization algebras. The quantum field theory in this volume is mostly limited to free field theories. In a moment, we give a detailed overview of this volume.

In the second volume, we focus on our core project: we develop the Batalin–Vilkovisky formalism for both classical and quantum field theories and show how it automatically produces a deformation quantization of factorization algebras of observables. In particular, Volume 2 will introduce the factorization algebras associated to interacting field theories. We also provide there a refinement of the Noether theorems in the setting of factorization algebras, in which, roughly speaking, local symmetries of a field theory lift to a map of factorization algebras,. This map realizes the symmetries as observables of the field theory. For a more detailed overview of Volume 2, see its introductory chapter.

1.5.1 Chapter by Chapter

Chapter 2 serves as a second introduction. In this chapter we explain, using informal language and without any background knowledge required, how the observables of a free scalar field theory naturally form a prefactorization algebra. Readers who want to understand the main ideas of this two-volume work with the minimum amount of technicalities should start there.

Chapter 3 gives a more careful definition of the concept of a prefactorization algebra, and analyzes some basic examples. In particular, the relationship between prefactorization algebras on $\mathbb{R}$ and associative algebras is developed in detail. We also introduce a construction that will play an important role in the rest of the book: the *factorization envelope* of a sheaf of Lie algebras on a manifold. This construction is the factorization version of the universal enveloping algebra of a Lie algebra.

In Chapter 4 we revisit the prefactorization algebras associated to a free field theory, but with more care and in greater generality than we used in Chapter 2. The methods developed in this chapter apply to gauge theories, treating with gauge symmetry with the method of Becchi, Rouet, Stora, and

Tyutin (BRST) and its extension by Batalin and Vilkovisky. We analyze in some detail the example of Abelian Chern–Simons theory, and verify that the expectation value of Wilson lines in this theory recovers the Gauss linking number.

Chapter 5 introduces the concept of a *holomorphic* prefactorization algebra on $\mathbb{C}^n$. The prefactorization algebra of observables of a field theory with a holomorphic origin – such as a holomorphic twist Costello (2013a) of a supersymmetric gauge theory – will be such a holomorphic factorization algebra. We prove that a holomorphic prefactorization algebra on $\mathbb{C}$ gives rise to a vertex algebra, thus linking our story with a more traditional point of view on the algebra of operators of a chiral conformal field theory.

The final chapters in this book develop the concept of factorization algebra, by adding a certain local-to-global axiom to the definition of prefactorization algebra. In Chapter 6, we provide the definition, discuss the relation between locally constant factorization algebras and E_n algebras, and explain how to construct several large classes of examples. In Chapter 7, we develop some formal properties of the theory of factorization algebras. Finally, in Chapter 8, we move beyond the formal and analyze some interesting explicit examples. For instance, we compute the factorization homology of the Kac–Moody enveloping factorization algebras, and we explain how Abelian Chern–Simons theory produces a quantum group.

1.5.2 A Comment on Functional Analysis and Algebra

This book uses an unusual array of mathematical techniques, including both homological algebra and functional analysis. The homological algebra appears because our factorization algebras live in the world of cochain complexes (ultimately, because they come from the BV formalism for field theory). The functional analysis appears because our factorization algebras are built from vector spaces of an analytic nature, such as the space of distributions on a manifold. We have included an expository introduction to the techniques we use from homological algebra, operads, and sheaf theory in Appendix A.

It is well known that it is hard to make homological algebra and functional analysis work well together. One reason is that, traditionally, the vector spaces that arise in analysis are viewed as topological vector spaces, and the category of topological vector spaces is not an Abelian category. In Appendix B, we introduce the concept of *differentiable vector spaces*. Differentiable vector spaces are more flexible than topological vector spaces, yet retain enough analytic structure for our purposes. We show that the category of differentiable vector spaces is an Abelian category, and indeed satisfies the strongest version of the axioms of an Abelian category: it is a locally presentable *AB*5 category. This means that homological algebra in the category of differentiable vector spaces works very nicely. We develop this in Appendix C.

A gentle introduction to differentiable vector spaces, containing more than enough to follow everything in both volumes, is contained in Chapter 3, Section 3.5.

1.6 Acknowledgments

The project of writing this book has stretched over many more years than we ever anticipated. In that course of time, we have benefited from conversations with many people, the chance to present this work at several workshops and conferences, and the feedback of various readers. The book and our understanding of this material is much better due to the interest and engagement of so many others. Thank you.

We would like to thank directly the following people, although this list is undoubtedly incomplete: David Ayala, David Ben-Zvi, Dan Berwick-Evans, Damien Calaque, Alberto Cattaneo, Lee Cohn, Ivan Contreras, Vivek Dhand, Chris Douglas, Chris Elliott, John Francis, Dennis Gaitsgory, Sachin Gautam, Ezra Getzler, Greg Ginot, Ryan Grady, André Henriques, Rune Haugseng, Theo Johnson-Freyd, David Kazhdan, Si Li, Jacob Lurie, Takuo Matsuoka, Aaron Mazel-Gee, Pavel Mnev, David Nadler, Thomas Nikolaus, Fred Paugam, Dmitri Pavlov, Toly Preygel, Kasia Rejzner, Kolya Reshetikhin, Nick Rozenblyum, Claudia Scheimbauer, Graeme Segal, Thel Seraphim, Yuan Shen, Jim Stasheff, Stephan Stolz, Dennis Sullivan, Matt Szczesny, Hiro Tanaka, Peter Teichner, David Treumann, Philsang Yoo, Brian Williams, and Eric Zaslow for helpful conversations. K. C. is particularly grateful to Graeme Segal for many illuminating conversations about quantum field theory over the years. O.G. would like to thank the participants in the Spring 2014 Berkeley seminar and Fall 2014 seminar at the Max Planck Institute for Mathematics for their interest and extensive feedback; he is also grateful to Rune Haugseng and Dmitri Pavlov for help at the intersection of functional analysis and category theory. Peter Teichner's feedback after teaching a Berkeley course in Spring 2016 on this material improved the book considerably. Finally, we are both grateful to John Francis and Jacob Lurie for introducing us to factorization algebras, in their topological incarnation, in 2008.

Our work has taken place at many institutions – Northwestern University, the University of California, Berkeley, the Max Planck Institute for Mathematics, and the Perimeter Institute for Theoretical Physics – which provided a supportive environment for writing and research. Research at the Perimeter Institute is supported by the Government of Canada through Industry Canada and by the Province of Ontario through the Ministry of Economic Development and Innovation, and K. C.'s research at the Perimeter Institute is supported by the Krembil Foundation. We have also benefitted from the support of the National Science Foundation (NSF): K. C. and O. G. were partially supported by NSF grants DMS 0706945 and DMS 1007168, and O. G. was supported by NSF postdoctoral fellowship DMS-1204826. K. C. was also partially supported by a Sloan fellowship.

Finally, during the period of this project, we have depended on the unrelenting support and love of our families. In fact, we have both gotten married and had children over these years! Thank you Josie, Dara, and Laszlo for being models of insatiable curiosity, ready sources of exuberant distraction, and endless fonts of joy. Thank you Lauren and Sophie—amidst all the tumult and the moves—for always encouraging us, for always sharing your wit and wisdom and warmth. It makes all the difference.

PART I

Prefactorization algebras

2

From Gaussian Measures to Factorization Algebras

This chapter serves as a kind of second introduction, demonstrating how the free scalar field theory on a manifold M produces a factorization algebra on M. We have already sketched in Chapter 1 why the observables of a field theory ought to form a factorization algebra, without making precise what we meant by a quantum field theory. Here we will meet the second main theme of the book – the Batalin–Vilkovisky (BV) formalism for field theory – and see how it naturally produces a factorization algebra.

Our approach to quantum field theory grows out of the idea of a path integral. Instead of trying to define such an integral directly, however, the BV formalism provides a homological approach to integration, similar in spirit to the de Rham complex. (As we will see in the next section, in the finite dimensional setting, the BV complex is isomorphic to the de Rham complex.) The philosophy goes like this. If the desired path integral were well defined mathematically, we could compute the expectation values of observables, and the expectation value map $\mathbb{E}$ is linear, so we obtain a linear equivalence relation between observables $O \sim O'$ whenever $\mathbb{E}(O - O') = 0$. We can reconstruct, in fact, the expectation value by describing the inclusion $\mathrm{Rel}_{\mathbb{E}} := \ker \mathbb{E} \hookrightarrow \mathrm{Obs}$ and taking the cokernel of this inclusion. In other words, we identify "integrands with the same integral." The BV formalism approaches the problem from the other direction: even though the desired path integral may not be well defined, we often know, from physical arguments, when two observables ought to have the same expectation value (e.g., via Ward identities), so that we can produce a subspace $\mathrm{Rel}_{\mathrm{BV}} \hookrightarrow \mathrm{Obs}$. The BV formalism produces a subspace $\mathrm{Rel}_{\mathrm{BV}}$ determined by the classical field theory, as we will see later, but this subspace is typically not of codimension 1. Further input, such as boundary conditions, is often necessary to get a number (i.e., to produce a codimension 1 subspace of relations). Nonetheless, the relations in $\mathrm{Rel}_{\mathrm{BV}}$ would hold for any such choice, so any expectation value map coming from physics would factor through Obs /Rel$_{BV}$.

In fact, the BV formalism produces a cochain complex, encoding relations between the relations and so on, whose zeroth cohomology group is the space Obs /Rel$_{BV}$. (Here, we are using Obs to denote the "naive" observables that one would first write down for the theory, not observables involving the auxiliary "anti-fields" introduced when applying the BV formalism.)

That description is quite abstract; the rest of this chapter is about making the idea concrete with examples. In physics, a free field theory is one in which the action functional is a purely quadratic function of the fields. A basic example is the free scalar field theory on a Riemannian manifold (M,g), where the space of fields is the space $C^\infty(M)$ of smooth functions on M, and the action funtional is

$$S(\phi) = \tfrac{1}{2} \int_M \phi(\triangle_g + m^2)\phi \, \mathrm{d}vol_g.$$

Here $\triangle_g$ refers to the Laplacian with the convention that its eigenvalues are nonnegative, and $\mathrm{d}vol_g$ denotes the Riemannian volume form associated to the metric. The positive real number m is the mass of the theory. The main quantities of interest in the free field theory are the correlation functions, defined by the heuristic expression

$$\langle \phi(x_1) \cdots \phi(x_n) \rangle = \int_{\phi \in C^\infty(M)} \phi(x_1) \dots \phi(x_n) \, e^{-S(\phi)} \mathrm{d}\phi,$$

where $x_1, \dots, x_n$ are points in M. Note that there is an observable

$$O(x_1, \dots, x_n) : \phi \mapsto \phi(x_1) \cdots \phi(x_n)$$

sending a field ϕ to the product of its values at those points. The support of this observable is precisely the set of points $\{x_1, \dots, x_n\}$, so this observable lives in Obs(U) for any open U containing all those points. The standard computations in quantum field theory tell us that this observable has the same expectation value as linear combinations of other correlation functions. For instance, Wick's lemma tells us how the two-point correlation function relates to the Green's function for $\triangle_g + m^2$.

Our task is to explain how the combination of the BV formalism and prefactorization algebras provides a simple and natural way to make sense of these relations. We will see that for a free field theory on a manifold M, there is a space of observables associated to any open subset $U \subset M$. We will see that the operations we can perform on these spaces of observables give us the structure of a prefactorization algebra on M. This example will serve as further motivation for the idea that observables of a field theory are described by a prefactorization algebra.

2.1 Gaussian Integrals in Finite Dimensions

As in many approaches to quantum field theory, we will motivate our definition of the prefactorization algebra of observables by studying finite dimensional Gaussian integrals. Thus, let A be an $n \times n$ symmetric positive-definite real matrix, and consider Gaussian integrals of the form

$$\int_{x \in \mathbb{R}^n} \exp\left(-\tfrac{1}{2} \sum x_i A_{ij} x_j\right) f(x) \, d^n x$$

where f is a polynomial function on $\mathbb{R}^n$. Note the formal analogy to the correlation function we wish to compute: here x replaces ϕ, $\tfrac{1}{2} \sum x_i A_{ij} x_j$ is quadratic in x as S is in ϕ, and $f(x)$ is polynomial in x as $O(\phi)$ is a polynomial in ϕ.

At this point, most textbooks on quantum field theory would explain Wick's lemma, which is a combinatorial expression for such integrals. It reduces the integral above to a sum involving the quadratic moments of the Gaussian measure. Then, such a textbook would go on to define similar infinite-dimensional Gaussian integrals using the analogous combinatorial expression. The key point is the simplicity of the moments of a Gaussian measure, which allows immediate generalization to infinite dimensions.

We will take a different approach, however. Instead of focusing on the combinatorial expression for the integral, we will focus on the *divergence operator* associated to the Gaussian measure. (This operator provides the inclusion map Rel $\hookrightarrow$ Obs discussed at the beginning of this chapter.)

Let $P(\mathbb{R}^n)$ denote the space of polynomial functions on $\mathbb{R}^n$. Let $\mathrm{Vect}(\mathbb{R}^n)$ denote the space of vector fields on $\mathbb{R}^n$ with polynomial coefficients. If $d^n x$ denotes the Lebesgue measure on $\mathbb{R}^n$, let ω_A denote the measure

$$\omega_A = \exp\left(-\tfrac{1}{2} \sum x_i A_{ij} x_j\right) d^n x.$$

Then, the divergence operator Div_{ω_A} associated to this measure is a linear map

$$\mathrm{Div}_{\omega_A} : \mathrm{Vect}(\mathbb{R}^n) \to P(\mathbb{R}^n),$$

defined abstractly by saying that if $V \in \mathrm{Vect}(\mathbb{R}^n)$, then

$$\mathcal{L}_V \omega_A = \left(\mathrm{Div}_{\omega_A} V\right) \omega_A$$

where $\mathcal{L}_V$ refers to the Lie derivative. Thus, the divergence of V measures the infinitesimal change in volume that arises when one applies the infinitesimal diffeomorphism V.

In coordinates, the divergence is given by the formula

$$\mathrm{Div}_{\omega_A} \left(\sum f_i \frac{\partial}{\partial x_i} \right) = - \sum_{i,j} f_i x_j A_{ij} + \sum_i \frac{\partial f_i}{\partial x_i}. \qquad (\dagger)$$

(This formula is an exercise in applying Cartan's magic formula: $\mathcal{L}_V = [\,, \iota_V]$. Note that this divergence operator is therefore a disguised version of the exterior derivative.)

By the definition of divergence, we see

$$\int \left(\mathrm{Div}_{\omega_A} V \right) \omega_A = 0$$

for all polynomial vector fields V, because

$$\int \left(\mathrm{Div}_{\omega_A} V \right) \omega_A = \int \mathcal{L}_V \omega_A = \int \left(\iota_V \omega_A \right)$$

and then we apply Stokes' lemma. By changing basis of $\mathbb{R}^n$ to diagonalize A, one sees that the image of the divergence map is a codimension 1 linear subspace of the space $P(\mathbb{R}^n)$ of polynomials on $\mathbb{R}^n$. (This statement is true as long as A is nondegenerate; positive-definiteness is not required).

Let us identify $P(\mathbb{R}^n)/\mathrm{Im}\,\mathrm{Div}_{\omega_A}$ with $\mathbb{R}$ by taking the basis of the quotient space to be the image of the polynomial function 1. What we have shown so far can be summarized in the following lemma.

Lemma 2.1.1 *The quotient map*

$$P(\mathbb{R}^n) \to P(\mathbb{R}^n)/\mathrm{Im}\,\mathrm{Div}_{\omega_A} \cong \mathbb{R}$$

is the map that sends a function f to its expected value

$$\langle f \rangle_A := \frac{\int_{\mathbb{R}^n} f \omega_A}{\int_{\mathbb{R}^n} \omega_A}.$$

This lemma plays a crucial motivational role for us. If we want to know expected values (which are the main interest in the physics setting), it suffices to describe a divergence operator. One does not need to produce the measure directly.

One nice feature of this approach to finite-dimensional Gaussian integrals is that it works over any ring in which $\det A$ is invertible (this follows from the explicit algebraic formula we wrote for the divergence of a polynomial vector field). This way of looking at finite-dimensional Gaussian integrals was analyzed further in Gwilliam and Johnson-Freyd (2012), where it was shown that one can derive the Feynman rules for finite-dimensional Gaussian integration from such considerations.

Remark: We should acknowledge here that our choice of *polynomial* functions and vector fields was important. Polynomial functions are integrable against a Gaussian measure, and the divergence of polynomial vector fields produces all the relations between these integrands. If we worked with all smooth functions and smooth vector fields, the cokernel would be zero. In the BV formalism, just as in ordinary integration, the choice of functions plays an important role. ◊

2.2 Divergence in Infinite Dimensions

So far, we have seen that finite-dimensional Gaussian integrals are entirely encoded in the divergence map from the Gaussian measure. In our approach to infinite-dimensional Gaussian integrals, the fundamental object we will define is such a divergence operator. We will recover the usual formulae for infinite-dimensional Gaussian integrals (in terms of the propagator or Green's function) from our divergence operator. Further, we will see that analyzing the cokernel of the divergence operator will lead naturally to the notion of prefactorization algebra.

For concreteness, we will work with the free scalar field theory on a Riemannian manifold (M, g), which need not be compact. We will define a divergence operator for the putative Gaussian measure on $C^\infty(M)$ associated to the quadratic form $\frac{1}{2} \int_M \phi(\triangle_g + m^2)\phi \, \mathrm{dvol}_g$. (Here $\triangle_g + m^2$ plays the role that the matrix A did in the preceding section.)

Before we define the divergence operator, we need to define spaces of polynomial functions and of polynomial vector fields. We will organize these spaces by their support in M. Namely, for each open subset $U \subset M$, we will define polynomial functions and vector fields on the space $C^\infty(U)$.

The space of all continuous linear functionals on $C^\infty(U)$ is the space $\mathcal{D}_c(U)$ of compactly supported distributions on U. To define the divergence operator, we need to restrict to functionals with more regularity. Hence we will work with $C_c^\infty(U)$, where every element $f \in C_c^\infty(U)$ defines a linear functional on $C^\infty(U)$ by the formula

$$\phi \mapsto \int_U f\phi \, \mathrm{dvol}_g.$$

ym (People sometimes call them "smeared" because these do not include beloved functionals such as delta functions, only smoothed-out approximations to them.)

As a first approximation to the algebra we wish to use, we define the space of polynomial functions on $C_c^\infty(U)$ to be

$$\widetilde{P}(C^\infty(U)) = \operatorname{Sym} C_c^\infty(U),$$

i.e., the symmetric algebra on $C_c^\infty(U)$. An element of $\widetilde{P}(C_c^\infty(U))$ that is homogeneous of degree n can be written as a finite sum of monomials $f_1 \cdots f_n$ where the $f_i \in C_c^\infty(U)$. Such a monomial defines a function on the space $C^\infty(U)$ of fields by the formula

$$\phi \mapsto \int_{(x_1,\dots,x_n)\in U^n} f_1(x_1)\phi(x_1)\dots f_n(x_n)\phi(x_n)\, dvol_g(x_1) \wedge \cdots \wedge dvol_g(x_n).$$

Note that because $C_c^\infty(U)$ is a topological vector space, it is more natural to use an appropriate completion of this purely algebraic symmetric power $\operatorname{Sym}^n C_c^\infty(U)$. Because this version of the algebra of polynomial functions is a little less natural than the completed version, which we will introduce shortly, we use the notation $\widetilde{P}$. The completed version is denoted P.

We define the space of polynomial vector fields in a similar way. Recall that if V is a finite-dimensional vector space, then the space of polynomial vector fields on V is isomorphic to $P(V) \otimes V$, where $P(V)$ is the space of polynomial functions on V. An element $X = f \otimes v$, with f a polynomial, acts on a polynomial g by the formula

$$X(g) = f\frac{\partial g}{\partial v}.$$

In particular, if g is homogeneous of degree n and we pick a representative $\widetilde{g} \in (V^*)^{\otimes n}$, then $\partial g/\partial v$ denotes the degree $n - 1$ polynomial

$$w \mapsto \widetilde{g}(v \otimes w \otimes \cdots \otimes w).$$

In other words, for polynomials, differentiation is a version of contraction.

In the same way, we would expect to work with

$$\widetilde{\operatorname{Vect}}(C^\infty(U)) = \widetilde{P}(C^\infty(U)) \otimes C^\infty(U).$$

We are interested, in fact, in a different class of vector fields. The space $C^\infty(U)$ has a foliation, coming from the linear subspace $C_c^\infty(U) \subset C^\infty(U)$. We are actually interested in vector fields along this foliation, owing to the role of variational calculus in field theory. This restriction along the foliation is clearest in terms of the divergence operator we describe later, so we explain it after Definition 2.2.1.

Thus, let

$$\widetilde{\operatorname{Vect}}_c(C^\infty(U)) = \widetilde{P}(C^\infty(U)) \otimes C_c^\infty(U).$$

Again, it is more natural to use a completion of this space that takes account of the topology on $C_c^\infty(U)$. We will discuss such completions shortly.

Any element of $\widetilde{\text{Vect}}_c(C^\infty(U))$ can be written as an finite sum of monomials of the form

$$f_1 \cdots f_n \frac{\partial}{\partial \phi}$$

for $f_i, \phi \in C_c^\infty(U)$. By $\frac{\partial}{\partial \phi}$ we mean the constant-coefficient vector field given by infinitesimal translation in the direction ϕ in $C^\infty(U)$.

Vector fields act on functions, in the same way as we described earlier: the formula is

$$f_1 \cdots f_n \frac{\partial}{\partial \phi}(g_1 \cdots g_m) = f_1 \cdots f_n \sum g_1 \cdots \widehat{g_i} \cdots g_m \int_U g_i(x)\phi(x)\,dvol_g$$

where $dvol_g$ is the Riemannian volume form on U.

Definition 2.2.1 The divergence operator associated to the quadratic form

$$S(\phi) = \int_U \phi(\Delta + m^2)\phi \, dvol_g$$

is the linear map

$$\widetilde{\text{Div}} : \widetilde{\text{Vect}}_c(C^\infty(U)) \to \widetilde{P}(C_c^\infty(U))$$

defined by

$$\widetilde{\text{Div}}\left(f_1 \cdots f_n \frac{\partial}{\partial \phi}\right) = -f_1 \cdots f_n(\Delta + m^2)\phi + \sum_{i=1}^n f_1 \cdots \widehat{f_i} \cdots f_n \int_U \phi(x)f_i(x)\,dvol_g.$$

$$(\ddagger)$$

Note that this formula is entirely parallel to the formula for divergence of a Gaussian measure in finite dimensions, given in formula ($\dagger$). Indeed, the formula makes sense even when ϕ is not compactly supported; however, the term

$$f_1 \cdots f_n(\Delta + m^2)\phi$$

need not be compactly supported if ϕ is not compactly supported. To ensure that the image of the divergence operator is in $\widetilde{P}(C_c^\infty(U))$, we work only with vector fields with compact support, namely $\widetilde{\text{Vect}}_c(C^\infty(U))$.

As we mentioned previously, it is more natural to use a completion of the spaces $\widetilde{P}(C^\infty(U))$ and $\text{Vect}_c(C^\infty(U))$ of polynomial functions and polynomial vector fields. We now explain a geometric approach to such a completion.

Let $dvol_g^n$ denote the Riemannian volume form on the product space U^n arising from the natural n-fold product metric induced by the metric g on U.

Any element $F \in C_c^\infty(U^n)$ then defines a polynomial function on $C^\infty(U)$ by

$$\phi \mapsto \int_{U^n} F(x_1, \ldots, x_n)\phi(x_1) \cdots \phi(x_n) \, dvol_g^n.$$

This functional does not change if we permute the arguments of F by an element of the symmetric group S_n, so that this function depends only on the image of F in the coinvariants of $C_c^\infty(U^n)$ by the symmetric group action. This quotient, of course, is isomorphic to invariants for the symmetric group action. Therefore we define

$$P(C^\infty(U)) = \bigoplus_{n \geq 0} C_c^\infty(U^n)_{S_n},$$

where the subscript indicates coinvariants. The (purely algebraic) symmetric power $\mathrm{Sym}^n \, C_c^\infty(U)$ provides a dense subspace of $C_c^\infty(U^n)^{S_n} \cong C_c^\infty(U^n)_{S_n}$. Thus, $\widetilde{P}(C^\infty(U))$ is a dense subspace of $P(C^\infty(U))$.

In a similar way, we define $\mathrm{Vect}_c(C^\infty(U))$ by

$$\mathrm{Vect}_c(C^\infty(U)) = \bigoplus_{n \geq 0} C_c^\infty(U^{n+1})_{S_n},$$

where the symmetric group S_n acts only on the first n factors. A dense subspace of $C_c^\infty(U^{n+1})^{S_n}$ is given by $\mathrm{Sym}^n \, C_c^\infty(U) \otimes C_c^\infty(U)$ so that $\widetilde{\mathrm{Vect}}_c(C^\infty(U))$ is a dense subspace of $\mathrm{Vect}_c(C^\infty(U))$.

Lemma 2.2.2 *The divergence map*

$$\widetilde{\mathrm{Div}} : \widetilde{\mathrm{Vect}}_c(C^\infty(U)) \to \widetilde{P}(C^\infty(U))$$

extends continuously to a map

$$\mathrm{Div} : \mathrm{Vect}_c(C^\infty(U)) \to P(C^\infty(U)).$$

Proof Suppose that

$$F(x_1, \ldots, x_{n+1}) \in C_c^\infty(U^{n+1})_{S_n} \subset \mathrm{Vect}_c(C^\infty(U)).$$

The divergence map in equation (‡) extends to a map that sends F to

$$-\Delta_{x_{n+1}} F(x_1, \ldots, x_{n+1}) + \sum_{i=1}^{n} \int_{x_i \in U} F(x_1, \ldots, x_i, \ldots, x_n, x_i) \, dvol_g.$$

Here, $\Delta_{x_{n+1}}$ denotes the Laplacian acting only on the $n + 1$st copy of U. Note that the integral produces a function on U^{n-1}. $\qquad\square$

With these objects in hand, we are able to define the *quantum observables* of a free field theory.

Definition 2.2.3 For an open subset $U \subset M$, let

$$H^0(\mathrm{Obs}^q(U)) = P(C^\infty(U))/\operatorname{Im}\operatorname{Div}.$$

In other words, $H^0(\mathrm{Obs}^q(U))$ be the cokernel of the operator Div. Later we will see that this linear map Div naturally extends to a cochain complex of quantum observables, which we will denote Obs^q, whose zeroth cohomology is what we just defined. This extension is why we write H^0.

Let us explain why we should interpret this space as the quantum observables. We expect that an observable in a field theory is a function on the space of fields. An observable on a field theory on an open subset $U \subset M$ is a function on the fields that depends only on the behavior of the fields inside U. Speaking conceptually, the expectation value of the observable is the integral of this function against the "functional measure" on the space of fields.

Our approach is that we do not try to define the functional measure, but instead we define the divergence operator. If we have some functional on $C^\infty(U)$ that is the divergence of a vector field, then the expectation value of the corresponding observable is zero. Thus, we would expect that the observable given by a divergence is not a physically interesting quantity, since its value is zero. Thus, we might as well identify it with zero.

The appropriate vector fields on $C^\infty(U)$ – the ones for which our divergence operator makes sense – are vector fields along the foliation of $C^\infty(U)$ by compactly supported fields. Thus, the quotient of functions on $C^\infty(U)$ by the subspace of divergences of such vector fields gives a definition of observables.

2.3 The Prefactorization Structure on Observables

Suppose that we have a Gaussian measure ω_A on $\mathbb{R}^n$. Then every function on $\mathbb{R}^n$ with polynomial growth is integrable, and this space of functions forms a commutative algebra. We showed that there is a short exact sequence

$$\mathrm{Vect}(\mathbb{R}^n) \xrightarrow{\operatorname{Div}_{\omega_A}} P(\mathbb{R}^n) \xrightarrow{\mathbb{E}_{\omega_A}} \mathbb{R} \to 0,$$

where $\mathbb{E}_{\omega_A}$ denotes the expectation value map for this measure. But the image of the divergence operator is *not* an ideal. (Indeed, usually an expectation value map is not an algebra map!) This fact suggests that, in the BV formalism, the quantum observables should *not* form a commutative algebra. One can check quickly that for our definition given earlier, $H^0(\mathrm{Obs}^q(U))$ is not an algebra.

However, we will find that some shadow of this commutative algebra structure exists, which allows us to combine observables on disjoint subsets. This residual structure will give the spaces $H^0(\mathrm{Obs}^q(U))$ of observables, viewed as a functor on the category of open subsets $U \subset M$, the structure of a *prefactorization algebra*.

Let us make these statements precise. Note that $P(C^\infty(U))$ is a commutative algebra, as it is a space of polynomial functions on $C^\infty(U)$. Further, if $U \subset V$ there is a map of commutative algebras ext : $P(C^\infty(U)) \to P(C^\infty(V))$, extending a polynomial map $F : C^\infty(U) \to \mathbb{R}$ to the polynomial map $F \circ \mathrm{res}$: $C^\infty(V) \to \mathbb{R}$ by precomposing with the restriction map res : $C^\infty(V) \to C^\infty(U)$. This map ext is injective. We will sometimes refer to an element of the subspace $P(C^\infty(U)) \subset P(C^\infty(V))$ as an element of $P(C^\infty(V))$ *with support in U*.

Lemma 2.3.1 *The product map*

$$P(C^\infty(V)) \otimes P(C^\infty(V)) \to P(C^\infty(V))$$

does not *descend to a map*

$$H^0(\mathrm{Obs}(V)) \otimes H^0(\mathrm{Obs}(V)) \to H^0(\mathrm{Obs}(V)).$$

If U_1, U_2 are disjoint open subsets of the open $V \subset M$, then we have a map

$$P(U_1) \otimes P(U_2) \to P(V)$$

obtained by combining the inclusion maps $P(U_i) \hookrightarrow P(V)$ with the product map on $P(V)$. This map does *descend to a map*

$$H^0(\mathrm{Obs}(U_1)) \otimes H^0(\mathrm{Obs}(U_2)) \to H^0(\mathrm{Obs}(V)).$$

In other words, although the product of general observables does not make sense, the product of observables with disjoint support does.

Proof Let U_1, U_2 be disjoint open subsets of M, both contained in an open V. Let us view the spaces $\mathrm{Vect}_c(C^\infty(U_i))$ and $P(C^\infty(U_i))$ as subspaces of $\mathrm{Vect}_c(C^\infty(V))$ and $P(C^\infty(V))$, respectively.

We denote by Div_{U_i}, the divergence operator on U_i, namely,

$$\mathrm{Div}_{U_i} : \mathrm{Vect}_c(C^\infty(U_i)) \to P(C^\infty(U_i)).$$

We use Div_V to denote the divergence operator on V.

Our situation is then described by the following diagram:

$$
\begin{array}{ccccc}
\ker q_{12} & \hookrightarrow & P(U_1) \otimes P(U_2) & \xrightarrow{q_{12}} & H^0(\mathrm{Obs}(U_1)) \otimes H^0(\mathrm{Obs}(U_2)) \\
 & & \downarrow & & \\
\mathrm{Im}\,\mathrm{Div}_V & \hookrightarrow & P(C^\infty(V)) & \xrightarrow{q} & H^0(\mathrm{Obs}(V)).
\end{array}
$$

The middle vertical arrow is multiplication map. We want to show there is a vertical arrow on the right that makes a commuting square. It suffices to show that the image of $\ker q_{12}$ in $H^0(\mathrm{Obs}(V))$ is zero.

Note that $H^0(\mathrm{Obs}(U_1)) \otimes H^0(\mathrm{Obs}(U_2))$ is the cokernel of the map

$$P(C^\infty(U_1)) \otimes \mathrm{Vect}_c(C^\infty(U_2)) \oplus \mathrm{Vect}_c(C^\infty(U_1)) \otimes P(C^\infty(U_2))$$

$$\xrightarrow{\ 1 \otimes \mathrm{Div}_{U_2} + \mathrm{Div}_{U_1} \otimes 1\ } P(C^\infty(U_1)) \otimes P(C^\infty(U_2)).$$

Hence, $\ker q_{12} = \mathrm{Im}\left(1 \otimes \mathrm{Div}_{U_2} + \mathrm{Div}_{U_2} \otimes 1\right)$.

We will show that the image of $1 \otimes \mathrm{Div}_{U_2} + \mathrm{Div}_{U_1} \otimes 1$ sits inside $\mathrm{Im}\,\mathrm{Div}_V$. This result ensures that we can produce the desired map.

Thus, it suffices to show that for any F is in $P(C^\infty(U_1))$ and X is in $\mathrm{Vect}_c(C^\infty(U_2))$, there exists $\widetilde{X}$ in $\mathrm{Vect}_c(C^\infty(V))$ such that $\mathrm{Div}_V(\widetilde{X}) = F\,\mathrm{Div}_{U_2}(X)$. (The same argument applies after switching the roles U_1 and U_2.) We will show, in fact, that

$$\mathrm{Div}(FX) = F\,\mathrm{Div}(X),$$

viewing F and X as living on the open V.

A priori, this assertion should be surprising. On an ordinary finite-dimensional manifold, the divergence Div with respect to any volume form has the following property: for any vector field X and any function f, we have

$$\mathrm{Div}(fX) - f\,\mathrm{Div}\,X = X(f),$$

where $X(f)$ denotes the action of X on f. Note that $X(f)$ is not necessarily in the image of Div, in which case $f\,\mathrm{Div}\,X$ is not in the image of Div, and so we see that $\mathrm{Im}\,\mathrm{Div}$ is not an ideal.

The same equation holds for the divergence operator we have defined in infinite dimensions. If $X \in \mathrm{Vect}_c(C^\infty(V))$ and $F \in P(C^\infty_c(V))$, then

$$\mathrm{Div}(FX) - F\,\mathrm{Div}X = X(F).$$

This computation tells us that the image of Div is not an ideal, as there exist $X(F)$ not in the image of Div.

When F is in $P(C^\infty(U_1))$ and X is in $\mathrm{Vect}_c(C^\infty(U_2))$, however, $X(F) = 0$, as their supports are disjoint. Thus, we have precisely the desired relation $F\,\mathrm{Div}(X) = \mathrm{Div}(FX)$.

We now prove that $X(F) = 0$ when X and F have disjoint support.

We are working with polynomial functions and vector fields, so that all computations can be done in a purely algebraic fashion; in other words, we will work with derivations as in algebraic geometry. Let ε satisfy $\varepsilon^2 = 0$. For a polynomial vector field X and a polynomial function F on a vector space V, we define the function $X(F)$ to assign to the vector $v \in V$, the ε component of $F(v + \varepsilon X_v) - F(v)$. Here X_v denotes the tangent vector at v that X produces.

In our situation, we know that for any $\phi \in C^\infty(V)$, we have the following properties:

- $F(\phi)$ depends only on the restriction of ϕ to U_1.
- X_ϕ is a function with support in U_2 and hence vanishes away from U_2.

Thus, $F(\phi + \varepsilon X(\phi)) = F(\phi)$ as the restriction of $\phi + \varepsilon X(\phi)$ to U_1 agrees with the restriction of ϕ. We see then that

$$X(F)(\phi) = \frac{d}{d\varepsilon} \left(F(\phi + \varepsilon X_\phi) - F(\phi) \right) = 0,$$

as asserted.   □

In a similar way, if $U_1, \ldots, U_n$ are disjoint opens all contained in V, then there is a map

$$H^0(\mathrm{Obs}^q(U_1)) \otimes \cdots \otimes H^0(\mathrm{Obs}^q(U_n)) \to H^0(\mathrm{Obs}^q(V))$$

descending from the map

$$P(U_1) \otimes \cdots \otimes P(U_n) \to P(V)$$

given by inclusion followed by multiplication.

Thus, we see that the spaces $H^0(\mathrm{Obs}^q(U))$ for open sets $U \subset M$ are naturally equipped with the structure maps necessary to define a prefactorization algebra. (See Section 1.2 in Chapter 1 for a sketch of the definition of a factorization algebra, and Section 3.1 in Chapter 3 for more details on the definition). It is straightforward to check, using the arguments from the preceding proof, that these structure maps satisfy the necessary compatibility conditions to define a prefactorization algebra.

2.4 From Quantum to Classical

Our general philosophy is that the quantum observables of a field theory are a factorization algebra given by deforming the classical observables. The classical observables are defined to be functions on the space of solutions to the equations of motion. We now examine how our construction can be seen as just such a deformation.

Let us see first why this holds for a class of measures on finite dimensional vector spaces. Let S be a polynomial function on $\mathbb{R}^n$. Let $d^n x$ denote the Lebesgue measure on $\mathbb{R}^n$, and consider the measure

$$\omega = e^{-S/\hbar}\,\mathrm{d}^n x,$$

where $\hbar$ is a small parameter. The divergence with respect to ω is given by the formula

$$\mathrm{Div}_\omega\left(\sum f_i \frac{\partial}{\partial x_i}\right) = -\frac{1}{\hbar}\sum f_i \frac{\partial S}{\partial x_i} + \sum \frac{\partial f_i}{\partial x_i}.$$

As before, let $P(\mathbb{R}^n)$ denote the space of polynomial functions on $\mathbb{R}^n$ and let $\mathrm{Vect}(\mathbb{R}^n)$ denote the space of polynomial vector fields. The divergence operator Div_ω is a linear map map $\mathrm{Vect}(\mathbb{R}^n) \to P(\mathbb{R}^n)$. Note that the operators Div_ω and $\hbar\,\mathrm{Div}_\omega$ have the same image so long as $\hbar \neq 0$. When $\hbar = 0$, the operator $\hbar\,\mathrm{Div}_\omega$ becomes the operator

$$\sum f_i \frac{\partial}{\partial x_i} \mapsto -\sum f_i \frac{\partial S}{\partial x_i}.$$

Therefore, the $\hbar \to 0$ limit of the image of Div_ω is the Jacobian ideal

$$\mathrm{Jac}(S) = \left(\frac{\partial S}{\partial x_i}\right) \subset P(\mathbb{R}^n),$$

which corresponds to the critical locus of S in $\mathbb{R}^n$. Hence, the $\hbar \to 0$ limit of the observables $P(\mathbb{R}^n)/\mathrm{Im}\,\mathrm{Div}_\omega$ is the commutative algebra $P(\mathbb{R}^n)/\mathrm{Jac}(S)$ that describes functions on the critical locus of S.

Let us now check the analogous property for the observables of a free scalar field theory on a manifold M. We will consider the divergence for the putative Gaussian measure

$$\exp\left(-\frac{1}{\hbar}\int_M \phi(\Delta + m^2)\phi\right)\mathrm{d}\phi$$

on $C^\infty(M)$. For any open subset $U \subset M$, this divergence operator gives us a map

$$\mathrm{Div}_\hbar : \mathrm{Vect}_c(C^\infty(U)) \to P(C^\infty(U))$$

with

$$f_1\cdots f_n \frac{\partial}{\partial\phi} \mapsto -\frac{1}{\hbar}f_1\cdots f_n(\Delta + m^2)\phi + \sum_i f_1\cdots \widehat{f_i}\int_M f\phi.$$

Note how $\hbar$ appears in this formula; it is just the same modification of the operator ($\ddagger$) as Div_ω is of the divergence operator for the measure $e^{-S}\mathrm{d}^n x$.

As in the finite dimensional case, the first term dominates in the $\hbar \to 0$ limit. The $\hbar \to 0$ limit of the image of $\mathrm{Div}_\hbar$ is the closed subspace of $P(C^\infty(U))$ spanned by functionals of the form $f_1\cdots f_n(\Delta + m^2)\phi$, where f_i and ϕ are in $C_c^\infty(U)$). This subspace is the topological ideal in $P(C^\infty(U))$ generated by linear functionals of the form $(\Delta + m^2)f$, where $f \in C_c^\infty(U)$. If $S(\phi) =$

$\int \phi(\triangle + m^2)\phi$ is the action functional of our theory, then this subspace is precisely the topological ideal generated by all the functional derivatives

$$\left\{ \frac{\partial S}{\partial \phi} \; : \; \phi \in C_c^\infty(U) \right\}.$$

In other words, it is the Jacobian ideal for S and hence is cut out by the Euler–Lagrange equations. Let us call this ideal $I_{EL}(U)$.

A more precise statement of what we have just sketched is the following. Define a prefactorization algebra $H^0(\mathrm{Obs}^{cl}(U))$ (the superscript cl stands for classical) that assigns to U the quotient algebra $P(C^\infty(U))/I_{EL}(U)$. Thus, $H^0(\mathrm{Obs}^{cl}(U))$ should be thought of as the polynomial functions on the space of solutions to the Euler–Lagrange equations. Note that each constituent space $H^0(\mathrm{Obs}^{cl}(U))$ in this prefactorization algebra has the structure of a commutative algebra, and the structure maps are all maps of commutative algebras. In short, $H^0(\mathrm{Obs}^{cl})$ forms a *commutative prefactorization algebra*. Heuristically, this terminology means that the product map defining the factorization structure is defined for all pairs of opens $U_1, U_2 \subset V$, and not just disjoint pairs.

Our work in the section is summarized as follows.

Lemma 2.4.1 *There is a prefactorization algebra $H^0(\mathrm{Obs}_\hbar^q)$ over $\mathbb{C}[\hbar]$ such that when specialized to $\hbar = 1$ is $H^0(\mathrm{Obs}^q)$ and to $\hbar = 0$ is $H^0(\mathrm{Obs}^{cl})$.*

This prefactorization algebra assigns to an open set the cokernel of the map

$$\hbar \, \mathrm{Div}_\hbar : \mathrm{Vect}_c(C^\infty(U))[\hbar] \to P(C^\infty(U))[\hbar],$$

where $\mathrm{Div}_\hbar$ is the map defined previously.

We will see later that $H^0(\mathrm{Obs}_\hbar^q(U))$ is free as an $\mathbb{R}[\hbar]$-module, although this is a special property of free theories and is not always true for an interacting theory.

2.5 Correlation Functions

We have seen that the observables of a free scalar field theory on a manifold M give rise to a factorization algebra. In this section, we explain how the structure of a factorization algebra is enough to define correlation functions of observables. We will calculate certain correlation functions explicitly and recover the standard answers.

Suppose now that M is a *compact* Riemannian manifold, and, as before, let us consider the observables of the free scalar field theory on M with mass $m > 0$. Then we have the following result.

Lemma 2.5.1 *If the mass m is positive, then $H^0(\mathrm{Obs}^q(M)) \cong \mathbb{R}$.*

Compare this result with the statement that for a Gaussian measure on $\mathbb{R}^n$, the image of the divergence map is of codimension 1 in the space of polynomial functions on $\mathbb{R}^n$. The assumption here that the mass is positive is necessary to ensure that the quadratic form $\int_M \phi(\triangle + m^2)\phi$ is nondegenerate.

This lemma will follow from our more detailed analysis of free theories in Chapter 4, but we can sketch the idea here. The main point is that there is a family of operators over $\mathbb{R}[\hbar]$ connecting $H^0(\mathrm{Obs}^q(M))$ to $H^0(\mathrm{Obs}^{cl}(M))$. It is straightforward to see that the algebra $H^0(\mathrm{Obs}^{cl}(M))$ is $\mathbb{R}$. Indeed, observe that the only solution to the equations of motion in the case that $m > 0$ is the function $\phi = 0$, since the assumption that $\triangle$ has nonnegative spectrum ensures that $\triangle\phi = -m^2\phi$ has no nontrivial solution. Functions on a point are precisely $\mathbb{R}$. To conclude that $H^0(\mathrm{Obs}^q(M))$ is also $\mathbb{R}$, we need to show that $H^0(\mathrm{Obs}^q_\hbar(M))$ is flat over $\mathbb{R}[\hbar]$, which will follow from a spectral sequence computation we will perform later in the book.

There is always a canonical observable $1 \in H^0(\mathrm{Obs}^q(U))$ for any open subset $U \subset M$. This element is defined to be the image of the function $1 \in P(C^\infty(U))$. We identify $H^0(\mathrm{Obs}^q(M))$ with $\mathbb{R}$ by taking this observable 1 to be a basis vector.

Definition 2.5.2 Let $U_1, \ldots, U_n \subset M$ be disjoint open subsets. The *correlator* is the prefactorization structure map

$$\langle - \rangle : H^0(\mathrm{Obs}^q(U_1)) \otimes \cdots \otimes H^0(\mathrm{Obs}^q(U_n)) \to H^0(\mathrm{Obs}^q(M)) \cong \mathbb{R}.$$

We should compare this definition with what happens in finite dimensions. If we have a Gaussian measure on $\mathbb{R}^n$, then the space of polynomial functions modulo divergences is one-dimensional. If we take the image of a function 1 to be a basis of this space, then we get a map

$$P(\mathbb{R}^n) \to \mathbb{R}$$

from polynomial functions to $\mathbb{R}$. This map is the integral against the Gaussian measure, normalized so that the integral of the function 1 is 1.

In our infinite dimensional situation, we are doing something very similar. Any reasonable definition of the correlation function of the observables $O_1, \ldots, O_n$, with O_i in $P(C^\infty(U_i))$, should depend only on the product function $O_1 \cdots O_n \in P(C^\infty(M))$. Thus, the correlation function map should be a linear map $P(C^\infty(M)) \to \mathbb{R}$. Further, it should send divergences to zero. We have seen that there is only one such map, up to an overall scale.

2.5.1 Comparison to Physics

Next we will check explicitly that this correlation function map really matches up with what physicists expect. Let $f_i \in C_c^\infty(U_i)$ be compactly supported smooth functions on the open sets $U_1, U_2 \subset M$. Let us view each f_i as a linear function on $C^\infty(U_i)$, and so as an element of the polynomial functions $P(C^\infty(U_i))$.

Let $G \in \mathcal{D}(M \times M)$ be the unique distribution on $M \times M$ with the property that

$$(\triangle_x + m^2)G(x, y) = \delta_{\text{Diag}}.$$

Here δ_{Diag} denotes the delta function supported on the diagonal copy of M inside $M \times M$. In other words, if we apply the operator $\triangle + m^2$ to the first factor of G, we find the delta function on the diagonal. Thus, G is the kernel for the operator $(\triangle + m^2)^{-1}$. In the physics literature, G is called the propagator; in the mathematics literature, it is called the Green's function for the operator $\triangle + m^2$. Note that G is smooth away from the diagonal.

Lemma 2.5.3 *Given* $f_i \in C_c^\infty(U_i)$, *there are classes* $[f_i] \in H^0(\text{Obs}^q(U_i))$. *Then*

$$\langle [f_1][f_2] \rangle = \int_{x,y \in M} f_1(x)G(x, y)f_2(y).$$

Note that this result is exactly the standard result in physics.

Proof Later we will give a slicker and more general proof of this kind of statement. Here we will give a simple proof to illustrate how the Green's function arises from our homological approach to defining functional integrals.

The operator $\triangle + m^2$ is surjective. Thus, there is a preimage

$$\phi = (\triangle + m^2)^{-1} f_2 \in C_c^\infty(M),$$

that is unique because $\triangle + m^2$ is also injective. Indeed, we know that

$$\phi(x) = \int_{y \in M} G(x, y)f_2(y).$$

Now consider the vector field

$$f_1 \frac{\partial}{\partial \phi} \in \text{Vect}_c(C^\infty(M)).$$

Note that

$$\mathrm{Div}\left(f_1\frac{\partial}{\partial\phi}\right) = \int_{x\in M} f_1(x)\phi(x) - f_1\left((\triangle + m^2)\phi\right)\mathrm{dvol}$$
$$= \left(\int_M f_1(x)G(x,y)f_2(y)\right)\cdot 1 - f_1f_2.$$

The element $f_1f_2 \in P(C^\infty(M))$ is a cocycle representing the factorization product $[f_1][f_2]$ in $H^0(\mathrm{Obs}^q(M))$ of the observables $[f_i] \in H^0(\mathrm{Obs}^q(U_i))$. The displayed equation tells us that

$$[f_1f_2] = \left(\int_M f_1(x)G(x,y)f_2(y)\right)\cdot 1 \in H^0(\mathrm{Obs}^q(M)).$$

Since the observable 1 is chosen to be the basis element identifying $H^0(\mathrm{Obs}^q(M))$ with $\mathbb{R}$, the result follows. □

With a little more work, the same arguments recover the usual Wick's formula for correlators of the form $\langle f_1\cdots f_n\rangle$.

Remark: These kinds of formulas are standard knowledge in physics, but not in mathematics. For a more extensive discussion in a mathematical style, see chapter 6 of Glimm and Jaffe (1987) or lecture 3 by Kazhdan in Deligne et al. (1999). ◊

2.6 Further Results on Free Field Theories

In this chapter, we showed that if we define the observables of a free field theory as the cokernel of a certain divergence operator, then these spaces of observables form a prefactorization algebra. We also showed that this prefactorization algebra contains enough information to allow us to define the correlation functions of observables, and that for linear observables we find the same formula that physicists would write.

In Chapter 3 we show that a certain class of factorization algebras on the real line are equivalent to associative algebras, together with a derivation. As Noether's theorem suggests, the derivation arises from infinitesimal translation on the real line, so that it encodes the Hamiltonian of the system.

In Chapter 4, we analyze the factorization algebra of free field theories in more detail. We will show that if we consider the free field theory on $\mathbb{R}$, the factorization algebra $H^0(\mathrm{Obs}^q)$ corresponds (under the relationship between factorization algebras on $\mathbb{R}$ and associative algebras) to the Weyl algebra. The Weyl algebra is generated by observables p, q corresponding to position and momentum satisfying $[p, q] = 1$. If we consider instead the family over $\mathbb{R}[\hbar]$ of factorization algebras $H^0(\mathrm{Obs}_\hbar^q)$ discussed earlier, then we find the commutation relation $[p, q] = \hbar$ of Heisenberg. This algebra, of course, is what is traditionally called the algebra of observables of quantum mechanics. In this case, we will further see that the derivation of this algebra (corresponding to infinitesimal time translation) is an inner derivation, given by bracketing with the Hamiltonian

$$H = p^2 - m^2 q^2,$$

which is the standard Hamiltonian for the quantum mechanics of the harmonic oscillator.

More generally, we recover canonical quantization of the free scalar field theory on higher dimensional manifolds as follows. Consider a free scalar theory on the product Riemannian manifold $N \times \mathbb{R}$, where N is a compact Riemannian manifold. This example gives rise to a factorization algebra on $\mathbb{R}$ that assigns to an open subset $U \subset \mathbb{R}$, the space $H^0(\mathrm{Obs}^q(N \times U))$. We will see that this factorization algebra on $\mathbb{R}$ has a dense sub-factorization algebra corresponding to an associative algebra. This associative algebra is the tensor product of Weyl algebras, where each each eigenspace of the operator $\triangle + m^2$ on $C^\infty(N)$ produces a Weyl algebra. In other words, we find quantum mechanics on $\mathbb{R}$ with values in the infinite-dimensional vector space $C^\infty(N)$. Since that space has a natural spectral decomposition for the operator $\triangle + m^2$, it is natural to interpret as the algebra of observables, the associative algebra given by tensoring together the Weyl algebra for each eigenspace. This result is entirely consistent with standard arguments in physics, being the factorization algebra analog of canonical quantization.

2.7 Interacting Theories

In any approach to quantum field theory, free field theories are easy to construct. The challenge is always to construct interacting theories. The core results of this two-volume work show how to construct the factorization algebra corresponding to interacting field theories, deforming the factorization algebra for free field theories discussed earlier.

Let us explain a little bit about the challenges we need to overcome to deal with interacting theories, and how we overcome these challenges.

Consider an interacting scalar field theory on a Riemannian manifold M. For instance, we could consider an action functional of the form

$$S(\phi) = -\tfrac{1}{2} \int_M \phi(\triangle + m^2)\phi + \int_M \phi^4.$$

In general, the action functional must be local: it must arise as the integral over M of some polynomial in ϕ and its derivatives. This condition is due to our interest in field theories, but it is also necessary to produce factorization algebras.

We will let $I(\phi)$ denote the interaction term in our field theory, which consists of the cubic and higher terms in S. In the above example, $I(\phi) = \int \phi^4$. We will always assume that the quadratic term in S is similar in form $-\tfrac{1}{2}\int_M \phi(\triangle + m^2)\phi$, i.e., we require an ellipticity condition. (Of course, the examples amenable to our techniques apply to a very general class of interacting theories, including many gauge theories.)

If $U \subset M$ is an open subset, we can consider, as before, the spaces $\mathrm{Vect}_c(C^\infty(M))$ and $P(C^\infty(M))$ of polynomial functions and vector fields on M. By analogy with the finite-dimensional situation, one can try to define the divergence for the putative measure $\exp S(\phi)/\hbar\, d\mu$ (where $d\mu$ refers to the "Lebesgue measure" on $C^\infty(M)$)) by the formula

$$\mathrm{Div}_\hbar\left(f_1 \dots f_n \frac{\partial}{\partial \phi}\right) = \tfrac{1}{\hbar}f_1 \dots f_n \frac{\partial S}{\partial \phi} + \sum f_1 \dots \widehat{f_i} \dots f_n \int_M f_i \phi.$$

This formula agrees with the formula we used when S was purely quadratic.

But now a problem arises. We defined $P(C^\infty(U))$ as the space of polynomial functions whose Taylor terms are given by integration against a smooth function on U^n. That is,

$$P(C^\infty(U)) = \bigoplus_n C_c^\infty(U^n)_{S_n}.$$

Unfortunately, if $\phi \in C_c^\infty(U))$, then $\frac{\partial S}{\partial \phi}$ is not necessarily in this space of functions. For instance, if $I(\phi) = \int \phi^4$ is the interaction term in the example of the ϕ^4 theory, then

$$\frac{\partial I}{\partial \phi}(\psi) = \int_M \phi \psi^3 \, d\text{vol}.$$

Thus $\frac{\partial I}{\partial \phi}$ provides a cubic function on the space $C^\infty(U)$, but it is not given by integration against an element in $C_c^\infty(U^3)$. Instead it is given by integrating against a distribution on U^3, namely, the delta distribution on the diagonal.

We can try to solve this issue by using a larger class of polynomial functions. Thus, we could let

$$\overline{P}(C^\infty(U)) = \bigoplus_n \mathcal{D}_c(U^n)_{S_n}$$

where $\mathcal{D}_c(U^n)$ is the space of compactly supported distributions on U. Similarly, we could let

$$\overline{\text{Vect}}_c(U) = \bigoplus_n \mathcal{D}_c(U^{n+1})_{S_n}.$$

The spaces $P(C^\infty(U))$ and $\text{Vect}_c(C^\infty(U))$ are dense subspaces of these spaces.

If $\phi_0 \in \mathcal{D}_c(U)$ is a compactly supported distribution, then for any local functional S, $\frac{\partial S}{\partial \phi}$ is a well-defined element of $\overline{P}(C^\infty(U))$. Thus, it looks like we have resolved our problem.

However, using this larger space of functions gives us a new problem: the second term in the divergence operator now fails to be well defined! For example, if $f, \phi \in \mathcal{D}_c(U)$, then we have

$$\text{Div}_\hbar\left(f\frac{\partial}{\partial \phi}\right) = \tfrac{1}{\hbar}f\frac{\partial S}{\partial \phi} + \int_M f\phi.$$

Now, $\int_M f\phi$ no longer make sense, because we are trying to multiply distributions. More explicitly, this term involves pairing the distribution $f \boxtimes \phi$ on the diagonal in M^2 with the delta function on the diagonal.

If we consider the $\hbar \to 0$ limit of this putative operator $\hbar \text{Div}_\hbar$, we nonetheless find a well-defined operator

$$\overline{\text{Vect}}_c(C^\infty(U)) \to \overline{P}(C^\infty(U))$$

$$X \to XS,$$

which sends a vector field X to its action on the local functional S.

The cokernel of this operator is the quotient of $\overline{P}(C^\infty(U))$ by the ideal $I_{EL}(U)$ generated by the Euler–Lagrange equations. We thus let

$$H^0(\overline{\text{Obs}}_S^{cl}(U)) = \overline{P}(C^\infty(U))/I_{\text{EL}}.$$

As U varies, this construction produces a factorization algebra on M, which we call the factorization algebra of classical observables associated to the action functional S.

Lemma 2.7.1 *If* $S = -\tfrac{1}{2}\int \phi(\Delta + m^2)\phi$, *then this definition of classical observables coincides with the one we discussed earlier:*

$$H^0(\overline{\text{Obs}}_S^{cl}(U)) \cong H^0(\text{Obs}^{cl}(U))$$

where $H^0(\text{Obs}^{cl}(U))$ *is defined, as earlier, to be the quotient of* $P(C^\infty(U))$ *by the ideal* $I_{\text{EL}}(U)$ *of the Euler–Lagrange equations.*

This result is a version of elliptic regularity, and we prove it later, in Chapter 4.

Now the challenge we face should be clear. If S is the action functional for the free field theory, then we have a factorization algebra of classical observables. This factorization algebra deforms in two ways. First, we can deform it into the factorization algebra of quantum observables for a free theory. Second, we can deform it into the factorization algebra of classical observables for an interacting field theory. The difficulty is to perform both of these deformations simultaneously.

To construct the observables of an interacting field theory, we use the renormalization technique of Costello (2011b). In Costello (2011b), the first author gives a definition of a quantum field theory and a cohomological method for constructing field theories. A field theory as defined in Costello (2011b) gives us (essentially from the definition) a family of divergence operators

$$\text{Div}[L] : \overline{\text{Vect}}_c(M) \to \overline{P}(C^\infty(M)),$$

one for every $L > 0$. These divergence operators, for varying L, can be conjugated to each other by continuous linear isomorphisms of $\overline{\text{Vect}}_c(M))$ and $\overline{P}(C^\infty(M))$. However, for a proper open subset U, these divergence operators do not map $\overline{\text{Vect}}_c(U))$ into $\overline{P}(C^\infty(U))$. However, roughly speaking, for L very small, the operator $\text{Div}[L]$ only increases the support of a vector field in $\overline{\text{Vect}}_c(U))$ by a small amount outside of U. For small enough L, it is almost support preserving. This property turns out to be enough to define the factorization algebra of quantum observables.

This construction of quantum observables for an interacting field theory is given in Volume 2, which is the most technically difficult part of this work. Before tackling it, we will develop more language and explore more examples. In particular, we will

- Develop some formal and structural aspects of the theory of factorization algebras.
- Analyze in more detail the factorization algebra associated to a free field theory.
- Construct and analyze factorization algebras associated to vertex algebras such as the Kac–Moody vertex algebra.
- Develop classical field theory using a homological approach arising from the BV formalism.
- Flesh out the description of the factorization algebra of classical observables we have sketched here.

The example of this chapter, however, already exhibits the central ideas.

3

Prefactorization Algebras and
Basic Examples

In this chapter we give a formal definition of the notion of prefactorization algebra. With the definition in hand, we proceed to examine several examples that arise naturally in mathematics. In particular, we explain how associative algebras can be viewed as prefactorization algebras on the real line, and when the converse holds.

We also explain how to construct a prefactorization algebra from a sheaf of Lie algebras on a manifold M. This construction is called the *factorization envelope*, and it is related to the universal enveloping algebra of a Lie algebra as well as to Beilinson–Drinfeld's notion of a chiral envelope. Although the factorization envelope construction is very simple, it plays an important role in field theory. For example, the factorization algebra for any free theories is a factorization envelope, as is the factorization algebra corresponding to the Kac–Moody vertex algebra. More generally, factorization envelopes play an important role in our formulation of Noether's theorem for quantum field theories.

Finally, when the manifold M is equipped with an action of a group G, we describe what a *G-equivariant prefactorization algebra* is. We will use this notion later in studying translation-invariant field theories (see Section 4.8 in Chapter 4) and holomorphically translation-invariant field theories (see Chapter 5).

3.1 Prefactorization Algebras

In this section we give a formal definition of the notion of a prefactorization algebra, starting concretely and then generalizing. In the first subsection, using plain language, we describe a prefactorization algebra taking values in vector spaces. Readers are free to generalize by replacing "vector space" and "linear map" with "object of a symmetric monoidal category C" and "morphism in

C." (Our favorite target category is cochain complexes.) The next subsections give a concise definition using the language of multicategories (also known as colored operads) and allow an arbitrary multicategory as the target. In the final subsections, we describe the category (and multicategory) of such prefactorization algebras.

3.1.1 The Definition in Explicit Terms

Let M be a topological space. A prefactorization algebra $\mathcal{F}$ on M, taking values in vector spaces, is a rule that assigns a vector space $\mathcal{F}(U)$ to each open set $U \subset M$ along with the following maps and compatibilities.

- There is a linear map $m_V^U : \mathcal{F}(U) \to \mathcal{F}(V)$ for each inclusion $U \subset V$.
- There is a linear map $m_V^{U_1,\ldots,U_n} : \mathcal{F}(U_1) \otimes \cdots \otimes \mathcal{F}(U_n) \to \mathcal{F}(V)$ for every finite collection of open sets where each $U_i \subset V$ and the U_i are pairwise disjoint. The following picture represents the situation.

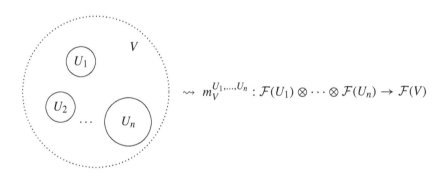

$$\rightsquigarrow m_V^{U_1,\ldots,U_n} : \mathcal{F}(U_1) \otimes \cdots \otimes \mathcal{F}(U_n) \to \mathcal{F}(V)$$

- The maps are compatible in the obvious way, so that if $U_{i,1} \sqcup \cdots \sqcup U_{i,n_i} \subseteq V_i$ and $V_1 \sqcup \cdots \sqcup V_k \subseteq W$, the following diagram commutes.

$$\bigotimes_{i=1}^{k} \bigotimes_{j=1}^{n_i} \mathcal{F}(U_j) \longrightarrow \bigotimes_{i=1}^{k} \mathcal{F}(V_i)$$
$$\mathcal{F}(W)$$

Thus $\mathcal{F}$ resembles a precosheaf, except that we tensor the vector spaces rather than take their direct sum.

Figure 3.1. The prefactorization algebra A^{fact} of an associative algebra A.

For an explicit example of the associativity, consider the following picture.

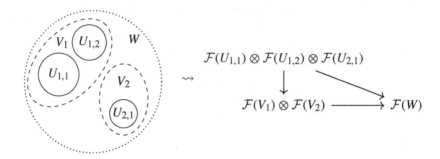

The case of $k = n_1 = 2, n_2 = 1$

These axioms imply that $\mathcal{F}(\emptyset)$ is a commutative algebra. We say that $\mathcal{F}$ is a *unital* prefactorization algebra if $\mathcal{F}(\emptyset)$ is a unital commutative algebra. In this case, $\mathcal{F}(U)$ is a pointed vector space by the image of the unit $1 \in \mathcal{F}(\emptyset)$ under the structure map $\mathcal{F}(\emptyset) \to \mathcal{F}(U)$. In practice, for our examples, $\mathcal{F}(\emptyset)$ is $\mathbb{C}$, $\mathbb{R}$, $\mathbb{C}[[\hbar]]$, or $\mathbb{R}[[\hbar]]$.

Example: The crucial example to bear in mind is an associative algebra. Every associative algebra A defines a prefactorization algebra A^{fact} on $\mathbb{R}$, as follows. To each open interval (a, b), we set $A^{\text{fact}}((a,b)) = A$. To any open set $U = \coprod_j I_j$, where each I_j is an open interval, we set $\mathcal{F}(U) = \bigotimes_j A$. The structure maps simply arise from the multiplication map for A. Figure 3.1 displays the structure of A^{fact}. Notice the resemblance to the notion of an E_1 or A_∞ algebra. (One takes an infinite tensor products of unital algebras, as follows. Given an infinite set I, consider the poset of finite subsets of I, ordered by inclusion. For each finite subset $J \subset I$, we can take the tensor product $A^J = \bigotimes_{j \in J} A$. For $J \hookrightarrow J'$, we define a map $A^J \to A^{J'}$ by tensoring with the identity $1 \in A$ for every $j \in J' \backslash J$. Then A^I is the colimit over this poset.) ◊

Example: Another important example for us is the symmetric algebra of a precosheaf. Let F be a precosheaf of vector spaces on a space X. (For example, consider $F = C_c^\infty$ the compactly supported smooth functions on a manifold.) The functor $\mathcal{F} = \operatorname{Sym} F : U \mapsto \operatorname{Sym}(F(U))$ defines a precosheaf of commutative algebras, but it also a prefactorization algebra. For instance, let U and V be disjoint opens. The structure maps $F(U) \to F(U \sqcup V)$ and $F(U) \to F(U \sqcup V)$ induce a canonical map

$$F(U) \oplus F(V) \to F(U \sqcup V),$$

and so we obtain a natural map $\operatorname{Sym}(F(U) \oplus F(V)) \to \operatorname{Sym} F(U \sqcup V)$. But

$$\mathcal{F}(U) \otimes \mathcal{F}(V) \cong \operatorname{Sym}(F(U)) \otimes \operatorname{Sym}(F(V)) \cong \operatorname{Sym}(F(U) \oplus F(V)),$$

so there is a natural map

$$\mathcal{F}(U) \otimes \mathcal{F}(V) \to \mathcal{F}(U \sqcup V).$$

In a similar way, one can provide all the structure maps to make $\mathcal{F}$ a prefactorization algebra. ◇

In the remainder of this section, we describe two other ways of phrasing this idea, but the reader who is content with this definition and eager to see examples should feel free to jump ahead, referring back as needed.

3.1.2 Prefactorization Algebras as Algebras over an Operad

We now provide a succinct and general definition of a prefactorization algebra using the efficient language of multicategories. (See Section A.2.3 in Appendix A for a quick overview of the notion of a multicategory, also known as a colored operad. Note that we mean the symmetric version of such definitions.)

Definition 3.1.1 Let Disj_M denote the following multicategory associated to M.

- The objects consist of all connected open subsets of M.
- For every (possibly empty) finite collection of open sets $\{U_\alpha\}_{\alpha \in A}$ and open set V, there is a set of maps $\operatorname{Disj}_M(\{U_\alpha\}_{\alpha \in A} \mid V)$. If the U_α are pairwise disjoint and all are contained in V, then the set of maps is a single point. Otherwise, the set of maps is empty.
- The composition of maps is defined in the obvious way.

A prefactorization algebra just *is* an algebra over this colored operad Disj_M.

Definition 3.1.2 Let $\mathcal{C}$ be a multicategory. A *prefactorization algebra* on M taking values in $\mathcal{C}$ is a functor (of multicategories) from Disj_M to $\mathcal{C}$.

Since symmetric monoidal categories are special kinds of multicategories, this definition makes sense for C any symmetric monoidal category.

Remark: If C is a symmetric monoidal category under coproduct, then a pre-cosheaf on M with values in C defines a prefactorization algebra valued in C. Hence, our definition broadens the idea of "inclusion of open sets leads to inclusion of sections" by allowing more general monoidal structures to "combine" the sections on disjoint open sets.   ◊

Note that if $\mathcal{F}$ is any prefactorization algebra, then $\mathcal{F}(\emptyset)$ is a commutative algebra object of C.

Definition 3.1.3 We say a prefactorization algebra $\mathcal{F}$ is *unital* if the commutative algebra $\mathcal{F}(\emptyset)$ is unital.

Remark: There is an important variation on this definition where one weakens the requirement that the composition of structure maps holds "on the nose" and instead requires homotopy coherence. For example, given disjoint opens U_1 and U_2 contained in V, which is then contained in W, we do not require that $m_W^{U_1,U_2} = m_W^V \circ m_V^{U_1,U_2}$ but that there is a "homotopy" between these maps. This kind of situation arises naturally whenever the target category is best viewed as an ∞-category, such as the category of cochain complexes. We do not develop here the formalism necessary to treat homotopy-coherent prefactorization algebras because our examples and constructions always satisfy the strictest version of composition. Readers interested in seeing this variant developed should see the treatment in Lurie (2016). (We remark that one typically has "strictification" results that ensure that a homotopy-coherent algebra over a colored operad can be replaced by a weakly equivalent strict algebra over a colored operad, so that working with strict algebras is sufficient for many purposes.)   ◊

3.1.3 Prefactorization Algebras in the Style of Precosheaves

Any multicategory C has an associated symmetric monoidal category SC, which is defined to be the universal symmetric monoidal category equipped with a functor of multicategories $C \to SC$. We call it the *symmetric monoidal envelope* of C. Concretely, an object of SC is a formal tensor product $a_1 \otimes \cdots \otimes a_n$ of objects of C. Morphisms in SC are characterized by the property that for any object b in C, the set of maps $SC(a_1 \otimes \cdots \otimes a_n, b)$ in the symmetric monoidal category is exactly the set of maps $C(\{a_1, \ldots, a_n\} \mid b)$ in the multicategory C.

We can give an alternative definition of prefactorization algebra by working with the symmetric monoidal category $S\,\mathrm{Disj}_M$ rather than the multicategory Disj_M.

Definition 3.1.4 Let $S\,\mathrm{Disj}_M$ denote the following symmetric monoidal category.

- An object of $S\,\mathrm{Disj}_M$ is a formal finite sequence $[V_1, \ldots, V_m]$ of opens V_i in M.
- A morphism $F : [V_1, \ldots, V_m] \to [W_1, \ldots, W_n]$ consists of a surjective function $\phi : \{1, \ldots, m\} \to \{1, \ldots, n\}$ and a morphism $f_j \in \mathrm{Disj}_M(\{V_k\}_{k \in \phi^{-1}(j)} \mid W_j)$ for each $1 \leq j \leq n$.
- The symmetric monoidal structure on $S\,\mathrm{Disj}_M$ is given by concatenation.

The alternative definition of prefactorization algebra is as follows. It resembles the notion of a precosheaf (i.e., a functor out of some category of opens) with the extra condition that it is symmetric monoidal.

Definition 3.1.5 A *prefactorization algebra* with values in a symmetric monoidal category $\mathscr{C}^{\otimes}$ is a symmetric monoidal functor $S\,\mathrm{Disj}_M \to \mathscr{C}$.

Remark: Although "algebra" appears in its name, a prefactorization algebra only allows one to "multiply" elements that live on disjoint open sets. The category of prefactorization algebras (taking values in some fixed target category) has a symmetric monoidal product, so we can study commutative algebra objects in that category. As an example, we will consider the observables for a classical field theory. ◇

3.1.4 Morphisms and the Category Structure

We now explain how prefactorization algebras form a category.

Definition 3.1.6 A *morphism of prefactorization algebras* $\phi : F \to G$ consists of a map $\phi_U : F(U) \to G(U)$ for each open $U \subset M$, compatible with the structure maps. That is, for any open V and any finite collection $U_1, \ldots, U_k$ of pairwise disjoint open sets, each contained in V, the following diagram commutes:

$$
\begin{array}{ccc}
F(U_1) \otimes \cdots \otimes F(U_k) & \xrightarrow{\phi_{U_1} \otimes \cdots \otimes \phi_{U_k}} & G(U_1) \otimes \cdots \otimes G(U_k) \\
\downarrow & & \downarrow \\
F(V) & \xrightarrow{\phi_V} & G(V)
\end{array} .
$$

Likewise, all the obvious associativity relations are respected.

Remark: When our prefactorization algebras take values in cochain complexes, we require the ϕ_U to be cochain maps, i.e., they each have degree 0 and commute with the differentials.   ◊

Definition 3.1.7 On a space X, we denote the category of prefactorization algebras on X taking values in the multicategory $\mathcal{C}$ by $\mathrm{PreFA}(X, \mathcal{C})$.

Remark: When the target multicategory $\mathcal{C}$ is a model category or dg category or some other kind of higher category, the category of prefactorization algebras naturally forms a higher category as well. Given the nature of our constructions and examples in the next few chapters, such aspects do not play a prominent role. When we define factorization algebras in Chapter 6, however, we will discuss such issues.   ◊

3.1.5 The Multicategory Structure

Let $\mathbf{S}\mathcal{C}$ denote the enveloping symmetric monoidal category of the multicategory $\mathcal{C}$ (see Section A.3.3 in Appendix A). Let $\otimes$ denote the symmetric monoidal product in $\mathbf{S}\mathcal{C}$. Any prefactorization algebra valued in $\mathcal{C}$ gives rise to one valued in $\mathbf{S}\mathcal{C}$.

There is a natural tensor product on $\mathrm{PreFA}(X, \mathbf{S}\mathcal{C})$, as follows. Let F, G be prefactorization algebras. We define $F \otimes G$ by

$$(F \otimes G)(U) = F(U) \otimes G(U),$$

and we simply define the structure maps as the tensor product of the structure maps. For instance, if $U \subset V$, then the structure map is

$$m(F)_V^U \otimes m(G)_V^U : (F \otimes G)(U) = F(U) \otimes G(U) \to F(V) \otimes G(V) = (F \otimes G)(V).$$

Definition 3.1.8 Let $\mathrm{PreFA}_{mc}(X, \mathcal{C})$ denote the multicategory arising from the symmetric monoidal product on $\mathrm{PreFA}(X, \mathbf{S}\mathcal{C})$. That is, if $F_1, \ldots F_n, G$ are prefactorization algebras valued in $\mathcal{C}$, we define the set of multimorphisms to be

$$\mathrm{PreFA}_{mc}(F_1, \cdots, F_n \mid G) = \mathrm{PreFA}(F_1 \otimes \cdots \otimes F_n, G),$$

the set of maps of $\mathbf{S}\mathcal{C}$-valued prefactorization algebras from $F_1 \otimes \cdots \otimes F_n$ to G.

3.2 Associative Algebras from Prefactorization Algebras on $\mathbb{R}$

We explained in the preceding text how an associative algebra provides a prefactorization algebra on the real line. There are, however, prefactorization algebras on $\mathbb{R}$ that do *not* come from associative algebras. Here we characterize those that do arise from associative algebras.

Definition 3.2.1 Let $\mathcal{F}$ be a prefactorization algebra on $\mathbb{R}$ taking values in the category of vector spaces (without any grading). We say $\mathcal{F}$ is *locally constant* if the map $\mathcal{F}(U) \to \mathcal{F}(V)$ is an isomorphism for every inclusion of intervals $U \subset V$.

Lemma 3.2.2 *Let $\mathcal{F}$ be a locally constant, unital prefactorization algebra on $\mathbb{R}$ taking values in vector spaces. Let $A = \mathcal{F}(\mathbb{R})$. Then A has a natural structure of an associative algebra.*

Remark: Recall that $\mathcal{F}$ being unital means that the commutative algebra $\mathcal{F}(\emptyset)$ is equipped with a unit. We will find that A is an associative algebra over $\mathcal{F}(\emptyset)$. $\Diamond$

Proof For any interval $(a, b) \subset \mathbb{R}$, the map

$$\mathcal{F}((a, b)) \to \mathcal{F}(\mathbb{R}) = A$$

is an isomorphism. Thus, we have a canonical isomorphism

$$A = \mathcal{F}((a, b))$$

for all intervals (a, b).

Notice that if $(a, b) \subset (c, d)$ then the diagram

$$
\begin{array}{ccc}
A & \xrightarrow{\;\cong\;} & \mathcal{F}((a, b)) \\
{\scriptstyle \text{Id}} \downarrow & & \downarrow {\scriptstyle i^{(a,b)}_{(c,d)}} \\
A & \xrightarrow{\;\cong\;} & \mathcal{F}((c, d))
\end{array}
$$

commutes.

The product map $m : A \otimes A \to A$ is defined as follows. Let $a < b < c < d$. Then, the prefactorization structure on $\mathcal{F}$ gives a map

$$\mathcal{F}((a, b)) \otimes \mathcal{F}((c, d)) \to \mathcal{F}((a, d)),$$

and so, after identifying $\mathcal{F}((a, b))$, $\mathcal{F}((c, d))$ and $\mathcal{F}((a, d))$ with A, we get a map

$$A \otimes A \to A.$$

This is the multiplication in our algebra.

It remains to check the following.

(i) This multiplication doesn't depend on the intervals $(a, b) \sqcup (c, d) \subset (a, d)$ we chose, as long as $(a, b) < (c, d)$.

(ii) This multiplication is associative and unital.

This is an easy (and instructive) exercise. $\square$

3.3 Modules as Defects

We want to explain another simple but illuminating class of examples, and then we apply this perspective in the context of quantum mechanics. We will work with prefactorization algebras taking values in vector spaces with the tensor product as symmetric monoidal structure.

3.3.1 Modules as Living at Points

We described already how to associate a prefactorization algebra $\mathcal{F}_A$ on $\mathbb{R}$ to an associative algebra A. It is easy to modify this construction to describe bimodules, as follows. Let A and B be associative algebras and M an A-B-bimodule, i.e., M is a left A-module and a right B-module and these structures are compatible in that $(am)b = a(mb)$ for all $a \in A$, $b \in B$, and $m \in M$. We now construct a prefactorization algebra $\mathcal{F}_M$ on $\mathbb{R}$ that encodes this bimodule structure.

Pick a point $p \in \mathbb{R}$. On the half-line $\{x \in \mathbb{R} \mid x < p\}$, $\mathcal{F}_M$ is given by $\mathcal{F}_A$: to an interval $I = (t_0, t_1)$ with $t_1 < p$, $\mathcal{F}_M(I) = A$, and the structure map for inclusion of finitely many disjoint intervals into a bigger interval is determined by multiplication in A. Likewise, on the half-line $\{x \in \mathbb{R} \mid p < x\}$, $\mathcal{F}_M$ is given by $\mathcal{F}_B$. Intervals containing p are determined, though, by M. When we consider an interval $I = (t_0, t_1)$ such that $t_0 < p < t_1$, we set $\mathcal{F}_M(I) = M$. The structure maps are also determined by the bimodule structure. For example, given

$$s_0 < s_1 < t_0 < p < t_1 < u_0 < u_1,$$

we have

$$\mathcal{F}_M((s_0, s_1) \sqcup (t_0, t_1) \sqcup (u_0, u_1)) = \mathcal{F}_M((s_0, s_1)) \otimes \mathcal{F}_M((t_0, t_1)) \otimes \mathcal{F}_M(u_0, u_1))$$
$$= A \otimes M \otimes B$$

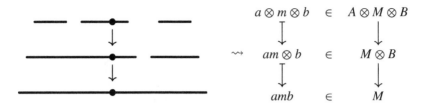

Figure 3.2. The prefactorization algebra $\mathcal{F}_M$ for the A-B-bimodule M.

and the inclusion of these three intervals into (s_0, u_1) is the map

$$\mathcal{F}_M((s_0, s_1)) \otimes \mathcal{F}_M((t_0, t_1)) \otimes \mathcal{F}_M)(u_0, u_1)) \quad \rightarrow \quad \mathcal{F}_M((s_0, u_1))$$
$$a \otimes m \otimes b \qquad\qquad \mapsto \qquad\qquad amb$$

The definition of a bimodule ensures that we have a prefactorization algebra. See Figure 3.2.

There is a structure map that we have not discussed yet, though. The inclusion of the empty set into an interval I containing p means that we need to pick an element m_I of M for each interval. The simplest case is to fix one element $m \in M$ and simply use it for every interval. If we are assigning the unit of A as the distinguished element for $\mathcal{F}_M$ on every interval to the left of p and the unit of B for every interval to the right of p, then the distinguished elements, then the structure maps we have given clearly respect these distinguished elements.

These distinguished elements, however, can change with the intervals, so long as they are preserved by the structure maps. For an interesting example, see the discussion of quantum mechanics that follows.

Let us examine one more interesting case. Suppose we have algebras A, B, and C, and an A-B-bimodule M and a B-C-bimodule N. There is a prefactorization algebra on $\mathbb{R}$ describing the natural algebra for this situation.

Fix points $p < q$. Let $\mathcal{F}_{M,N}$ be the prefactorization algebra on $\mathbb{R}$ such that

- On $\{x \in \mathbb{R} \mid x < p\}$, it agrees with $\mathcal{F}_A$.
- On $\{x \in \mathbb{R} \mid p < x < q\}$, it agrees with $\mathcal{F}_B$.
- On $\{x \in \mathbb{R} \mid q < x\}$, it agrees with $\mathcal{F}_C$.

and

- To intervals (t_0, t_1) with $t_0 < p < t_1 < q$, it assigns M.
- To intervals (t_0, t_1) with $p < t_0 < q < t_1$, it assigns N.

In other words, on $\{x \in \mathbb{R} \mid x < q\}$, this prefactorization algebra $\mathcal{F}_{M,N}$ behaves like $\mathcal{F}_M$, and like $\mathcal{F}_N$ on $\{x \in \mathbb{R} \mid p < x\}$.

We still need to describe what it does on an interval I of the form (T_0, T_1) with $T_0 < p < q < T_1$. There is a natural choice, dictated by the requirement that we produce a prefactorization algebra.

We know that $\mathcal{F}_{M,N}(I)$ must receive maps from M and N, by considering smaller intervals that contain either p or q, respectively. It also receives maps from A, B, and C from intervals not hitting these marked points, but these factor through intervals containing one of the marked points. Finally, we must have a structure map

$$\mu : M \otimes N \to \mathcal{F}_{M,N}(I)$$

for each pair of disjoint intervals hitting both marked points.

Note, in particular, what the associativity condition requires in the situation where we have three disjoint intervals given by

$$s_0 < p < s_1 < t_0 < t_1 < u_0 < q < u_1,$$

contained in I. We can factor the inclusion of these three intervals through the pair of intervals $(s_0, t_1) \sqcup (u_0, u_1)$ or the pair of intervals $(s_0, s_1) \sqcup (t_0, u_1)$ Thus, our structure map

$$M \otimes B \otimes N \to \mathcal{F}_{M,N}(I)$$

must satisfy that $\mu((mb) \otimes n) = \mu(m \otimes (bn))$ for every $m \in M$, $b \in B$, and $n \in N$. This condition means that $\mathcal{F}_{M,N}(I)$ receives a canonical map from the tensor product $M \otimes_B N$.

Hence, the most natural choice is to set $\mathcal{F}_{M,N}(I) = M \otimes_B N$. One can make other choices for how to extend to these longer intervals, but such a prefactorization algebra will receive a map from this one. The local-to-global principle satisfied by a factorization algebra is motivated by this kind of reasoning.

Remark: We have shown how thinking about prefactorization algebras on a real line "decorated" with points (i.e., with a kind of stratification) reflects familiar algebraic objects like bimodules. By moving into higher dimensions and allowing more interesting submanifolds and stratifications, one generalizes this familiar algebra into new, largely unexplored directions. See Ayala et al. (2015) for an extensive development of these ideas in the setting of locally constant factorization algebras. ◇

3.3.2 Standard Quantum Mechanics as a Prefactorization Algebra

We will now explain how to express the standard formalism of quantum mechanics in the language of prefactorization algebras, using the kind of construction just described. As our goal is to emphasize the formal structure, we will work with a finite-dimensional complex Hilbert space and avoid discussions of functional analysis.

Remark: In a sense, this section is a digression from the central theme of the book. Throughout this book we take the path integral formalism as fundamental, and hence we do not focus on the Hamiltonian, or operator, approach to quantum physics. Hopefully, juxtaposed with our work in Section 4.3 in Chapter 4, this example clarifies how to connect our methods with others. ◊

Let V denote a finite-dimensional complex Hilbert space V. That is, there is a nondegenerate symmetric sesquilinear form $(-, -) : V \times V \to \mathbb{C}$ so that

$$(\lambda v, \lambda' v') = \overline{\lambda} \lambda' (v, v')$$

where λ, λ' are complex numbers and v, v' are vectors in V. Let $A = \text{End}(V)$ denote the algebra of endomorphisms, which has a $*$-structure via $M^* = \overline{M}^T$, the conjugate-transpose. The space V is a representation of A, and the $*$-structure is characterized by the property that

$$(M^* v, v') = (v, M v').$$

It should be clear that one could work more generally with a Hilbert space equipped with the action of a $*$-algebra of operators, *aka* observables.

Now that we have fixed the kinematics of the situation, let's turn to the dynamics. Let $(U_t)_{t \in \mathbb{R}}$ be a one-parameter group of unitary operators on V. Since we are in the finite-dimensional setting, there is no problem identifying

$$U_t = e^{itH}$$

for some Hermitian operator H that we call the Hamiltonian. We view V as a state space for our system, A as where the observables live, and H as determining the time evolution of our system.

We now rephrase this structure to make it easier to articulate via the factorization picture. Let $\overline{V}$ denote V with the conjugate complex structure. We will denote elements of V by "kets" $| v' \rangle$ and elements of $\overline{V}$ by "bras" $\langle v |$, and we provide a bilinear pairing between them by

$$\langle v \mid v' \rangle = (v, v').$$

We equip $\overline{V}$ with the right A-module structure by

$$\langle v \mid M = \langle M^* v \mid.$$

We will write $\langle v \mid M \mid v' \rangle$ as can think of M acting on v' from the left or on v from the right and it will produce the same number.

Our goal is to describe a scattering-type experiment.

- At time $t = 0$, we prepare our system in the initial state $\langle v_{\text{in}} \mid$.
- We modify the governing Hamiltonian over some finite time interval (i.e., apply an operator, or equivalently, an observable).
- At time $t = T$, we measure whether our system is in the final state $\mid v_{\text{out}}\rangle$.

If we run this experiment many times, with the same initial and final states and the same operator, we should find a statistical pattern in our data. If an operator $\mathcal{O}$ acts during a time interval (t, t'), then we are trying to compute the number

$$\langle v_{\text{in}} \mid e^{itH} \mathcal{O} e^{i(T-t')H} \mid v_{\text{out}}\rangle.$$

Our formalism does not include the idealized situation of an operator $\mathcal{O}$ acting at a single moment t_0 in time, but in the appropriate limit of shorter and shorter time intervals around t_0, we would compute

$$\langle v_{\text{in}} \mid e^{it_0 H} \mathcal{O} e^{i(T-t_0)H} \mid v_{\text{out}}\rangle,$$

which agrees with the usual prescription.

Note that we use a bra $\langle v_{\text{in}} \mid$ for the "incoming" state and a ket $\mid v_{\text{out}}\rangle$ for the "outgoing" state so that the left-to-right ordering will agree with the left-to-right ordering of the real line viewed as a timeline. The prefactorization algebra F on the interval $[0, T]$ describing this situation has the following structure. Interior open intervals describe moments when operators can act on our system. An interval that contains 0 (but not the other end) should describe possible "incoming" states of the system; dually, an interval containing the other endpoint should describe "outgoing" states. Let us now spell things out explicitly.

To open subintervals, our prefactorization algebra F assigns the following vector spaces:

- $[0, t) \mapsto \overline{V}$
- $(s, t) \mapsto A$
- $(t, T] \mapsto V$
- $[0, T] \mapsto \mathbb{C}$.

In light of our discussion about modules in the preceding section, note that the natural choice for the value $F([0, T])$ is the vector space $\overline{V} \otimes_A V$. However, $\overline{V} \otimes_{\text{End}(V)} V$ is isomorphic to the ground field $\mathbb{C}$ due to the compatibility of the left and right actions with the inner product. We must now describe the structure maps coming from inclusion of intervals; we will describe enough so that the mechanism is clear.

The case that determines the rest is that the inclusion

$$[0, t_0) \sqcup (t_1, t_2) \sqcup (t_3, T] \subset [0, T]$$

corresponds to the structure map

$$\begin{array}{ccc} \overline{V} \otimes A \otimes V & \to & \mathbb{C} \\ \langle v_0 \mid \otimes \mathcal{O} \otimes \mid v_1 \rangle & \mapsto & \langle v_0 \mid e^{i(t_1-t_0)H} \mathcal{O} e^{i(t_3-t_2)H} \mid v_1 \rangle \end{array}.$$

In other words, the system evolves according the Hamiltonian during the closed intervals in the complement of opens during which we specify the incoming and outgoing states and the operator.

Note that if we set $\mathcal{O} = e^{i(t_2-t_1)H}$, then we obtain

$$\langle v_0 \mid e^{i(t_1-t_3)H} \mid v_1 \rangle$$

and so recover the expected value of being in state $\langle v_0 \mid$ at time t_0 and going to state $\mid v_1 \rangle$ at time t_3. Setting $t_3 = t_0$, we recover the inner product on V.

For another example of structure maps, the inclusion $(t_0, t_1) \subset (t_0', t_1')$ goes to $\mathcal{O} \mapsto e^{i(t_0-t_0')H} \mathcal{O} e^{i(t_1'-t_1)H}$. More generally, for k disjoint open intervals inside a big open interval

$$(t_0, t_1) \sqcup \cdots \sqcup (t_{2k-2}, t_{2k-1}) \subset (t_0', t_1'),$$

we again time-order the operators and then multiply with evolution operators inserted for the closed intervals between them: the structure map for F is

$$\mathcal{O}_1 \otimes \cdots \otimes \mathcal{O}_k \mapsto e^{i(t_0-t_0')H} \mathcal{O}_1 e^{i(t_2-t_1)H} \cdots \mathcal{O}_k e^{i(t_1'-t_{2k-1})H}.$$

Note that these structure maps reduce to those for an arbitrary associative algebra if we set $H = 0$ and hence have the identity map as the evolution operator. (In general, a one-parameter semigroup of algebra automorphisms can be used like this to "twist" the prefactorization algebra associated to an associative algebra.)

As further examples, we have

- The inclusion $[0, t) \subset [0, t')$ has structure map $\langle v_0 \mid \mapsto \langle v_0 \mid e^{i(t'-t)H}$.
- The inclusion $(t, T] \subset (t', T]$ has structure map $\mid v_1 \rangle \mapsto e^{i(t-t')H} \mid v_1 \rangle$.
- The inclusion $[0, t_0) \sqcup (t_1, t_2) \subset [0, t')$ has structure map $\langle v_0 \mid \otimes \mathcal{O} \mapsto \langle v_0 \mid e^{i(t_1-t_0)H} \mathcal{O} e^{i(t'-t_2)H}$.

All these choices ensure that we are free to choose what happens during open intervals but that the system evolves according to H during their closed complements. In this way, the prefactorization algebra encodes the basic abstract structure of quantum mechanics. The open interior $(0, T)$ encodes the algebra of observables, and the boundaries encode the state spaces.

Attentive readers might notice that we have not discussed, e.g., the structure map associated to $(t_0, t_1) \subset [0, t')$. Here we need to use the distinguished elements in $F((t_0, t_1)) = A$ and $F([0, t')) = \overline{V}$ arising from the inclusion of the empty set into these opens. If we fix $\langle v_{\text{in}} \mid$ as the "idealized" initial state at time 0, then we set $\langle v_{\text{in}} \mid e^{it'H}$ to be the distinguished element in $F([0, t')) = \overline{V}$. The distinguished element of $F((t_0, t_1))$ is the evolution operator $e^{i(t_1 - t_0)H}$. Thus, the structure map for $(t_0, t_1) \subset [0, t')$ is naturally

$$\mathcal{O} \mapsto \langle v_{\text{in}} \mid e^{it_0 H} \mathcal{O} e^{i(t' - t_1)H}.$$

We now describe the dual situation for intervals containing the other endpoint. Here we specify an "idealized" final state $\mid v_{\text{out}} \rangle$ so that the distinguished element of $F((t, T])$ is $e^{i(T-t)H} \mid v_{\text{out}} \rangle$. We do not need these initial or final states to recover the quantum mechanical formalism from the prefactorization algebra, so it is interesting that the prefactorization perspective pushes toward fixing these boundary states in the form of the distinguished elements.

Remark: Although we have explained here how to start with the standard ingredients of quantum mechanics and encode them as a prefactorization algebra, one can also turn the situation around and motivate (or interpret), via the factorization perspective, aspects of the quantum mechanical formalism. For instance, time-reversal amounts to reflection across a point in $\mathbb{R}$. Requiring a locally constant prefactorization algebra on $\mathbb{R}$ to be equivariant under time-reversal corresponds to equipping the corresponding associative algebra A with an involutive algebra antiautomorphism. In other words, the prefactorization algebra corresponds to a $*$-algebra A. Likewise, suppose we want a prefactorization algebra on a closed interval $[0, T]$ such that it corresponds to A on the open interior. The right end point corresponds to a left A-module V, viewed as the "outgoing" states. It is natural to want the "incoming" states – given by some right A-module $\overline{V}$ – to be "of the same type" as V (e.g., abstractly isomorphic) and to want the global sections to be $\mathbb{C}$. This forces us to have a map

$$\overline{V} \otimes A \otimes V \to \mathbb{C}.$$

This map induces a pairing between $\overline{V}$ and V, which provides a pre-Hilbert structure. $\Diamond$

Remark: Our construction above captures much of the standard formalism of quantum mechanics, but there are a few loose ends to address.

First, in standard quantum mechanics, a state is not a vector in V but a line. Above, however, we fixed vectors $\langle v_{\text{in}} \mid$ and $\mid v_{\text{out}} \rangle$, so there seems to be a discrepancy. The observation that rescues us is a natural one, from the mathematical viewpoint. Consider scaling the initial and final states by elements of $\mathbb{C}^\times$. This defines a new factorization algebra, but it is isomorphic to what we described earlier, and the expected value "$\langle v_0 | \mathcal{O} v_1 \rangle$" of an operator depends linearly in the rescaling of the input and output vectors. More precisely, there is a natural equivalence relation we can place on the factorization algebras described earlier that corresponds to the usual notion of state in quantum mechanics. In other words, we could make a groupoid of factorization algebras where the underlying vector spaces and structure maps are all the same, but the distinguished elements are allowed to change.

Another issue that might bother readers is that our formalism matches nicely only with experiments that resemble scattering experiments. It does not seem well suited to descriptions of systems such as bound states (e.g., an atom sitting quietly, minding its own business). For such systems, we might consider running over the whole space of states, which is described as a groupoid in the previous paragraph. Alternatively, we might drop the endpoints and simply work with the factorization algebra on the open interval, which focuses on the algebra of operators. ◊

3.4 A Construction of the Universal Enveloping Algebra

Let $\mathfrak{g}$ be a Lie algebra. In this section we describe a procedure that produces the universal enveloping algebra $U\mathfrak{g}$ as a prefactorization algebra on $\mathbb{R}$. This construction is useful both because it is a model for a more general result about E_n algebras (see Section 6.4 in Chapter 6) and because it appears in several of our examples (such as the Kac–Moody factorization algebra). For a mathematician, it may be useful to see techniques similar to those we use in Section 4.2 in Chapter 4 shorn of any connection with physics, so that the underlying process is clearer.

Let $\mathfrak{g}^{\mathbb{R}}$ denote the cosheaf on $\mathbb{R}$ that assigns $(\Omega_c^*(U) \otimes \mathfrak{g}, d_{dR})$ to each open U, with d_{dR} the exterior derivative. This is a cosheaf of cochain complexes, but it is only a *pre*cosheaf of dg Lie algebras. Note that the cosheaf axiom involves the use of coproducts, and the coproduct in the category of dg Lie algebras is not given by direct sum of the underlying cochain complexes. In fact, $\mathfrak{g}^{\mathbb{R}}$ is a prefactorization algebra taking value in dg Lie algebras equipped with direct sum $\oplus$ as the symmetric monoidal product (here direct sum means the sum of the underlying cochain complexes, which inherits a natural Lie bracket).

Let $C_* \mathfrak{h}$ denote the Chevalley–Eilenberg complex for Lie algebra *homology*, written as a cochain complex. In other words, $C_* \mathfrak{h}$ is the graded vector space $\mathrm{Sym}(\mathfrak{h}[1])$ with a differential determined by the bracket of $\mathfrak{h}$. See Section A.4 in Appendix A for the definition and further discussion of this construction.

Our main result shows how to construct the universal enveloping algebra Ug using $C_*(g^{\mathbb{R}})$.

Proposition 3.4.1 *Let $\mathcal{H}$ denote the cohomology prefactorization algebra of $C_*(g^{\mathbb{R}})$. That is, we take the cohomology of every open and every structure map, so*

$$\mathcal{H}(U) = H^*(C_*(g^{\mathbb{R}}(U)))$$

for any open U. Then $\mathcal{H}$ is locally constant, and the corresponding associative algebra is isomorphic to Ug, the universal enveloping algebra of g.

Remark: Recall Lemma 3.2.2, which says that every locally constant prefactorization algebra on $\mathbb{R}$ corresponds to an associative algebra. This proposition above provides a homological mechanism for recovering the universal enveloping algebra of a Lie algebra, but readers have probably noticed that we could apply the same construction with $\mathbb{R}$ replaced by any smooth manifold M. In Chapter 6, we will explain how to understand what this general procedure means.   ◇

Proof Local constancy of $\mathcal{H}$ is immediate from the fact that, if $I \subset J$ is the inclusion of one interval into another, the map of dg Lie algebras

$$\Omega_c^*(I) \otimes g \to \Omega_c^*(J) \otimes g$$

is a quasi-isomorphism. We let A_g be the associative algebra constructed from $\mathcal{H}$ by Lemma 3.2.2.

The underlying vector space of A_g is the space $\mathcal{H}(I)$ for any interval I. To be concrete, we will use the interval $\mathbb{R}$, so that we identify

$$A_g = \mathcal{H}(\mathbb{R}) = H^*(C_*(\Omega_c^*(\mathbb{R}) \otimes g)).$$

We now identify that vector space.

The dg Lie algebra $\Omega_c^*(\mathbb{R}) \otimes g$ is concentrated in degrees 0 and 1 and maps quasi-isomorphically to its cohomology $H_c^*(\mathbb{R}) \otimes g = g[-1]$, which is concentrated in degree 1 by the Poincaré lemma. This cohomology is an Abelian Lie algebra because the cup product on $H_c^*(I)$ is zero. It follows that $C_*(\Omega_c^*(\mathbb{R}) \otimes g)$ is quasi-isomorphic to the Chevalley–Eilenberg chains of the Abelian Lie algebra $g[-1]$, which is simply $\operatorname{Sym} g$. Thus, as a vector space, A_g is isomorphic to the symmetric algebra $\operatorname{Sym} g$.

There is a map

$$\Phi : g \to A_g$$

that sends an element $X \in g$ to $\varepsilon \otimes X$ where $\varepsilon \in H_c^1(I)$ is a basis element for the compactly supported cohomology of the interval I; we require the integral of ε

to be 1. We will show that Φ is a map of Lie algebras, where $A_{\mathfrak{g}}$ is given the Lie bracket coming from the associative structure. This result immediately implies the theorem, as we then have an map of associative algebra $\Phi : U\mathfrak{g} \to A_{\mathfrak{g}}$ that is an isomorphism of vector spaces.

Let us check explicitly that Φ is a map of Lie algebras. Let $\delta > 0$ be small number, and let $f_0 \in C_c^\infty(-\delta, \delta)$ be a compactly supported smooth function with $\int f_0 \, dx = 1$. Let $f_t(x) = f_0(x - t)$. Note that f_t is supported on the interval $(t - \delta, t + \delta)$. If $X \in \mathfrak{g}$, a cochain representative for $\Phi(X) \in A_{\mathfrak{g}}$ is provided by

$$f_0 \, dx \otimes X \in \Omega_c^1((-\delta, \delta)) \otimes \mathfrak{g}.$$

Indeed, because every $f_t \, dx$ is cohomologous to $f_0 \, dx$ in $\Omega_c^1(\mathbb{R})$, the element $f_t \, dx \otimes X$ is a cochain representative of $\Phi(X)$ for any t.

Given elements $\alpha, \beta \in A_g$, the product $\alpha \cdot \beta$ is defined as follows.

(i) We choose intervals I, J with $I < J$.
(ii) We regard α as an element of $\mathcal{H}(I)$ and β as an element of $\mathcal{H}(J)$ using the inverses to the isomorphisms $\mathcal{H}(I) \to \mathcal{H}(\mathbb{R})$ and $\mathcal{H}(J) \to \mathcal{H}(\mathbb{R})$ coming from the inclusions of I and J into $\mathbb{R}$.
(iii) The product $\alpha \cdot \beta$ is defined by taking the image of $\alpha \otimes \beta$ under the factorization structure map

$$\mathcal{H}(I) \otimes \mathcal{H}(J) \to \mathcal{H}(\mathbb{R}) = A_g.$$

Let us see how this works with our representatives.

The cohomology class $[f_t \, dx \otimes X] \in \mathcal{H}(t - \delta, t + \delta)$ becomes $\Phi(X)$ under the natural map from $\mathcal{H}(t - \delta, t + \delta)$ to $\mathcal{H}(\mathbb{R})$. If we take δ to be sufficiently small, the intervals $(-\delta, \delta)$ and $(1 - \delta, 1 + \delta)$ are disjoint. It follows that the product $\Phi(X)\Phi(Y)$ is represented by the cocycle

$$(f_0 \, dx \otimes X)(f_1 \, dx \otimes Y) \in \mathrm{Sym}^2(\Omega_c^1(\mathbb{R}) \otimes \mathfrak{g}) \subset C_*(\Omega_c^*(\mathbb{R}) \otimes \mathfrak{g}).$$

Similarly, the commutator $[\Phi(X), \Phi(Y)]$ is represented by the expression

$$(f_0 \, dx \otimes X)(f_1 \, dx \otimes Y) - (f_0 \, dx \otimes Y)(f_{-1} \, dx \otimes X).$$

It suffices to show that this cocycle in $C_*(\Omega_c^*(\mathbb{R}) \otimes \mathfrak{g})$ is cohomologous to $\Phi([X, Y])$.

Note that the 1-form $f_1 \, dx - f_{-1} \, dx$ has integral 0. It follows that there exists a compactly supported function $h \in C_c^\infty(\mathbb{R})$ with

$$d_{dR} h = f_{-1} \, dx - f_1 \, dx.$$

We can assume that h takes value 1 in the interval $(-\delta, \delta)$.

We now calculate the differential of the element

$$(f_0 \, dx \otimes X)(h \otimes Y) \in C_*(\Omega_c^*(\mathbb{R}) \otimes \mathfrak{g}).$$

We have

$$d\left((f_0 \, dx \otimes X)(h \otimes Y)\right) = (f_0 \, dx \otimes X)(dh \otimes Y) + f_0 h \, dx \otimes [X, Y]$$
$$= (f_0 \, dx \otimes X)((f_{-1} - f_1)dx \otimes Y) + f_0 h \, dx \otimes [X, Y].$$

Since h takes value 1 on the interval $(-\delta, \delta)$, we know $f_0 h = f_0$. This equation tells us that a representative for $[\Phi(X), \Phi(Y)]$ is cohomologous to $\Phi([X, Y])$.

$\square$

3.5 Some Functional Analysis

Nearly all of the examples of factorization algebras that we will consider in this book will assign to an open subset, a cochain complex built from vector spaces of analytical provenance: for example, the vector space of smooth sections of a vector bundle or the vector space of distributions on a manifold. Such vector spaces are best viewed as being equipped with an extra structure, such as a topology, reflecting their analytical origin. In this section we briefly sketch a flexible multicategory of vector spaces equipped with an extra "analytic" structure. More details are contained in Appendix B.

3.5.1 Differentiable Vector Spaces

The most common way to encode the analytic structure on a vector space such as the space of smooth functions on a manifold is to endow it with a topology. Homological algebra with topological vector spaces is not easy, however. (For instance, topological vector spaces do not form an Abelian category.) To get around this issue, we will work with *differentiable vector spaces*. Let us first define the slightly weaker notion of a C^∞-module.

Definition 3.5.1 Let Mfld be the site of smooth manifolds, i.e., the category of smooth manifolds and smooth maps between them, where a cover is a surjective local diffeomorphism.

Let C^∞ denote the sheaf of rings on Mfld that assigns to any manifold M the commutative algebra $C^\infty(M)$. A C^∞-*module* is a module sheaf over C^∞ on Mfld.

In other words, to each manifold M, a C^∞-module $\mathcal{F}$ assigns a module $\mathcal{F}(M)$ over the algebra $C^\infty(M)$, and for any map of manifolds $f : M \to N$, the pullback map $f^* : \mathcal{F}(N) \to \mathcal{F}(M)$ is a map of $C^\infty(N)$-modules. For example, if V is any topological vector space, then there is a natural notion of smooth map from any manifold M to V (see, e.g., Kriegl and Michor (1997)). The space $C^\infty(M, V)$ of such smooth maps is a module over the algebra $C^\infty(M)$. Since smoothness is a local condition on M, sending M to $C^\infty(M, V)$ gives a sheaf of C^∞-modules on the site Mfld.

As an example of this construction, let us consider the case when V is the space of smooth functions on a manifold N, equipped with its usual Fréchet topology. One can show that for each manifold M, the space $C^\infty(M, C^\infty(N))$ is naturally isomorphic to the space $C^\infty(M \times N)$ of smooth functions on $M \times N$.

As we will see shortly, we lose very little information when we view a topological vector space as a C^∞-module.

Sheaves of C^∞-modules on Mfld that arise from topological vector spaces are endowed with an extra structure: we can always differentiate smooth maps from a manifold M to a topological vector space V. Differentiation can be viewed as an action of the vector fields on M on the vector space $C^\infty(M, V)$. Dually, it can be viewed as coming from a connection

$$\nabla : C^\infty(M, V) \to \Omega^1(M, V),$$

where $\Omega^1(M, V)$ is defined to be the tensor product

$$\Omega^1(M, V) = \Omega^1(M) \otimes_{C^\infty(M)} C^\infty(M, V).$$

This tensor product is just the algebraic one, which is well behaved because $\Omega^1(M)$ is a projective $C^\infty(M)$-module of finite rank.

This connection is flat, in that the curvature

$$F(\nabla) = \nabla \circ \nabla : C^\infty(M, V) \to \Omega^2(M, V)$$

is zero. Flatness ensures that the action of vector fields is actually a Lie algebra action, so that we get an action of all differential operators on M and not just vector fields.

This flatness property suggests the following definition.

Definition 3.5.2 Let Ω^1 denote the C^∞-module that assigns $\Omega^1(M)$ to a manifold M. For F a C^∞-module, the C^∞-*module of k-forms valued in F* is the C^∞-module $\Omega^k(F)$ that assigns to a manifold M, the tensor product $\Omega^k(M, F) = \Omega^k(M) \otimes_{C^\infty(M)} F(M)$.

A *connection* on a C^∞-module F is a map of sheaves on the site Mfld,

$$\nabla : F \to \Omega^1(F),$$

that satisfies the Leibniz rule on every manifold M. A connection is *flat* if it is flat on every manifold M.

A *differentiable vector space* is a C^∞-module equipped with a flat connection. A *map of differentiable vector spaces* $f : F \to G$ is a map of C^∞-modules that intertwines with the flat connections. We denote the set of all such maps by DVS(F, G).

Almost all of the differentiable vector spaces we will consider are concrete in nature. Indeed, most satisfy the formal definition of a being a *concrete* sheaf, which we now explain. Let Set denote the category of sets and let Set(S, T) denote the collection of functions from the set S to the set T. For any sheaf $\mathcal{F}$ on Mfld taking values in Set, there is a natural map $\mathcal{F}(M) \to \text{Set}(M, \mathcal{F}(*))$: each element of the set $\mathcal{F}(M)$ has an associated function from the underlying set M to the set $\mathcal{F}(*)$, the value of the sheaf on a point. We say a sheaf $\mathcal{F}$ is *concrete* if this map $\mathcal{F}(M) \to \text{Set}(M, \mathcal{F}(*))$ is injective. Hence, for a concrete sheaf, one can think of any section on M as just a particular function from M to $\mathcal{F}(*)$; a section is just a "smooth" function on M with values in the set $\mathcal{F}(*)$. As an example of a concrete sheaf, consider the sheaf $\underline{X}$ associated to a smooth manifold X, where $\underline{X}(M) = C^\infty(M, X)$. In this case, a section of the sheaf $\underline{X}$ really is just a function to X. This sheaf just identifies which set-theoretic functions are smooth.

We often work with the differentiable vector space arising from a topological vector space V, which, just like the example $\underline{X}$, simply records which set-theoretic maps are smooth. For this reason, we will normally think of a differentiable vector space $\mathcal{F}$ as being an ordinary vector space, given by its value on a point $\mathcal{F}(*)$, together with extra structure. We will often refer to the sections $\mathcal{F}(M)$ (i.e., the value of the sheaf $\mathcal{F}$ on the manifold M) as the space of smooth maps to the value on a point. If V is a differentiable vector space, we often write $C^\infty(M, V)$ for the space of smooth maps from a manifold M. Abusively, we will also often call a map of differentiable vector spaces simply a *smooth map*.

3.5.2 Differentiable Cochain Complexes

Although we typically work with differentiable vector spaces coming from topological vector spaces, the category DVS is much easier to use – for our purposes – than that of topological vector spaces. The key reason is that they are (essentially) just sheaves on a site, and homological algebra for such objects is very well developed, as we explain in Appendix C.

Definition 3.5.3 A *differentiable cochain complex* is a cochain complex in the category of differentiable vector spaces.

A cochain map $f : V \to W$ of differentiable cochain complexes is a *quasi-isomorphism* if the map $C^\infty(M, V) \to C^\infty(M, W)$ is a quasi-isomorphism for all manifolds M. This condition is equivalent to asking that the map be a quasi-isomorphism at the level of stalks.

We use Ch(DVS) to denote the category of differentiable cochain complexes and cochain maps. It can be enriched over cochain complexes of vector spaces in the usual way.

3.5.3 Differentiable Prefactorization Algebras

We have defined the notion of prefactorization algebra with values in any multicategory. To discuss factorization algebras valued in differentiable vector spaces, we need to define a multicategory structure on differentiable vector spaces. Let us first discuss the multicategory structure on the C^∞-modules.

Definition 3.5.4 Let $V_1, \ldots, V_n, W$ be differentiable vector spaces. A *smooth multilinear map*

$$\Phi : V_1 \times \cdots \times V_n \to W$$

is a C^∞-multilinear map Φ of sheaves that satisfies the following Leibniz rule with respect to the connections on the V_i and W. For every manifold M, and for every $v_i \in V_i(M)$, we require that

$$\nabla \Phi(v_1, \ldots, v_n) = \sum_{i=1}^{n} \Phi(v_1, \ldots, \nabla v_i, \ldots, v_n) \in \Omega^1(M, W).$$

We let $\mathrm{DVS}(V_1, \ldots, V_n \mid W)$ denote this space of smooth multilinear maps.

In more down-to-earth terms, such a Φ is a $C^\infty(M)$-multilinear map $\Phi(M) : V_1(M) \times \cdots \times V_n(M) \to W(M)$ for every manifold M, in a way compatible with the connections and with the maps $V_i(M) \to V_i(N)$ associated to a map $f : N \to M$ of manifolds.

The category of differentiable cochain complexes acquires a multicategory structure from that on differentiable vector spaces, where the multimaps are smooth multilinear maps that are compatible with the differentials. (Here "compatible" means precisely the same thing as it does with ordinary vector spaces and cochain complexes. It means, in the case of unary maps, that we work with cochain maps and not arbitrary linear maps of graded vector spaces.)

Definition 3.5.5 A *differentiable prefactorization algebra* is a prefactorization algebra valued in the multicategory of differentiable cochain complexes.

In words, a differentiable prefactorization algebra $\mathcal{F}$ on a space X assigns a differentiable cochain complex $\mathcal{F}(U)$ to every open subset $U \subset X$, and a smooth multilinear cochain map

$$\mathcal{F}(U_1) \times \cdots \times \mathcal{F}(U_n) \to \mathcal{F}(V),$$

whenever $U_1, \ldots, U_n$ are disjoint opens contained in V.

Note the very different roles played here by the space X, on which the prefactorization algebra lives, and the site Mfld of manifolds, on which the differentiable cochain complexes live. The topology of the space X organizes the algebraic structure we are interested in. By contrast, the geometry of manifolds encoded in Mfld organizes the structure of the vector spaces (or cochain complexes) we work with. Said succinctly, as a substitute for a topology on these vector spaces, we use a sheaf structure over Mfld.

3.5.4 Relationship with Topological Vector Spaces

As we have seen, every topological vector space gives rise to a differentiable vector space. There is a beautiful theory developed in Kriegl and Michor (1997) concerning the precise relationship between topological vector spaces and differentiable vector spaces. These results are discussed in much more detail in Appendix B; we briefly summarize them now.

Let LCTVS denote the category of locally convex Hausdorff topological vector spaces, and continuous linear maps. Let BVS denote the category with the same objects, but whose morphisms are *bounded* linear maps. Every continuous linear map is bounded, but not conversely. These categories have natural enrichments to multicategories, where the multimaps are continuous (respectively, bounded) multilinear maps. The category BVS is equivalent to a full subcategory of LCTVS whose objects are called bornological vector spaces.

Theorem 3.5.6 *The functor dif_t : LCTVS $\to$ DVS restricts to a functor dif_β : BVS $\to$ DVS, which embeds BVS as a full sub-multicategory of DVS.*

In other words: if V, W are topological vector spaces, and if $dif_t(V), dif_t(W)$ denote the corresponding differentiable vector spaces, then the maps in DVS from $dif_t(V)$ to $dif_t(W)$ are the same as bounded linear maps from V to W. More generally, if $V_1, \ldots, V_n$ and W are topological vector spaces, the bounded multilinear maps $\mathrm{BVS}(V_1, \ldots, V_n \mid W)$ are the same as smooth multilinear maps $\mathrm{DVS}(dif_t(V_1), \ldots, dif_t(V_n) \mid dif_t(W))$.

This theorem tells us that we lose very little information if we think of a topological vector space as being a differentiable vector space. We just end up thinking about bounded maps rather than continuous maps.

Remark: On occasion, we will exploit this relationship by using a concept from topological or bornological vector spaces without proffering a differentiable analog. For example, in a few places we talk about a *dense inclusion* and use it to show that a map out of the bigger space is determined by the subspace. In such cases, we are working only with bornological vector spaces and depending upon the fact that dif_β is a full and faithful functor. ◊

So far, however, we have not discussed how completeness of topological vector spaces appears in this context. We need a notion of completeness for a topological vector space that depends only on smooth maps to that vector space. The relevant concept was developed in Kriegl and Michor (1997). We will view the category BVS as being a full subcategory of DVS.

Definition 3.5.7 A topological vector space $V \in$ BVS is c^∞-*complete*, or *convenient*, if every smooth map $c : \mathbb{R} \to V$ has an antiderivative.

We denote the category of convenient vector spaces and bounded linear maps by CVS.

This completeness condition is a little weaker than the one normally studied for topological vector spaces. That is, every complete topological vector space is c^∞-complete.

Proposition 3.5.8 *The full subcategory* dif_c : CVS $\subset$ DVS *is closed under the formation of limits, countable coproducts, and sequential colimits of closed embeddings.*

We give CVS the multicategory structure inherited from BVS. Since BVS is a full sub-multicategory of DVS, so is CVS.

Theorem 3.5.9 *The multicategory structure on* CVS *is represented by a symmetric monoidal structure* $\widehat{\otimes}_\beta$.

This symmetric monoidal structure is called the completed bornological tensor product. If $E, F \in$ CVS, this completed bornological tensor product is written as $E \widehat{\otimes}_\beta F$. The statement that it represents the multicategory structure means that smooth (equivalently, bounded) bilinear maps $f : E_1 \times E_2 \to F$ are the same as bounded linear maps $f' : E_1 \widehat{\otimes}_\beta E_2 \to F$, for objects E_1, E_2, F of CVS.

When it should cause no confusion, we may use the symbol $\otimes$ instead of $\widehat{\otimes}_\beta$ for this tensor product.

3.5.5 Examples from Differential Geometry

Let us now give some examples of differentiable vector spaces. These examples will include the basic building blocks for most of the factorization algebras we will consider.

Let E be a vector bundle on a manifold X. We let $\mathscr{E}(X)$ denote the vector space of smooth sections of E on X, and we let $\mathscr{E}_c(X)$ denote the vector space of compactly supported sections of E on X.

Let us give these vector spaces the structure of differentiable vector spaces, as follows. If M is a manifold, we say a smooth map from M to $\mathscr{E}$ is a smooth section of the bundle $\pi_X^* E$ on $M \times X$. We denote this set of smooth maps $C^\infty(M, \mathscr{E}(X))$. Sending M to $C^\infty(M, \mathscr{E}(X))$ defines a sheaf of C^∞-modules on the site of smooth manifolds with a flat connection, and so a differentiable vector space.

Similarly, we say a smooth map from M to $\mathscr{E}_c(X)$ is a smooth section of the bundle $\pi_X^* E$ on $M \times X$, whose support maps properly to M. Let us denote this set by $C^\infty(M, \mathscr{E}_c(X))$; this defines, again, a sheaf of C^∞-modules on the site of smooth manifolds with a flat connection.

Theorem 3.5.10 *With this differentiable structure, the spaces $\mathscr{E}(X)$ and $\mathscr{E}_c(X)$ are in the full subcategory* CVS *of convenient vector spaces. Further, this differentiable structure is the same as the one that arises from the natural topologies on $\mathscr{E}(X)$ and $\mathscr{E}_c(X)$.*

The proof (like the proofs of all results in this section) are contained in Appendix B, and based heavily on the book Kriegl and Michor (1997).

Let E be a vector bundle on a manifold X. Throughout this book, we will often use the notation $\overline{\mathscr{E}}(X)$ to denote the distributional sections on X, defined by

$$\overline{\mathscr{E}}(X) = \mathscr{E}(X) \otimes_{C^\infty(X)} \mathcal{D}(X),$$

where $\mathcal{D}(X)$ is the space of distributions on X. Similarly, let $\overline{\mathscr{E}}_c(X)$ denote the compactly supported distributional sections of E on X. There are natural inclusions

$$\mathscr{E}_c(X) \hookrightarrow \overline{\mathscr{E}}_c(X) \hookrightarrow \overline{\mathscr{E}}(X),$$

$$\mathscr{E}_c(X) \hookrightarrow \mathscr{E}(X) \hookrightarrow \overline{\mathscr{E}}(X),$$

by viewing smooth functions as distributions.

Let $E^!$ denote the vector bundle $E^\vee \otimes \mathrm{Dens}_X$, which possesses a natural vector bundle pairing ev : $E \otimes E^! \to \mathrm{Dens}_X$. In other words, a smooth section of E and a smooth section of $E^!$ can be paired to produce a smooth density on X. If this smooth density has compact support, it can certainly be integrated to produce a real number. With a little more work, one can show that $\overline{\mathscr{E}}_c(X)$ is the continuous dual to $\mathscr{E}^!(X)$, the space of smooth sections of $E^!$ on X. Likewise, one can show that $\mathscr{E}_c(X)$ is the continuous dual to $\overline{\mathscr{E}}^!(X)$, the distributional sections of $E^!$.

These topological vector spaces $\overline{\mathscr{E}}(X)$ and $\overline{\mathscr{E}}_c(X)$ thus obtain natural differentiable structures via the functor dif_t. We have the following description of the differentiable structure.

Theorem 3.5.11 *For any manifold M, a smooth map from M to $\overline{\mathscr{E}}(X)$ is the same as a smooth linear map*

$$\mathscr{E}^!_c(X) \to C^\infty(M).$$

Similarly, a smooth map from M to $\overline{\mathscr{E}}_c(X)$ is the same as a smooth linear map $\mathscr{E}^!(X) \to C^\infty(M)$.

This is a consequence of Lemma B.5.2.2. See also Section B.2.1 in Appendix B for more discussion.

3.5.6 Multilinear Maps and Enriched Spaces of Maps

The category of differentiable vector spaces has a natural tensor product. In other words, it is a symmetric monoidal category. The tensor product in DVS is very simple: if V, W are differentiable vector spaces, then they are C^∞-modules (by forgetting the flat connections) so $V \otimes_{C^\infty} W$ is another C^∞-module, but it inherits a natural flat connection $\nabla_V \otimes \mathrm{Id}_W + \mathrm{Id}_V \otimes \nabla_W$, so we obtain a new differentiable vector space.

One must be careful with this tensor product, however. From the point of view of analysis, this tensor product is not very meaningful: it is somewhat similar to an uncompleted tensor product of topological vector spaces. For example, if M and N are manifolds, then $C^\infty(M)$ and $C^\infty(N)$ naturally have the structure of differentiable vector spaces. It is *not* true that

$$C^\infty(M) \otimes_{C^\infty} C^\infty(N) = C^\infty(M \times N). \tag{$\dagger$}$$

This issue means that our examples will *not* assign to a disjoint union of opens the tensor product of values on the components.

The multicategory structure on DVS that we use coincides with this symmetric monoidal structure. The multicategory structure is better behaved than the symmetric monoidal structure: when restricted to the full subcategory CVS the multicategory becomes the one associated to the symmetric monoidal structure on CVS, which has good analytical properties (and in particular satisfies the equality in (†)).

Similarly, there is an internal hom in the category of C^∞-modules, and this sheaf hom likewise inherits a natural flat connection. We denote this sheaf hom by $\mathcal{H}om_{DVS}(V, W)$. This sheaf hom is not as well behaved as one would hope, however, and does not capture the concept of "smooth families of maps." In particular, it is *not* true that the value of the sheaf $\mathcal{H}om_{DVS}(V, W)$ on a point is the vector space $DVS(V, W)$ of smooth maps from V to W. For any reasonable definition of the notion of smooth family of maps parametrized by a manifold, a smooth family of maps parametrized by a point should be simply a map, and the self-enrichment given by $\mathcal{H}om_{DVS}(V, W)$ does not satisfy this.

There is, however, another way to enrich the category DVS over itself that better captures the notion of smooth family of maps. (For a careful treatment, see Section B.6 in Appendix B.) Before we define this enrichment, we need the following definition.

Definition 3.5.12 For V a differentiable vector space and M a manifold, let $C^\infty(M, V)$ denote the differentiable vector space whose value on a manifold N is $C^\infty(N \times M, V)$. The flat connection map

$$\nabla_{N, C^\infty(M, V)} : C^\infty(N, C^\infty(M, V)) \to \Omega^1(N, C^\infty(M, V))$$

is the composition of the flat connection

$$\nabla_{N \times M} : C^\infty(M \times N, V) \to \Omega^1(M \times N, V)$$
$$= \Omega^1(M, C^\infty(N, V)) \oplus \Omega^1(N, C^\infty(M, V))$$

with the projection onto $\Omega^1(N, C^\infty(M, V))$.

Here the direct sum decomposition of 1-forms is a consequence of the fact that $T^*_{M \times N}$ splits as the direct sum $\pi_M^* T_M^* \oplus \pi_N^* T_N^*$, where $\pi_M : M \times N \to M$ and $\pi_N : M \times N \to N$ denote the projection maps.

This definition makes the category of differentiable vector spaces into a category cotensored over the category of smooth manifolds. Note that $C^\infty(M)$ defines a ring object in the category DVS, and for any differentiable vector space V, the mapping space $C^\infty(M, V)$ is a $C^\infty(M)$-module.

This construction generalizes in a natural way. If E is a vector bundle on a manifold M and V is a differentiable vector space, then we can define

$$C^\infty(M, E \otimes V) := \mathscr{E}(M) \otimes_{C^\infty(M)} C^\infty(M, V)$$

where on the right-hand side we are taking tensor products in the category DVS over the differentiable ring $C^\infty(M)$. Although, in general, tensor products in DVS are not analytically well behaved, in this case there are no problems because $\mathscr{E}(M)$ is a projective $C^\infty(M)$ module of finite rank. We will denote $C^\infty(M, T^*M \otimes E)$ by $\Omega^1(M, E)$.

Definition 3.5.13 Let $V_1, \ldots, V_n, W$ be differentiable vector spaces. Given a manifold M, a *smooth family of multilinear maps* $V_1 \times \cdots \times V_n \to W$ parametrized by M is an element of

$$\mathrm{DVS}(V_1, \ldots, V_n \mid C^\infty(M, W))$$

where we regard $C^\infty(M, W)$ as being a differentiable vector space using the definition given earlier.

In Section B.6 of Appendix B we shown that there is a differentiable vector space

$$\mathbb{H}\mathrm{om}_{\mathrm{DVS}}(V_1, \ldots, V_n \mid W)$$

defined by saying that its value on a manifold M is $\mathrm{DVS}(V_1, \ldots, V_n \mid C^\infty(M, W))$. The flat connection comes from the natural map

$$\nabla_{C^\infty(M,W)} : C^\infty(M, W) \to \Omega^1(M, W),$$

given by applying the projection in the M-direction to the connection $\nabla_{M \times N, W}$ (compare to the connection on $C^\infty(M, W)$).

This definition makes the category DVS into a multicategory enriched in itself, in such a way that if we evaluate the differentiable Hom-space on a point, we recover the original vector space of smooth maps.

In this text, whenever we consider a smooth family of maps between differentiable vector spaces, we are always referring to this self-enrichment.

3.5.7 Algebras of Observables

The prefactorization algebras we will use for most of the book are built as algebras of functions on the convenient vector spaces $\mathscr{E}(X)$, which for us will mean symmetric algebras on their dual spaces. Recall that $\mathscr{E}(X)$ and $\overline{\mathscr{E}}_c(X)$ are both convenient vector spaces, and that, in the full subcategory CVS $\subset$ DVS of convenient vector spaces, the multicategory is represented by a symmetric monoidal structure $\widehat{\otimes}_\beta$. Recall as well that $\widehat{\otimes}_\pi$ denotes the completed projective tensor product in LCTVS.

Proposition 3.5.14 *Let E be a vector bundle on a manifold X, and let F be a vector bundle on a manifold Y. Let $\mathscr{E}(X)$ denote the convenient vector space of* *smooth sections of E on X, and let $\mathscr{F}(Y)$ denote the convenient vector space of smooth sections of F on Y. Then*

$$\mathscr{E}(X) \widehat{\otimes}_\beta \mathscr{F}(Y) \cong \Gamma(X \times Y, E \boxtimes F) \cong \mathscr{E}(X) \widehat{\otimes}_\pi \mathscr{F}(Y),$$

where $E \boxtimes F$ denotes the external tensor product of vector bundles and Γ denotes smooth sections.

Remark: An alternative approach to the one we've taken is to use the category of nuclear topological vector spaces, with the completed projective tensor product, instead of the category of convenient (or differentiable) vector spaces. Using nuclear spaces raises a number of technical issues, but one immediate issue is the following: although it is true that $C^\infty(X) \widehat{\otimes}_\pi C^\infty(Y) = C^\infty(X \times Y)$, it seems not to be true that the same statement holds if we use compactly supported smooth functions. The problem stems from the fact that the projective tensor product does not commute with colimits, whereas the bornological tensor product does.   ◊

We can define symmetric powers of convenient vector spaces using the symmetric monoidal structure we have described. If, as before, E is a vector bundle on X and U is an open subset of X, this proposition allows us to identify

$$\mathrm{Sym}^n(\mathscr{E}_c(U)) = C_c^\infty(U^n, E^{\boxtimes n})_{S_n}.$$

The symmetric algebra $\mathrm{Sym}\,\mathscr{E}_c(U)$ is defined as usual to be the direct sum of the symmetric powers. It is an algebra in the symmetric monoidal category of convenient vector spaces.

A related construction is the algebra of functions on a differentiable vector space. If V is a differentiable vector space, we can define, as we have seen, the space of linear functionals on V to be the space of maps $\mathbb{Hom}_{\mathrm{DVS}}(V, \mathbb{R})$. Because the category DVS is self-enriched, this is again a differentiable vector space. In a similar way, we can define the space of polynomial functions on V homogeneous of degree n to be the space

$$P_n(V) = \mathbb{Hom}_{\mathrm{DVS}}(\underbrace{V, \ldots, V}_{n \text{ times}} \mid C^\infty)_{S_n}.$$

In other words, we take smooth multilinar maps from n copies of V to $\mathbb{R}$, and then take the S_n coinvariants. The self-enrichment of DVS gives this the structure of differentiable vector space. Concretely, a smoooth map from a manifold M to $P_n(V)$ is

$$C^\infty(M, P_n(V)) = \mathbb{Hom}_{\mathrm{DVS}}(\underbrace{V, \ldots, V}_{n \text{ times}} \mid C^\infty(M))_{S_n}.$$

One can then define the algebra of functions on V by

$$\mathscr{O}(V) = \prod_n P_n(V).$$

(In this formula, we take the product rather than the direct sum, so that $\mathscr{O}(V)$ should be thought of as a space of formal power series on V. One can, of course, also consider the version using the sum.) The space $\mathscr{O}(V)$ is a commutative algebra object of the category DVS in a natural way.

This construction is a very general one, of course: one can define the algebra of functions on any object in any multicategory in the same way.

An important example is the following.

Lemma 3.5.15 *Let E be a vector bundle on a manifold X. Then, $P_n(\mathcal{E}(X))$ is isomorphic as a differentiable vector space to the S_n covariants of the space of compactly supported distributional sections of $E^!$ on X^n.*

Proof We know that the multicategory structure on the full subcategory CVS $\subset$ DVS is represented by a symmetric monoidal category, and that in this symmetric monoidal category,

$$\mathcal{E}(X)^{\widehat{\otimes}_\beta n} = \Gamma(X^n, E^{\boxtimes n}).$$

It follows from this that, for any manifold M, we can identify $C^\infty(M, P_n(\mathcal{E}(X)))$ with the S_n covariants of the space of smooth linear maps

$$\Gamma(X^n, E^{\boxtimes n}) \to C^\infty(M).$$

In fact, because the spaces in this equation are bornological (see Appendix B), such smooth linear maps are the same as continuous linear maps.

We have seen that this space of smooth linear maps is – with its differentiable structure – the same as the space $\mathcal{D}_c(X^n, (E^!)^{\boxtimes n})$ of compactly supported distributional sections of the bundle $(E^!)^{\boxtimes n}$. $\square$

Note that $\mathscr{O}(\mathcal{E}(U))$ is naturally the same as $\mathbb{Hom}_{\mathrm{DVS}}(\mathrm{Sym}\,\mathcal{E}(U), \mathbb{R})$, i.e., it is the dual of the symmetric algebra of $\mathcal{E}(U)$.

3.6 The Factorization Envelope of a Sheaf of Lie Algebras

In this section, we will introduce an important class of examples of factorization algebras. We will show how to construct, for every fine sheaf of Lie algebras $\mathcal{L}$ on a manifold M, a factorization algebra that we call the *factorization envelope*. This construction is our version of the chiral envelope introduced in

Beilinson and Drinfeld (2004). The construction can also be viewed as a natural generalization of the universal enveloping algebra of a Lie algebra. Indeed, we have shown in Section 3.4 that the factorization envelope of the constant sheaf of Lie algebras $\mathfrak{g}$ on $\mathbb{R}$ is the universal enveloping algebra of $\mathfrak{g}$, viewed as a factorization algebra on $\mathbb{R}$.

The factorization envelope plays an important role in our story.

(i) The factorization algebra of observables for a free field theory is an example of a factorization envelope.

(ii) In Section 5.5 in Chapter 5, we will show, following Beilinson and Drinfeld, that the Kac–Moody vertex algebra arises as a (twisted) factorization envelope.

The most important appearance of factorization envelopes appears in our treatment of Noether's theorem at the quantum level, which is covered in Volume 2. We show there that if a sheaf of Lie algebras $\mathcal{L}$ acts on a quantum field theory on a manifold M, then there is a morphism from a twisted factorization envelope of $\mathcal{L}$ to the quantum observables of the field theory. This construction is very useful. For example, for a chiral conformal field theory, it allows one to construct a map of factorization algebras from a Virasoro factorization algebra to the quantum observables. This map induces a map of vertex algebras via our construction in Section 5.3 in Chapter 5.

3.6.1 The Key Idea

Thus, let M be a manifold. Let $\mathcal{L}$ be a fine sheaf of dg Lie algebras on M. Let $\mathcal{L}_c$ denote the cosheaf of compactly supported sections of $\mathcal{L}$. (We restrict to fine sheaves so that taking compactly supported sections is a straightforward operation.)

Remark: Note that, although $\mathcal{L}_c$ is a cosheaf of cochain complexes, and a precosheaf of dg Lie algebras, it is *not* a cosheaf of dg Lie algebras. This is because colimits of dg Lie algebras are not the same as colimits of cochain complexes. ◇

We can view $\mathcal{L}_c$ as a prefactorization algebra valued in the category of dg Lie algebras with symmetric monoidal structure given by direct sum. Indeed, if $\{U_i\}$ is a finite collection of disjoint opens in M contained in the open V, there is a natural map of dg Lie algebras

$$\bigoplus_i \mathcal{L}_c(U_i) = \mathcal{L}_c(\sqcup_i U_i) \to \mathcal{L}_c(V)$$

giving the factorization product.

Taking Chevalley–Eilenberg chains is a symmetric monoidal functor from dg Lie algebras, equipped with the direct sum monoidal structure, to cochain complexes.

Definition 3.6.1 If $\mathcal{L}$ is a sheaf of dg Lie algebras on M, the *factorization envelope* $\mathbb{U}\mathcal{L}$ is the prefactorization algebra obtained by applying the Chevalley–Eilenberg chain functor to $\mathcal{L}_c$, viewed as a prefactorization algebra valued in dg Lie algebras.

Concretely, $\mathbb{U}\mathcal{L}$ assigns to an open subset $V \subset M$ the cochain complex

$$\mathbb{U}\mathcal{L}(V) = C_*(\mathcal{L}_c(V)),$$

where C_* is the Chevalley–Eilenberg chain complex. The factorization structure maps are defined as follows: given a finite collection of disjoint opens $\{U_i\}$ in V, we have

$$\bigotimes_i C_*\mathcal{L}_c(U_i) \cong C_*\left(\bigoplus_i \mathcal{L}_c(U_i)\right) \to C_*\mathcal{L}_c(V).$$

This construction is parallel to the example in Section 3.1.1.

We will see later (see Theorem 6.5.3 in Chapter 6) that this prefactorization algebra is a factorization algebra.

3.6.2 Local Lie Algebras

In practice, we will need an elaboration of this construction that involves a small amount of analysis.

Definition 3.6.2 Let M be a manifold. A *local dg Lie algebra* on M consists of the following data:

(i) A graded vector bundle L on M, whose sheaf of smooth sections will be denoted $\mathcal{L}$.
(ii) A differential operator $d : \mathcal{L} \to \mathcal{L}$, of cohomological degree 1 and square 0.
(iii) An alternating bi-differential operator

$$[-,-] : \mathcal{L}^{\otimes 2} \to \mathcal{L}$$

that endows $\mathcal{L}$ with the structure of a sheaf of dg Lie algebras.

Remark: This definition will play an important role in our approach to interacting classical field theories, developed in Volume 2. ◇

If $U \subset M$, then $\mathcal{L}(U)$ is a topological vector space, because it is the space of smooth sections of a graded vector bundle on U. We would like to form, as above, the Chevalley–Eilenberg chain complex $C_*(\mathcal{L}_c(U))$. The underlying vector space of $C_*(\mathcal{L}_c(U))$ is the (graded) symmetric algebra on $\mathcal{L}_c(U)[1]$. We need to take account of the topological structure on $\mathcal{L}_c(U)$ when we take the tensor powers of $\mathcal{L}_c(U)$.

We explained how to do this in Section 3.5: we define $(\mathcal{L}_c(U))^{\otimes n}$ to be the tensor power defined using the completed projective tensor product on the topological vector space $\mathcal{L}_c(U)$. Concretely, if $L^{\boxtimes n}$ denotes the vector bundle on M^n obtained as the external tensor product, then

$$(\mathcal{L}_c(U))^{\otimes n} = \Gamma_c(U^n, L^{\boxtimes n})$$

is the space of compactly supported smooth sections of $L^{\boxtimes n}$ on U^n. Symmetric (or exterior) powers of $\mathcal{L}_c(U)$ are defined by taking coinvariants of $\mathcal{L}_c(U)^{\otimes n}$ with respect to the action of the symmetric group S_n. The completed symmetric algebra on $\mathcal{L}_c(U)[-1]$ that is the underlying graded vector space of $C_*(\mathcal{L}_c(U))$ is defined using these completed symmetric powers. The Chevalley–Eilenberg differential is continuous, and therefore defines a differential on the completed symmetric algebra of $\mathcal{L}_c(U)[-1]$, giving us the cochain complex $C_*(\mathcal{L}_c(U))$.

Example: Let $\mathfrak{g}$ be a Lie algebra. Consider the sheaf $\Omega^*_{\mathbb{R}} \otimes \mathfrak{g}$ of dg Lie algebras on $\mathbb{R}$, which assigns to the open U the dg Lie algebra $\Omega^*(U) \otimes \mathfrak{g}$. Then, as we saw in detail in Section 3.4, the factorization envelope of this sheaf of Lie algebras encodes the universal enveloping algebra $U\mathfrak{g}$ of $\mathfrak{g}$. Indeed, factorization algebras on $\mathbb{R}$ with an additional "locally constant" property give rise to associative algebras, and the associative algebra associated to $\mathbb{U}(\Omega^*_{\mathbb{R}} \otimes \mathfrak{g})$ recovers the ordinary universal enveloping algebra.

In the same way, for any Lie algebra $\mathfrak{g}$ we can construct a factorization algebra on $\mathbb{R}^n$ as the factorization envelope of $\Omega^*_{\mathbb{R}^n} \otimes \mathfrak{g}$. The resulting factorization algebra is locally constant: it has the property that the inclusion map from one disc to another is a quasi-isomorphism. A theorem of Lurie (2016) tells us that locally constant factorization algebras on $\mathbb{R}^n$ are the same as E_n algebras. The E_n algebra we have constructed is the E_n-enveloping algebra of $\mathfrak{g}$. (See the discussion in Section 6.4 in Chapter 6 for more about locally constant factorization algebras, E_n algebras, and the E_n enveloping functor.) $\Diamond$

3.6.3 Shifted Central Extensions and the Twisted Envelope

Many interesting factorization algebras – such as the Kac–Moody factorization algebra, and the factorization algebra associated to a free field theory – can be constructed from a variant of the factorization envelope construction, which we call the *twisted* factorization envelope.

Definition 3.6.3 Let $\mathcal{L}$ be a sheaf of dg Lie algebras on M. A *k-shifted central extension* of $\mathcal{L}_c$ is a precosheaf of dg Lie algebras $\tilde{\mathcal{L}}_c$ fitting into an exact sequence

$$0 \to \underline{\mathbb{C}}[k] \to \tilde{\mathcal{L}}_c \to \mathcal{L}_c \to 0$$

of precosheaves, where $\underline{\mathbb{C}}[k]$ is the constant presheaf that assigns the one-dimensional vector space $\mathbb{C}[k]$ in degree $-k$ to every open.

If $\mathcal{L}$ is a local Lie algebra, we require in addition that locally there is a splitting

$$\tilde{\mathcal{L}}_c(U) = \mathbb{C}[k] \oplus \mathcal{L}_c(U)$$

such that the differential and bracket maps from $\mathcal{L}_c(U) \to \mathbb{C}[k]$ and $\mathcal{L}_c(U)^{\otimes 2} \to \mathbb{C}[k]$ are continuous.

Definition 3.6.4 In this situation, the *twisted factorization envelope* is the prefactorization algebra $\mathbb{U}\tilde{L}$ that sends an open set U to $C_*(\tilde{\mathcal{L}}_c(U))$. (In the case that $\mathcal{L}$ is a local dg Lie algebra, we use the completed tensor product as above.)

The chain complex $C_*(\tilde{\mathcal{L}}_c(U))$ is a module over chains on the Abelian Lie algebra $\mathbb{C}[k]$ for every k. Thus, we will view the twisted factorization envelope as a prefactorization algebra in modules for $\mathbb{C}[\mathbf{c}]$ where $\mathbf{c}$ has degree $-k-1$.

Under the assumption that $\mathcal{L}$ is a homotopy sheaf, Theorem 6.6.1 shows that the twisted factorization envelope $\mathbb{U}\tilde{L}$ is a factorization algebra over the base ring $\mathbb{C}[\mathbf{c}]$. Of particular interest is the case when $k = -1$, so that the central parameter c is of degree 0.

Let us now introduce some important examples of this construction.

Example: Let $\mathfrak{g}$ be a Lie algebra and consider the local Lie algebra $\Omega_{\mathbb{R}}^* \otimes \mathfrak{g}$ on the real line $\mathbb{R}$, which we will denote $\mathfrak{g}^{\mathbb{R}}$. Given a skew-symmetric, invariant bilinear form ω on $\mathfrak{g}$, there is a natural shifted extension of $\mathfrak{g}_c^{\mathbb{R}}$ where

$$[\alpha \otimes X, \beta \otimes Y]_\omega = \alpha \wedge \beta \otimes [X, Y] + \int_{\mathbb{R}} \alpha \wedge \beta \, \omega(X, Y) \, \mathbf{c},$$

where we use $\mathbf{c}$ to denote the generator in degree 1 of the central extension. Let $\mathbb{U}_\omega \mathfrak{g}^{\mathbb{R}}$ denote the twisted factorization envelope for this central extension. By mimicking the proof of Proposition 3.4.1, one can see that the cohomology of this twisted factorization envelope recovers the enveloping algebra $U\hat{\mathfrak{g}}$ of the central extension of $\mathfrak{g}$ given by ω. ◇

Example: Let $\mathfrak{g}$ be a simple Lie algebra, and let $\langle -, - \rangle_\mathfrak{g}$ denote a symmetric invariant pairing on $\mathfrak{g}$. We define the *Kac–Moody* factorization algebra as follows.

Let Σ be a Riemann surface, and consider the local Lie algebra $\Omega^{0,*} \otimes \mathfrak{g}$ on Σ, which sends an open subset U to the dg Lie algebra $\Omega^{0,*}(U) \otimes \mathfrak{g}$. The differential here is $\bar{\partial}$, so we are describing the Dolbeault analog of the de Rham construction in Section 3.4.

There is a -1-shifted central extension of $\Omega_c^{0,*} \otimes \mathfrak{g}$ defined by the cocycle

$$\omega(\alpha, \beta) = \int_U \langle \alpha, \partial \beta \rangle_\mathfrak{g}$$

where $\alpha, \beta \in \Omega_c^{0,*}(U) \otimes \mathfrak{g}$ and $\partial : \Omega^{0,*} \to \Omega^{1,*}$ is the holomorphic de Rham operator. Note that this is a -1-shifted cocycle because $\omega(\alpha, \beta)$ is zero unless $\deg(\alpha) + \deg(\beta) = 1$.

The twisted factorization envelope $\mathbb{U}_\omega(\Omega^{0,*} \otimes \mathfrak{g})$ is the Kac–Moody factorization algebra. Locally, it recovers the Kac–Moody vertex algebra, just as the real one-dimensional case recovers the enveloping algebra. (We show this in Section 5.5 in Chapter 5.) $\qquad\qquad\Diamond$

Example: In this example we will define a higher-dimensional analog of the Kac–Moody vertex algebra.

Let X be a complex manifold of dimension n. Let $\Omega^*(X)$ denote the de Rham complex. Let $\phi \in \Omega^{n-1,n-1}(X)$ be a closed form.

Then, given any Lie algebra $\mathfrak{g}$ equipped with an invariant pairing $\langle -, - \rangle_\mathfrak{g}$, we can construct a -1-shifted central extension $\widetilde{\mathcal{L}}$ of $\mathcal{L} = \Omega_c^{0,*} \otimes \mathfrak{g}$, defined as above by the cocycle

$$\omega : \alpha \otimes \beta \mapsto \int_X \langle \alpha, \partial \beta \rangle_\mathfrak{g} \wedge \phi.$$

It is easy to verify that ω is a cocycle. (The case of the Kac–Moody extension is when $n = 1$ and ϕ is a constant.) The cohomology class of this cocycle is unchanged if we change ϕ to $\phi + \bar{\partial}\psi$, where $\psi \in \Omega^{n-1,n-2}(X)$ satisfies $\partial\psi = 0$.

The twisted factorization envelope $\mathbb{U}\widetilde{\mathcal{L}} = \mathbb{U}_\omega\mathcal{L}$ is closely related to the Kac–Moody algebra. For instance, if $X = \Sigma \times \mathbb{P}^{n-1}$ where Σ is a Riemann surface, and the form ϕ is a volume form on $\mathbb{P}^{n-1}$, then the pushforward of this factorization algebra to Σ is quasi-isomorphic to the Kac–Moody factorization algebra described earlier. (Let $p : X \to \Sigma$ denote the projection map. The pushforward is defined by

$$(p_*\mathcal{F})(U) = \mathcal{F}(p^{-1}(U))$$

for each $U \subset \Sigma$.)

There is an important special case of this construction. Let X be a complex surface (i.e., $\dim_\mathbb{C}(X) = 2$) and the form ϕ is the curvature of a connection on the canonical bundle of X (i.e., ϕ represents $c_1(X)$). As we will show when we discuss Noether's theorem at the quantum level, if we have a field theory with an action of a local dg Lie algebra $\mathcal{L}$ then a twisted factorization envelopes of Ł will map to the factorization algebra of observables of the theory. One can show that the local dg Lie algebra $\Omega_X^{0,*} \otimes \mathfrak{g}$ acts on a twisted $N = 1$ gauge theory with matter, and that the twisted factorization envelope – with central extension determined by $c_1(X)$ – maps to observables of this theory (following Johansen (1995)). $\diamond$

3.7 Equivariant Prefactorization Algebras

Let M be a topological space with an action of a group G by homeomorphisms. For $g \in G$ and $U \subset M$, we use gU to denote the subset $\{gx \mid x \in U\} \subset M$. We will formulate what it means to have a G-equivariant prefactorization algebra on M. When M is a manifold and G is a Lie group acting smoothly on M, we will formulate the notion of a smoothly G-equivariant prefactorization algebra on M.

3.7.1 Discrete Equivariance

We begin with the case where G is viewed simply as a group (i.e., we do not require any compatibility between the action and a possible smooth structure on G). We give a concrete definition.

Definition 3.7.1 Let $\mathcal{F}$ be a prefactorization algebra on M. We say $\mathcal{F}$ is G-*equivariant* if for each $g \in G$ and each open subset $U \subset M$ we are given isomorphisms

$$\sigma_g : \mathcal{F}(U) \xrightarrow{\cong} \mathcal{F}(gU)$$

satisfying the following conditions.

(i) For all $g, h \in G$ and all opens U, $\sigma_g \circ \sigma_h = \sigma_{gh} : \mathcal{F}(U) \to \mathcal{F}(ghU)$.

(ii) Every σ_g respects the factorization product. That is, for any finite tuple $U_1, \ldots, U_k$ of disjoint opens contained in an open V, the diagram

$$\begin{array}{ccc} \mathcal{F}(U_1) \otimes \cdots \otimes \mathcal{F}(U_n) & \longrightarrow & \mathcal{F}(gU_1) \otimes \cdots \otimes \mathcal{F}(gU_n) \\ \downarrow & & \downarrow \\ \mathcal{F}(V) & \longrightarrow & \mathcal{F}(gV) \end{array}$$

commutes.

There is a more categorical way to phrase this definition. As G acts continuously, each element $g \in G$ provides an endofunctor of the category Opens_M of opens in M. Indeed, g provides an endofunctor of Disj_M and hence, via precomposition, provides an endofunctor on the category of algebras over Disj_M: it sends a prefactorization algebra $\mathcal{A}$ to a prefactorization algebra $g\mathcal{A}$, where $g\mathcal{A}(U) = \mathcal{A}(gU)$. An equivariant prefactorization algebra is then a collection of isomorphisms $\sigma_g : \mathcal{A} \to g\mathcal{A}$ of prefactorization algebras such that $(h\sigma_g) \circ \sigma_h = \sigma_{gh}$.

3.7.2 A Useful Reinterpretation

We want to be able to talk about a prefactorization algebra $\mathcal{F}$ that is equivariant with respect to the smooth action on a manifold M of a Lie group G. In particular, we need to formulate what it means to give a "smooth" action of G on a prefactorization algebra.

As a first step, we will rework the notion of equivariant prefactorization algebra. We will start by constructing a colored operad for each group G. An algebra over this colored operad will recover the preceding definition of a G-equivariant prefactorization algebra.

Definition 3.7.2 Let Disj_M^G denote the colored operad where the set of colors is the set of opens in M and where the operations $\mathrm{Disj}_M^G(U_1, \ldots, U_n \mid V)$ are given by the set

$$\left\{ (g_1, \ldots, g_n) \in G^n \mid \forall i, \ g_i U_i \subset V \text{ and } \forall i \neq j, \ g_i U_i \cap g_j U_j = \emptyset \right\}$$

for any choice of inputs $U_1, \ldots, U_n$ and output V.

There is a map of colored operads $\mathrm{Disj}_M \to \mathrm{Disj}_M^G$, sending each open to itself and sending a nonempty operation $U_1, \ldots, U_n \to V$ to the identity in G^n. Hence, given an algebra $\mathcal{A}$ over Disj_M^G, there is an "underlying" prefactorization algebra on M.

Next, observe that for an open set U, the set $\mathrm{Disj}_M^G(U \mid gU)$ is the coset $g \cdot \mathrm{Stab}(U)$, where the stabilizer subgroup $\mathrm{Stab}(U) \subset G$ consists of elements in G that preserve U. Hence for any algebra $\mathcal{A}$ over Disj_M^G, we see that there is an isomorphism $\sigma_g : \mathcal{A}(U) \xrightarrow{\cong} \mathcal{A}(gU)$ for every $g \in G$ and that they compose in the natural way. Hence the "underlying" prefactorization algebra is G-equivariant.

Now observe that every operation in Disj_M^G factors as a collection of unary operations σ_g arising from the G-action followed by a operation from the "underlying" prefactorization algebra. (If the input opens U_i already sit in the output open V, we are done. Otherwise, pick elements g_i of G that move the input opens inside V and keep them pairwise disjoint.) Hence we obtain the following lemma.

Lemma 3.7.3 *For G a group, every G-equivariant prefactorization algebra on M produces a unique algebra over Disj_M^G, and conversely.*

Some notation will make it easier to understand how the G-action intertwines with the structure of the prefactorization algebra $\mathcal{A}$. If $(g_1, \ldots, g_n) \in \mathrm{Disj}_M^G(U_1, \ldots, U_n \mid V)$, then we denote the associated operation by

$$m_{(g_1, \ldots, g_k)} : \mathcal{A}(U_1) \otimes \cdots \otimes \mathcal{A}(U_k) \to \mathcal{A}(V).$$

It can understood as the composition

$$\bigotimes_i \mathcal{A}(U_i) \xrightarrow{\otimes \sigma_{g_i}} \bigotimes_i \mathcal{A}(g_i U_i) \to \mathcal{A}(V),$$

where the second map is the structure map of the prefactorization algebra and the first map is given by the unary operations arising from the action of G on M.

3.7.3 Smooth Equivariance

From now on, we focus on the situation where G is a Lie group acting smoothly on a smooth manifold M. In this situation, an algebra $\mathcal{A}$ over Disj_M^G has an underlying prefactorization algebra on M but now we want the operations to vary smoothly as the input opens are moved by elements of G. To accomplish this, we need to equip with sets of operations with a "smooth structure" and we need the target category, in which an algebra takes values, to admit a notion of "smoothly varying family of multilinear maps."

Throughout this book, our prefactorization algebras take values in the category DVS of differentiable vector spaces or in the category Ch(DVS) of cochain complexes of differentiable vector spaces. We view these categories as enriched over themselves. In the case of DVS, the self-enrichment is discussed in Section 3.5.6. If V, W are differentiable vector spaces, we denote by $\mathbb{H}\mathrm{om}_{\mathrm{DVS}}(V, W)$ the differentiable vector space of maps from V to W. The key feature of $\mathbb{H}om_{DVS}(V, W)$ is that its value on a point is $DVS(V, W)$ and, more generally, its value on a manifold X is smooth families over X of maps of differentiable vector spaces from V to W. The self-enrichment of DVS leads, in an obvious way, to a self-enrichment of Ch(DVS).

The Colored Operad

Equipping each set of operations $\mathrm{Disj}_M^G(U_1, \ldots, U_n \mid V)$ with some kind of smooth structure is a little subtle. We might hope such a set, which is a subset of G^n, inherits a manifold structure from G^n, but it is easy to produce examples of opens U_i and V where the set of operations is far from being a submanifold of G^n. We take the following approach instead. Given a subset $S \subset X$ of a smooth manifold, there is a set of smooth maps $Y \to X$ that factor through S for any smooth manifold Y. In other words, S provides a sheaf of sets on the site Mfld of smooth manifolds, sometimes known as a "generalized manifold." (See Definition B.2.1 in Appendix B for the site.)

Let Shv(Mfld) denote the category of sheaves (of sets). Then every collection of operations

$$\mathrm{Disj}_M^G(U_1, \ldots, U_n \mid V)$$

provides such a sheaf, so we naturally obtain a colored operad enriched in "generalized manifolds," i.e., Shv(Mfld).

Definition 3.7.4 Let $\widetilde{\mathrm{Disj}}_M^G$ denote the colored operad where the set of colors is the set of opens in M and where the operations $\widetilde{\mathrm{Disj}}_M^G(U_1, \ldots, U_n \mid V)$ are the sheaves on Mfld determined by the subset

$$\mathrm{Disj}_M^G(U_1, \ldots, U_n \mid V) \subset G^n.$$

Let us unpack what this definition means. An algebra $\mathcal{F}$ over $\widetilde{\mathrm{Disj}}_M^G$ in DVS associates to each open $U \subset M$, a differentiable vector space $\mathcal{F}(U)$. To each finite collection of opens $U_1, \ldots, U_n, V$, we have a map of sheaves

$$\widetilde{\mathrm{Disj}}_M^G(U_1, \ldots, U_n \mid V) \to \mathbb{H}om_{DVS}(\mathcal{F}(U_1), \ldots, \mathcal{F}(U_n) \mid \mathcal{F}(V)).$$

Thus, for each smooth manifold Y and for each smooth map $Y \to G^n$ factoring through $\mathrm{Disj}_M^G(U_1, \ldots, U_n \mid V)$, we obtain a section in

$$C^\infty(Y, \mathbb{H}om_{DVS}(\mathcal{F}(U_1), \ldots, \mathcal{F}(U_n) \mid \mathcal{F}(V))),$$

which encodes a Y-family of multilinear morphisms from the $\mathcal{F}(U_i)$ to $\mathcal{F}(V)$.

Note that we can evaluate each sheaf $\widetilde{\mathrm{Disj}}_M^G(U_1, \ldots, U_n \mid V)$ at the point $* \in$ Mfld to obtain the underlying set $\mathrm{Disj}_M^G(U_1, \ldots, U_n \mid V)$. Thus, each algebra over $\widetilde{\mathrm{Disj}}_M^G$ has an underlying G-equivariant prefactorization algebra.

An algebra over this new colored operad is very close to what we need for our purposes in this book. What's missing so far is the ability to differentiate the action of G on the prefactorization algebra to obtain an action of the Lie algebra $\mathfrak{g}$.

Ideally, this action of $\mathfrak{g}$ would simply exist. Instead, we will put it in by hand, as data, since that suffices for our purposes. After giving our definition, however, we will indicate a condition on an algebra over Disj_M^{iG} that provides the desired action automatically.

Derivations

First, we need to introduce the notion of a *derivation* of a prefactorization algebra on a manifold M. We will construct a differential graded Lie algebra of derivations of any prefactorization algebra.

Recall that a derivation of an associative algebra A is a linear map $D : A \to A$ that intertwines in an interesting way with the multiplication map:

$$D(m(a, b)) = m(Da, b) + m(a, Db),$$

a relation known as the Leibniz rule. We simply write down the analog for an algebra over the colored operad Disj_M taking values in cochain complexes.

Definition 3.7.5 A *degree k derivation* of a prefactorization algebra $\mathcal{F}$ is a collection of maps $D_U : \mathcal{F}(U) \to \mathcal{F}(U)$ of cohomological degree k for each open subset $U \subset M$, with the property that for any finite collection of pairwise disjoint opens $U_1, \dots, U_n$, all contained in an open V, and an element $\alpha_i \in \mathcal{F}(U_i)$ for each open, the derivation acts by a Leibniz rule on the structure maps:

$$D_V m_V^{U_1,\dots,U_n}(\alpha_1, \dots, \alpha_n) = \sum_i (-1)^{k(|\alpha_1| + \dots + |\alpha_{i-1}|)} m_V^{U_1,\dots,U_n}$$
$$\times (\alpha_1, \dots, D_{U_i}\alpha_i, \dots, \alpha_n),$$

where the sign is determined by the usual Koszul rule of signs.

Let $\mathrm{Der}^k(\mathcal{F})$ denote the derivations of degree k. It is easy to verify that the derivations of all degrees $\mathrm{Der}^*(\mathcal{F})$ forms a differential graded Lie algebra. The differential is defined by $(dD)_U = [d_U, D_U]$, where d_U is the differential on $\mathcal{F}(U)$. The Lie bracket is defined by $[D, D']_U = [D_U, D'_U]$.

The concept of derivation allows us to talk about the action of a dg Lie algebra $\mathfrak{g}$ on a prefactorization algebra $\mathcal{F}$. Such an action is simply a homomorphism of differential graded Lie algebras from $\mathfrak{g}$ to $\mathrm{Der}^*(\mathcal{F})$.

The Main Definition

We now provide the main definition. As our prefactorization algebra $\mathcal{F}$ takes values in the category of differentiable cochain complexes, it makes sense to differentiate an element of $\mathcal{F}(U)$ for any open U.

Definition 3.7.6 A *smoothly G-equivariant prefactorization algebra* on M is an algebra $\mathcal{F}$ over $\widetilde{\mathrm{Disj}}_M^{.G}$ and an action of the Lie algebra $\mathfrak{g}$ of G on the underlying prefactorization algebra of $\mathcal{F}$ such that for every $X \in \mathfrak{g}$, every operation $(g_1, \ldots, g_n) \in \widetilde{\mathrm{Disj}}_M^{.G}(U_1, \ldots, U_n \mid V)$, and every $1 \leq i \leq n$, we have

$$\frac{\partial}{\partial X_i} m_{g_1, \ldots, g_n}(\alpha_1, \ldots, \alpha_n) = m_{g_1, \ldots, g_n}(\alpha_1, \ldots, X(\alpha_i), \ldots, \alpha_n).$$

On the left-hand side, $\frac{\partial}{\partial X_i}$ indicates the action of the left-invariant vector field on G^k associated to X in the ith factor of G^k and zero in the remaining factors.

Remark: In some cases, an algebra $\mathcal{A}$ over $\widetilde{\mathrm{Disj}}_M^{.G}$ should possess a natural action of $\mathfrak{g}$. We want to recover how $\mathfrak{g}$ acts on each value $\mathcal{A}(U)$ from the action of G on $\mathcal{A}$. Suppose V is an open such that $\overline{gU} \subset V$ for every g in some neighborhood of the identity in G. Then we can differentiate the structure maps $m_g : \mathcal{A}(U) \to \mathcal{A}(V)$ to obtain a map $X : \mathcal{A}(U) \to \mathcal{A}(V)$ for every $X \in \mathfrak{g}$. If $\mathcal{A}(U) = \lim_{V \supset \overline{U}} \mathcal{A}(V)$, we obtain via this limit, an action of $\mathfrak{g}$ on $\mathcal{A}(U)$. Hence, if we have $\mathcal{A}(U) = \lim_{V \supset \overline{U}} \mathcal{A}(V)$ for every open U, we obtain an action of $\mathfrak{g}$ on the underlying prefactorization algebra of $\mathcal{A}$. (The compatibility of the G-action with the structure ensures that we get derivations of $\mathcal{A}$.) $\Diamond$

Here is an example we will revisit.

Example: Let $\mathcal{F}$ be a locally constant, smoothly translation invariant factorization algebra on $\mathbb{R}$, valued in vector spaces. Hence, $A = \mathcal{F}((0, 1))$ has the structure of an associative algebra.

Being locally constant means that for any two intervals $(0, 1)$ and $(t, t + 1)$, there is an isomorphism $\mathcal{F}((0, 1)) \cong \mathcal{F}((t, t + 1))$ coming from the isomorphism $\mathcal{F}((a, b)) \to \mathcal{F}(\mathbb{R})$ associated to inclusion of an interval into $\mathbb{R}$. As $\mathcal{F}$ is translation invariant, there is *another* isomorphism $\mathcal{F}((0, 1)) \to \mathcal{F}((t, t+1))$ for any $t \in \mathbb{R}$. Composing these two isomorphism yields an action of the group $\mathbb{R}$ on $A = \mathcal{F}((0, 1))$. One can check that this is an action on associative algebras, not just vector spaces.

The fact that $\mathcal{F}$ is both smoothly translation-invariant and locally constant means that the action of $\mathbb{R}$ on A is smooth, and thus it differentiates to an infinitesimal action of the Lie algebra $\mathbb{R}$ on A by derivations. The basis element $\frac{\partial}{\partial x}$ of $\mathbb{R}$ becomes a derivation H of A, called the *Hamiltonian*.

In the case that $\mathcal{F}$ is the cohomology of the factorization algebra of observables of the free scalar field theory on $\mathbb{R}$ with mass m, we will see in Section 4.3 in Chapter 4 that the algebra A is the Weyl algebra, generated by $p, q, \hbar$ with commutation relation $[p, q] = \hbar$. The Hamiltonian is given by $H(a) = \frac{1}{2\hbar}[p^2 - m^2 q^2, a]$. $\diamond$

PART II

First examples of field theories and
their observables

4

Free Field Theories

4.1 The Divergence Complex of a Measure

In this section, we revisit the ideas and constructions from Chapter 2. Recall that in that chapter, we studied the divergence operator associated to a Gaussian measure on a finite-dimensional vector space and then generalized this construction to the infinite-dimensional vector spaces that occur in field theory. We used these ideas to define a vector space $H^0(\text{Obs}^q)$ of quantum observables.

Our goal here is to lift these ideas to the level of cochain complexes. We will find a cochain complex Obs^q such that $H^0(\text{Obs}^q)$ is the vector space we constructed in Chapter 2. To make the narrative as clear as possible, we will recapitulate our approach there.

4.1.1 The Divergence Complex of a Finite-Dimensional Measure and Its Classical Limit

We start by considering again Gaussian integrals in finite dimensions. Let M be a smooth manifold of dimension n, and let ω_0 be a smooth measure on M. Let f be a function on M. (For example, M could be a vector space, ω_0 the Lebesgue measure and f a quadratic form.) The divergence operator for the measure $e^{f/\hbar}\omega_0$ is a map

$$\text{Div}_\hbar : \text{Vect}(M) \to C^\infty(M)$$
$$X \mapsto \hbar^{-1}(Xf) + \text{Div}_{\omega_0} X.$$

One way to describe the divergence operator is to contract with the volume form $e^{f/\hbar}\omega_0$ to identify $\text{Vect}(M)$ with $\Omega^{n-1}(M)$ and $C^\infty(M)$ with $\Omega^n(M)$. Explicitly, the contraction of a function ϕ with $e^{f/\hbar}\omega_0$ is the volume form $\phi e^{f/\hbar}\omega_0$, and the contraction of a vector field X with $e^{f/\hbar}\omega_0$ is the $n-1$-form

$\iota_X e^{f/\hbar}\omega_0$. Under this identification, the divergence operator is simply the de Rham operator from $\Omega^{n-1}(M)$ to $\Omega^n(M)$.

The de Rham operator, of course, is part of the de Rham complex. In a similar way, we can define the *divergence complex*, as follows. Let

$$\mathrm{PV}^i(M) = C^\infty(M, \wedge^i TM)$$

denote the space of polyvector fields on M. The divergence complex is the complex

$$\cdots \to \mathrm{PV}^i(M) \xrightarrow{\mathrm{Div}_\hbar} \mathrm{PV}^1(M) \xrightarrow{\mathrm{Div}_\hbar} \mathrm{PV}^0(M)$$

where the differential

$$\mathrm{Div}_\hbar : \mathrm{PV}^i(M) \to \mathrm{PV}^{i-1}(M)$$

is defined so that the diagram

$$
\begin{array}{ccc}
\mathrm{PV}^i(M) & \xrightarrow{\vee e^{f/\hbar}\omega_0} & \Omega^{n-i}(M) \\
\downarrow{\scriptstyle\mathrm{Div}_\hbar} & & \downarrow{\scriptstyle d_{dR}} \\
\mathrm{PV}^{i-1}(M) & \xrightarrow{\vee e^{f/\hbar}\omega_0} & \Omega^{n-i+1}(M)
\end{array}
$$

commutes. Here we use $\vee e^{f/\hbar}\omega_0$ to denote the contraction map; it is the natural extension to polyvector fields of the contractions we defined for functions and vector fields.

In summary, after contracting with the volume $e^{f/\hbar}\omega_0$, the divergence complex becomes the de Rham complex. It is easy to check that, as maps from $\mathrm{PV}^i(M)$ to $\mathrm{PV}^{i-1}(M)$, we have

$$\mathrm{Div}_\hbar = \vee \hbar^{-1} df + \mathrm{Div}_{\omega_0}$$

where $\vee df$ is the operator of contracting with the 1-form df. In the $\hbar \to 0$ limit, the dominant term is $\vee \hbar^{-1} df$.

More precisely, by multiplying the differential by $\hbar$, we see that there is a flat family of cochain complexes over the algebra $\mathbb{C}[\hbar]$ with the properties that

(i) The family is isomorphic to the divergence complex when $\hbar \neq 0$.
(ii) At $\hbar = 0$, the cochain complex is

$$\cdots \to \mathrm{PV}^2(M) \xrightarrow{\vee df} \mathrm{PV}^1(M) \xrightarrow{\vee df} \mathrm{PV}^0(M).$$

Note that this second cochain complex is a differential graded algebra, which is not the case for the divergence complex.

The image of the map $\vee df : PV^1(M) \to PV^0(M)$ is the ideal cutting out the critical locus. Indeed, this whole complex is the Koszul complex for the equations cutting out the critical locus. This observation leads to the following definition.

Definition 4.1.1 The *derived critical locus of f* is the locally dg ringed space whose underlying manifold is M and whose dg commutative algebra of functions is the complex $PV^*(M)$ with differential $\vee df$.

Remark: We will call a manifold with a nice sheaf of dg commutative algebras a *dg manifold*. Since the purpose of this section is motivational, we will not develop a theory of such dg manifolds. We are using the concrete object here as a way to think about the derived geometry of this situation. (More details on derived geometry, from a different point of view, are discussed in Volume 2.) The crucial point is that the dg algebra keeps track of the behavior of df, including higher homological data. ◊

Here is a bit more motivation for our terminology. Let $\Gamma(df) \subset T^*M$ denote the graph of df. The ordinary critical locus of f is the intersection of $\Gamma(df)$ with the zero-section $M \subset T^*M$. The derived critical locus is defined to be the derived intersection. In derived geometry, functions on derived intersections are defined by derived tensor products:

$$C^\infty(\mathrm{Crit}^h(f)) = C^\infty(\Gamma(df)) \otimes^{\mathbb{L}}_{C^\infty(T^*M)} C^\infty(M).$$

By using a Koszul resolution of $C^\infty(M)$ as a module for $C^\infty(T^*M)$, one finds a quasi-isomorphism of dg commutative algebras between this derived intersection and the complex $PV^*(M)$ with differential $\vee df$.

In short, we find that the $\hbar \to 0$ limit of the divergence complex is the dg commutative algebra of functions on the derived critical locus of f.

An important special case of this relationship is when the function f is zero. In that case, the derived critical locus of f has as functions the algebra $PV^*(M)$ with zero differential. We view this algebra as the functions on the graded manifold $T^*[-1]M$. The derived critical locus for a general function f can be viewed as a deformation of $T^*[-1]M$ obtained by introducing a differential $\vee df$.

4.1.2 A Different Construction

We will define the prefactorization algebra of observables of a free scalar field theory as a divergence complex, just like we defined H^0 of observables to be given by functions modulo divergences in Chapter 2. It turns out that there is a slick way to write this prefactorization algebra as a twisted factorization envelope of a certain sheaf of Heisenberg Lie algebras. We will explain this point in a finite-dimensional toy model, and then use the factorization envelope picture to define the prefactorization algebra of observables of the field theory in the next section.

Let V be a vector space, and let $q : V \to \mathbb{R}$ be a quadratic function on V. Let ω_0 be the Lebesgue measure on V. We want to understand the divergence complex for the measure $e^{q/\hbar}\omega_0$. The construction is quite general: we do not need to assume that q is nondegenerate.

The derived critical locus of the function q is a *linear* dg manifold. (The Jacobi ideal is generated by linear equations.) Linear dg manifolds are equivalent to cochain complexes: any cochain complex B gives rise to the linear dg manifold whose functions are the symmetric algebra on the dual of B.

The derived critical locus of q is described by the cochain complex W given by

$$V \xrightarrow{\partial/\partial q} V^*[-1],$$

where the differential sends $v \in V$ to the linear functional $\frac{\partial}{\partial q}v = q(v, -)$.

Note that W is equipped with a graded antisymmetric pairing $\langle -, - \rangle$ of cohomological degree -1, defined by pairing V and V^*. In other words, W has a symplectic pairing of cohomological degree -1. We let

$$\mathcal{H}_W = \mathbb{C} \cdot \hbar[-1] \oplus W,$$

where $\mathbb{C} \cdot \hbar$ indicates a one-dimensional vector space with basis $\hbar$. We give the cochain complex $\mathcal{H}_W$ a Lie bracket by saying that

$$[w, w'] = \hbar \langle w, w' \rangle.$$

Thus, $\mathcal{H}_W$ is a shifted-symplectic version of the Heisenberg Lie algebra of an ordinary symplectic vector space.

Consider the Chevalley–Eilenberg chain complex $C_* \mathcal{H}_W$. This cochain complex is defined to be the symmetric algebra of the underlying graded vector space $\mathcal{H}_W[1] = W[1] \oplus \mathbb{C} \cdot \hbar$ but equipped with a differential determined by the Lie bracket. Since the pairing on W identifies $W[1] = W^*$, we can identify

$$C_*(\mathcal{H}_W) = (\mathrm{Sym}\,(W^*)\,[\hbar], d_{\mathrm{CE}})$$

where d_{CE} denotes the differential. Since, as a graded vector space, $W = V \oplus V^*[-1]$, we have a natural identification

$$C_n(\mathcal{H}_W) = \mathrm{PV}^n(V)[\hbar]$$

where $\mathrm{PV}^*(V)$ refers to polyvector fields on V with polynomial coefficients and where, as before, we place $\mathrm{PV}^n(V)$ in degree $-n$.

Lemma 4.1.2 *The differential on* $C_*(\mathcal{H}_W)$ *is, under this identification, the operator*

$$\hbar \operatorname{Div}_{e^{q/\hbar}\omega_0} : \mathrm{PV}^i(V)[\hbar] \to \mathrm{PV}^{i-1}(V)[\hbar],$$

where ω_0 *is the Lebesgue measure on* V, *and* q *is the quadratic function on* V *used to define the differential on the complex* W.

Proof The proof is an explicit calculation, which we leave to interested readers. The calculation is facilitated by choosing a basis of V in which we can explicitly write both the divergence operator and the differential on the Chevalley–Eilenberg complex $C_*(\mathcal{H}_W)$. □

In what follows, we define the prefactorization algebra of observables of a free field theory as a Chevalley–Eilenberg chain complex of a certain Heisenberg Lie algebra, constructed as in this lemma.

4.2 The Prefactorization Algebra of a Free Field Theory

In this section, we construct the prefactorization algebra associated to any free field theory. We will concentrate, however, on the free scalar field theory on a Riemannian manifold. We will show that, for one-dimensional manifolds, this prefactorization algebra recovers the familiar Weyl algebra, the algebra of observables for quantum mechanics. In general, we will show how to construct correlation functions of observables of a free field theory and check that these agree with how physicists define correlation functions.

4.2.1 The Classical Observables of the Free Scalar Field

We start with a standard example: the free scalar field. Let M be a Riemannian manifold with metric g, so M is equipped with a natural density. We will use this natural density both to integrate functions and also to provide an isomorphism between functions and densities, and we will use this isomorphism implicitly from hereon.

The scalar field theory has smooth functions as fields, and we use the notation $\phi \in C^\infty(M)$ for an arbitrary field. The action functional of the theory is

$$S(\phi) = \int_M \phi \triangle \phi,$$

where $\triangle$ is the Laplacian on M. (Normally we will reserve the symbol $\triangle$ for the Batalin–Vilkovisky Laplacian, but that's not necessary in this section.)

This functional is not well defined on a field ϕ unless $\phi \triangle \phi$ is integrable, but it is helpful to bear in mind that for classical field theory, what is crucial is the Euler–Lagrange equation, which in this case is $\triangle \phi = 0$ and which is thus well-defined on all smooth functions. The action can be viewed here as a device for producing these partial differential equations.

If $U \subset M$ is an open subset, then the space of solutions to the equation of motion on U is the space of harmonic functions on U. In this book, we always consider the *derived* space of solutions of the equation of motion. (For more details about the derived philosophy, readers should consult Volume 2.) In this simple situation, the derived space of solutions to the free field equations, on an open subset $U \subset M$, is the two-term complex

$$\mathscr{E}(U) = \left(C^\infty(U) \xrightarrow{\triangle} C^\infty(U)[-1] \right),$$

where the complex is concentrated in cohomological degrees 0 and 1. (The bracket $[-1]$ denotes "shift up by 1.")

The observables of this classical field theory are simply the functions on this derived space of solutions to the equations of motion. As this derived space is a cochain complex (and hence linear in nature), it is natural to work with the polynomial functions. (One could work with more complicated types of functions, but the polynomial functions provide a concrete and useful collection of observables.) To be explicit, the classical observables are the symmetric algebra on the dual space to the fields.

Let's make this idea precise, using the technology we introduced in Section 3.5 of Chapter 3. The space $\mathscr{E}(M)$ has the structure of differentiable cochain complex (essentially, it is a sheaf of vector spaces on the site of smooth manifolds). We define the space of polynomial functions homogeneous of degree n on $\mathscr{E}(M)$ to be the space

$$P_n(\mathscr{E}(M)) = \mathbb{H}\mathrm{om}_{\mathrm{DVS}}(\mathscr{E}(M), \dots, \mathscr{E}(M)|\mathbb{R})_{S_n}.$$

In other words, we consider smooth multilinear maps from n copies of $\mathscr{E}(M)$, and then we take the S_n-coinvariants. The algebra of all polynomial functions on $\mathscr{E}(M)$ is the space $P(\mathscr{E}(M)) = \bigoplus_n P_n(\mathscr{E}(M))$.

As we discussed in Section 3.5 of Chapter 3, we can identify

$$P_n(\mathscr{E}(M)) = \mathcal{D}_c(M^n, (E^!)^{\boxtimes n})_{S_n}$$

as the S_n-coinvariants of the space of compactly supported distributional sections of the bundle $(E^!)^{\boxtimes n}$ on M^n. In general, if $\mathscr{E}(M)$ is sections of a graded bundle E, then $E^!$ is $E^\vee \otimes \mathrm{Dens}$. In the case at hand, the bundle $E^!$ is two copies of the trivial bundle, one in degree -1 and one in degree 0.

For example, the space $P_1(\mathscr{E}(M)) = \mathscr{E}(M)^\vee$ of smooth linear functionals on $\mathscr{E}(M)$ is the space

$$\mathscr{E}(M)^\vee = \left(\mathcal{D}_c(M)^{-1} \xrightarrow{\triangle} \mathcal{D}_c(M)^0 \right),$$

where $\mathcal{D}_c(M)$ indicates the space of compactly supported distributions on M.

We also want to keep track of where measurements are taking place on the manifold M, so we will organize the observables by where they are supported. The classical observables with *support in* $U \subset M$ are then the symmetric algebra of

$$\mathscr{E}(U)^\vee = \left(\mathcal{D}_c(U)[1] \xrightarrow{\triangle} \mathcal{D}_c(U) \right),$$

where $\mathcal{D}_c(U)$ indicates the space of compactly supported distributions on U. The complex $\mathscr{E}(U)^\vee$ is thus the graded, smooth linear dual to the two-term complex $\mathscr{E}(U)$ given earlier. Note that, as the graded dual to $\mathscr{E}$, this complex is concentrated in cohomological degrees 0 and -1. These are precisely the observables that depend only on the behavior of the field ϕ on the open set U.

Thus, as a first pass, one would want to define the classical observables as the symmetric algebra on $\mathscr{E}(U)^\vee$. This choice leads, however, to difficulties defining the quantum observables. When we work with an interacting theory, these difficulties can be surmounted only by using the techniques of renormalization. For a free field theory, though, there is a much simpler solution, as we now explain.

Recall from Section 3.5 in Chapter 3 that $E^!$ denotes the vector bundle $E^\vee \otimes \mathrm{Dens}_M$. Our identification between densities and functions then produces an isomorphism

$$\mathscr{E}_c^!(U) \cong \left(C_c^\infty(U)[1] \to C_c^\infty(U) \right),$$

for compactly supported sections of $\mathscr{E}^!$. Note that there is a natural map of cochain complexes $\mathscr{E}_c^!(U) \to \mathscr{E}(U)^\vee$, given by viewing a compactly supported function as a distribution.

Lemma 4.2.1 *The inclusion map* $\mathscr{E}_c^!(U) \to \mathscr{E}(U)^\vee$ *is a cochain homotopy equivalence of differentiable cochain complexes.*

Proof This assertion is a special case of a general result proved in Appendix D. Note that by differentiable homotopy equivalence we mean that there is an "inverse" map $\mathscr{E}(U)^\vee \to \mathscr{E}_c^!(U)$, and differentiable cochain homotopies between the two composed maps and the identity maps. "Differentiable" here means that all maps are in the category DVS of differentiable vector spaces.

As these are convenient cochain complexes, suffices to construct a continuous homotopy equivalence.   □

This lemma says that, since we are working homotopically, we can replace a distributional linear observable by a smooth linear observable. In other words, any distributional observable that is closed in the cochain complex $\mathscr{E}(U)^{\vee}$ is chain homotopy equivalent to a closed smooth observable.We think of the smooth linear observables as "smeared." For example, we can replace a delta function δ_x by some bump function supported near the point x.

The observables we will work with is the space of "smeared observables," defined by

$$\mathrm{Obs}^{cl}(U) = \mathrm{Sym}(\mathscr{E}_c^!(U)) = \mathrm{Sym}(C_c^{\infty}(U)[1] \xrightarrow{\triangle} C_c^{\infty}(U)),$$

the symmetric algebra on $\mathscr{E}_c^!(U)$. As we explained in Section 3.5 in Chapter 3, this symmetric algebra is defined using the natural symmetric monoidal structure on the full subcategory CVS $\subset$ DVS of convenient vector spaces. Concretely, we can identify

$$\mathrm{Sym}^n(\mathscr{E}_c^!(U)) = C_c^{\infty}(U^n, [E^!]^{\otimes n})_{S_n}.$$

In other words, $\mathrm{Sym}^n \mathscr{E}_c^!(U)$ is the subspace of $P_n(\mathscr{E}(U))$ defined by taking all distributions to be smooth functions with compact support.

Lemma 4.2.2 *The map* $\mathrm{Obs}^{cl}(U) \to P(\mathscr{E}(U))$ *is a homotopy equivalence of cochain complexes of differentiable vector spaces.*

Proof It suffices to show that the map $\mathrm{Sym}^n \mathscr{E}_c^!(U) \to P_n(\mathscr{E}(U))$ is a differentiable homotopy equivalence for each n. But for this, it suffices to observe that the map

$$C_c^{\infty}(U, [E^!]^{\otimes n}) \to \mathcal{D}_c(U, [E^!]^{\otimes n})$$

is an S_n-equivariant differentiable homotopy equivalence.   □

In parallel to the previous lemma, this lemma says that, since we are working homotopically, we can replace a distributional polynomial observable (given by integration against some distribution on U^n) by a smooth polynomial observable (given by integration against a smooth function on U^n).

4.2.2 Interpreting this Construction

Let us describe the cochain complex $\mathrm{Obs}^{cl}(U)$ more explicitly, to clarify the relationship with what we discussed in Chapter 2. The complex $\mathrm{Obs}^{cl}(U)$ looks like

$$\cdots \rightarrow \wedge^2 C_c^\infty(U) \otimes \text{Sym}\, C_c^\infty(U) \rightarrow C_c^\infty(U) \otimes \text{Sym}\, C_c^\infty(U) \rightarrow \text{Sym}\, C_c^\infty(U).$$

All tensor products appearing in this expression are completed tensor products in the category of convenient vector spaces.

We should interpret $\text{Sym}\, C_c^\infty(U)$ as being an algebra of polynomial functions on $C^\infty(U)$, using (as we explained previously) the Riemannian volume form on U to identify $C_c^\infty(U)$ with a subspace of the dual of $C^\infty(U)$. There is a similar interpretation of the other terms in this complex using the geometry of the space $C^\infty(U)$ of fields. Let $T_c C^\infty(U)$ refer to the sub-bundle of the tangent bundle of $C^\infty(U)$ given by the subspace $C_c^\infty(U) \subset C^\infty(U)$. An element of a fiber of $T_c C^\infty(U)$ is a first-order variation of a field that is zero outside of a compact set. This subspace $T_c C^\infty(U)$ defines an integrable foliation on $C^\infty(U)$, and this foliation can be defined if we replace $C^\infty(U)$ by any sheaf of spaces.

Then, we can interpret $C_c^\infty(U) \otimes \text{Sym}\, C_c^\infty(U)$ as a space of polynomial sections of $T_c C^\infty(U)$. Similarly, $\wedge^k C_c^\infty(U) \otimes \text{Sym}\, C_c^\infty(U)$ should be interpreted as a space of polynomial sections of the bundle $\wedge^k T_c C^\infty(U)$.

That is, if $\text{PV}_c(C^\infty(U))$ refers to polynomial polyvector fields on $C^\infty(U)$ along the foliation given by $T_c C^\infty(U)$, we have

$$\text{Obs}^{cl}(U) = \text{PV}_c(C^\infty(U)).$$

So far, this is just an identification of graded vector spaces. We need to explain how to identify the differential. Roughly speaking, the differential on $\text{Obs}^{cl}(U)$ corresponds to the differential on $\text{PV}_c(C^\infty(U))$ obtain this complex is given by contracting with the 1-form dS for the function

$$S(\phi) = \tfrac{1}{2} \int \phi \triangle \phi,$$

the action functional. Before making this idea precise, we recall the finite-dimensional model of a quadratic function $Q(x) = (x, Ax)$ on a vector space V. The contraction of the 1-form dQ with a tangent vector $v_0 \in T_0 V$ gives the linear functional $x \mapsto (v_0, Ax)$. As S is quadratic in ϕ, we expect that the contraction of dS with a tangent vector $\phi_0 in C^\infty(U)$ is the linear functional

$$\phi \mapsto \tfrac{1}{2} \int \phi_0 \triangle \phi.$$

But we run into an issue here: the functional S is not well defined for all fields ϕ, because the integral may not converge, and similarly the linear functional above is not well-defined for arbitrary smooth functions ϕ_0 and ϕ. However, the expression

$$\frac{\partial S}{\partial \phi_0}(\phi) = \frac{1}{2} \int \phi_0 \triangle \phi$$

does make sense for any $\phi \in C^\infty(U)$ and $\phi_0 \in C_c^\infty(U)$. (Note that we now only consider tangent vectors ϕ_0 with compact support.) In other words, the desired 1-form dS does not make sense as a section of the cotangent bundle of $C^\infty(U)$, but it is well defined as a section of the space $T_c^* C^\infty(U)$, the dual of the subbundle $T_c C^\infty(U) \subset TC^\infty(U)$ describing vector fields along the leaves. This leafwise 1-form is closed. Such 1-forms are the kinds of things we can contract with elements of $PV_c(C^\infty(U))$. The differential on $PV_c(C^\infty(U))$ thus matches the differential on Obs^{cl} given by contracting with dS.

4.2.3 General Free Field Theories

With this example in mind, we introduce a general definition.

Definition 4.2.3 Let M be a manifold. A *free field theory* on M is the following data:

(i) A graded vector bundle E on M, whose sheaf of sections will be denoted $\mathcal{E}$, and whose compactly supported sections will be denoted $\mathcal{E}_c$.

(ii) A differential operator d $: \mathcal{E} \to \mathcal{E}$, of cohomological degree 1 and square zero, making $\mathcal{E}$ into an elliptic complex.

(iii) Let $E^! = E^\vee \otimes \text{Dens}_M$, and let $\mathcal{E}^!$ be the sections of $E^!$. Let d$^!$ be the differential on $\mathcal{E}^!$ which is the formal adjoint to the differential on $\mathcal{E}$. Note that there is a natural pairing between $\mathcal{E}_c(U)$ and $\mathcal{E}^!(U)$, and this pairing is compatible with differentials.

We require an isomorphism $E \to E^![-1]$ compatible with differentials, with the property that the induced pairing of cohomological degree -1 on each $\mathcal{E}_c(U)$ is graded antisymmetric.

The complex $\mathcal{E}(U)$ is the derived version of the solutions to the equations of motion of the theory on an open subset U. More motivation for this definition is presented in Volume 2.

Note that the equations of motion for a free theory are always linear, so that the space of solutions is a vector space. Similarly, the derived space of solutions of the equations of motion of a free field theory is a cochain complex, which is a linear derived stack. The cochain complex $\mathcal{E}(U)$ should be thought of as the derived space of solutions to the equations of motion on an open subset U. As we explain in Volume 2, the pairing on $\mathcal{E}_c(U)$ arises from the fact that the equations of motion of a field theory are not arbitrary differential equations, but describe the critical locus of an action functional.

For example, for the free scalar field theory on a manifold M with mass m, we have, as previously,

$$\mathscr{E} = C^\infty(U) \xrightarrow{\Delta+m^2} C^\infty(U)[-1].$$

Our convention is that Δ is a nonnegative operator, so that on $\mathbb{R}^n$, $\Delta = -\sum \frac{\partial}{\partial x_i}^2$. The pairing on $\mathscr{E}_c(U)$ is defined by

$$\left\langle \phi^0, \phi^1 \right\rangle = \int_M \phi^0 \phi^1$$

for ϕ^k in the graded piece $C^\infty(U)[k]$.

As another example, let us describe Abelian Yang–Mills theory (with gauge group $\mathbb{R}$) in this language. Let M be a manifold of dimension 4. If $A \in \Omega^1(M)$ is a connection on the trivial $\mathbb{R}$-bundle on a manifold M, then the Yang–Mills action functional applied to A is

$$S_{YM}(A) = -\tfrac{1}{2} \int_M dA \wedge *dA = \tfrac{1}{2} \int_M A(d * d)A.$$

The equations of motion are that $d * dA = 0$. There is also gauge symmetry, given by $X \in \Omega^0(M)$, which acts on A by $A \to A + dX$. The complex $\mathscr{E}$ describing this theory is

$$\mathscr{E} = \Omega^0(M)[1] \xrightarrow{d} \Omega^1(M) \xrightarrow{d*d} \Omega^3(M)[-1] \xrightarrow{d} \Omega^4(M)[-2].$$

We explain how to derive this statement in Volume 2. For now, note that $H^0(\mathscr{E})$ is the space of those $A \in \Omega^1(M)$ that satisfy the Yang–Mills equation $d * dA = 0$, modulo gauge symmetry.

For any free field theory with cochain complex of fields $\mathscr{E}$, we define the classical observables of the theory as

$$\mathrm{Obs}^{cl}(U) = \mathrm{Sym}(\mathscr{E}_c^!(U)) = \mathrm{Sym}(\mathscr{E}_c(U)[1]).$$

It is clear that classical observables form a prefactorization algebra (recall the example in Section 3.1.1). Indeed, $\mathrm{Obs}^{cl}(U)$ is a commutative differential graded algebra for every open U. If $U \subset V$, there is a natural algebra homomorphism

$$i_V^U : \mathrm{Obs}^{cl}(U) \to \mathrm{Obs}^{cl}(V),$$

which on generators is the extension-by-zero map $C_c^\infty(U) \to C_c^\infty(V)$.

If $U_1, \ldots, U_n \subset V$ are disjoint open subsets, the prefactorization structure map is the continuous multilinear map

$$
\begin{array}{ccc}
\mathrm{Obs}^{cl}(U_1) \times \cdots \times \mathrm{Obs}^{cl}(U_n) & \to & \mathrm{Obs}^{cl}(V) \\
(\alpha_1, \ldots, \alpha_n) & \mapsto & \prod_{i=1}^{n} i_V^{U_i} \alpha_i.
\end{array}
$$

The product denotes the product in the symmetric algebra on V. On a field ϕ, this observable takes the value

$$
\prod_{i=1}^{n} i_V^{U_i} \alpha_i(\phi) = \alpha_1(\phi|_{U_1}) \cdots \alpha_n(\phi|_{U_n}),
$$

which is the product of the value that each observable α_i takes on ϕ restricted to U_i.

4.2.4 The One-Dimensional Case, in Detail

This space is particularly simple in dimension 1. Indeed, we recover the usual answer at the level of cohomology.

Lemma 4.2.4 *If $U = (a, b) \subset \mathbb{R}$ is an interval in $\mathbb{R}$, then the algebra of classical observables for the free field with mass $m \geq 0$ has cohomology*

$$
H^*(\mathrm{Obs}^{cl}((a, b))) = \mathbb{R}[p, q],
$$

the polynomial algebra in two variables.

Proof The idea is the following. The equations of motion for free classical mechanics on the interval (a, b) are that the field ϕ satisfies $(\Delta + m^2)\phi = 0$. This space is two dimensional, spanned by the functions $\{e^{\pm mx}\}$, for $m > 0$, and by the functions $\{1, x\}$, for $m = 0$. Classical observables are functions on the space of solutions to the equations of motion. We would this expect that classical observables are a polynomial algebra in two generators.

We need to be careful, however, because we use the *derived* version of the space of solutions to the equations of motion. We will show that the complex

$$
\mathcal{E}_c^!((a, b)) = \left(C_c^{\infty}((a, b))^{-1} \xrightarrow{\Delta + m^2} C_c^{\infty}((a, b))^0 \right)
$$

is homotopy equivalent to the complex $\mathbb{R}^2$ situated in degree 0. Since the algebra $\mathrm{Obs}^{cl}((a, b))$ of observables is defined to be the symmetric algebra on this complex, this will imply the result. Without loss of generality, we can take $a = -1$ and $b = 1$.

First, let us introduce some notation. We start in the complex $\mathscr{E}$ of fields. If $m = 0$, let $\phi_q(x) = 1$ and $\phi_p(x) = x$. If $m > 0$, let

$$\phi_q(x) = \tfrac{1}{2} \left(e^{mx} + e^{-mx} \right),$$
$$\phi_p(x) = \tfrac{1}{2m} \left(e^{mx} - e^{-mx} \right).$$

For any value of m, the functions ϕ_p and ϕ_q are annihilated by the operator $-\partial_x^2 + m^2$, and they form a basis for the kernel of this operator. Further, $\phi_p(0) = 1$ and $\phi_q(0) = 0$, whereas $\phi_p'(0) = 0$ and $\phi_q'(0) = 1$. Finally, $\phi_q = \phi_p'$.

Define a map

$$\pi : \mathscr{E}_c^!((-1, 1)) \to \mathbb{R}\{p, q\}$$

by

$$g \mapsto \pi(g) = q \int g(x)\phi_q \, dx + p \int g(x)\phi_p \, dx.$$

It is a cochain map, because if $g = (\Delta + m^2)f$, where f has compact support, then $\pi(g) = 0$. This map is easily seen to be surjective.

This map π says how to identify a linear observable on the field ϕ over the time interval $(-1, 1)$ into a linear combination of the "position" and "momentum."

We need to construct a contracting homotopy on the kernel of π. That is, if $\operatorname{Ker} \pi^k \subset C_c^\infty((-1, 1))$ refers to the kernel of π in cohomological degree k, we need to construct an inverse to the differential

$$C_c^\infty((-1, 1)) = \operatorname{Ker} \pi^{-1} \xrightarrow{\Delta + m^2} \operatorname{Ker} \pi^0.$$

This inverse is defined as follows. Let $G(x) \in C^0(\mathbb{R})$ be the Green's function for the operator $\Delta + m^2$. Explicitly, we have

$$G(x) = \begin{cases} \frac{m}{2} e^{-m|x|} & \text{if } m > 0 \\ -\frac{1}{2} |x| & \text{if } m = 0. \end{cases}$$

Then $(\Delta + m^2)G$ is the delta function at 0. The inverse map sends a function

$$f \in \operatorname{Ker} \pi^0 \subset C_c^\infty((-1, 1))$$

to

$$G \star f = \int_y G(x - y)f(y) \, dy.$$

The fact that $\int f\phi_q = 0$ and $\int f\phi_p = 0$ implies that $G \star f$ has compact support. The fact that G is the Green's function implies that this operator is the inverse to $\triangle + m^2$. It is clear that the operator of convolution with G is smooth (and even continuous), so the result follows.   □

4.2.5 The Poisson Bracket

We now return to the general case and construct the P_0 algebra structure on classical observables.

Suppose we have any free field theory on a manifold M, with complex of fields $\mathscr{E}$. Classical observables are the symmetric algebra $\text{Sym}\,\mathscr{E}_c(U)[1]$. Recall that the complex $\mathscr{E}_c(U)$ is equipped with an antisymmetric pairing of cohomological degree -1. Thus, $\mathscr{E}_c(U)[1]$ is equipped with a symmetric pairing of degree 1.

Lemma 4.2.5 *There is a unique smooth Poisson bracket on* $\text{Obs}^{cl}(U)$ *of cohomological degree* 1, *with the property that*

$$\{\alpha, \beta\} = \langle \alpha, \beta \rangle$$

for any two linear *observables* $\alpha, \beta \in \mathscr{E}_c(U)[1]$.

Recall that "smooth" means that the Poisson bracket is a smooth bilinear map

$$\{-, -\} : \text{Obs}^{cl}(U) \times \text{Obs}^{cl}(U) \to \text{Obs}^{cl}(U)$$

as defined in Section 3.5 in Chapter 3.

Proof The argument we will give is very general, and it applies in any reasonable symmetric monoidal category. Recall that, as stated in Section 3.5 in Chapter 3, the category of convenient vector spaces is a symmetric monoidal category with internal Homs and a Hom-tensor adjunction.

Let A be a commutative algebra object in the category $\text{Ch}(\text{CVS})$ of convenient cochain complexes, and let M be a dg A-module. Then we define $\text{Der}(A, M)$ to be the space of algebra homomorphisms $\phi : A \to A \oplus M$ that are the identity on A modulo the ideal M. (Here $A \oplus M$ is given the "square-zero" algebra structure, where the product of any two elements in M is zero and the product of an element in A with one in M is via the module structure.)

Since the category of convenient cochain complexes has internal Homs, this cochain complex $\text{Der}(A, M)$ is again an A-module in $\text{Ch}(\text{CVS})$.

The commutative algebra

$$\text{Obs}^{cl}(U) = \text{Sym}\,\mathscr{E}_c^!(U)$$

is the initial commutative algebra in the category $\text{Ch}(\text{CVS})$ of convenient cochain complexes equipped with a smooth linear cochain map $\mathscr{E}_c^!(U) \to \text{Obs}^{cl}(U)$.

(We are simply stating the universal property characterizing Sym.) It follows that for any dg module M in Ch(CVS) over the algebra $\mathrm{Sym}\,\mathscr{E}_c^!(U)$,

$$\mathrm{Der}(\mathrm{Sym}\,\mathscr{E}_c^!(U), M) = \mathbb{H}\mathrm{om}_{\mathrm{DVS}}(\mathscr{E}_c^!(U), M).$$

A Poisson bracket on $\mathrm{Sym}\,\mathscr{E}_c^!(U)$ is, in particular, a biderivation. A biderivation is something that assigns to an element of $\mathrm{Sym}\,\mathscr{E}_c^!(U)$ a derivation of the algebra $\mathrm{Sym}\,\mathscr{E}_c^!(U)$. Thus, the space of biderivations is the space

$$\mathrm{Der}\left(\mathrm{Sym}\,\mathscr{E}_c^!(U), \mathrm{Der}\left(\mathrm{Sym}\,\mathscr{E}_c^!(U), \mathrm{Sym}\,\mathscr{E}_c^!(U)\right)\right).$$

What we have said so far thus identifies the space of biderivations with

$$\mathbb{H}\mathrm{om}_{\mathrm{DVS}}(\mathscr{E}_c^!(U), \mathbb{H}\mathrm{om}_{\mathrm{DVS}}(\mathscr{E}_c^!(U), \mathrm{Sym}\,\mathscr{E}_c^!(U))),$$

or equivalently with

$$\mathbb{H}\mathrm{om}_{\mathrm{DVS}}(\mathscr{E}_c^!(U) \otimes \mathscr{E}_c^!(U), \mathrm{Sym}\,\mathscr{E}_c^!(U))$$

via the Hom-tensor adjunction in the category Ch(CVS).

The Poisson bracket we are constructing corresponds to the biderivation which is the pairing on $\mathscr{E}_c^!(U)$ viewed as a map

$$\mathscr{E}_c^!(U) \otimes \mathscr{E}_c^!(U) \to \mathbb{R} = \mathrm{Sym}^0\,\mathscr{E}_c^!(U).$$

This biderivation is antisymmetric and satisfies the Jacobi identity. Since Poisson brackets are a subspace of biderivations, we have proved both the existence and uniqueness clauses. □

Note that for U_1, U_2 disjoint open subsets of V and for observables $\alpha_i \in \mathrm{Obs}^{cl}(U_i)$, we have

$$\{i_V^{U_1}\alpha_1, i_V^{U_2}\alpha_2\} = 0$$

in $\mathrm{Obs}^{cl}(V)$. That is, observables coming from disjoint open subsets commute with respect to the Poisson bracket. This property of $\mathrm{Obs}^{cl}(U)$ is the definition of a P_0 prefactorization algebra. (We will show in Section 6.5 in Chapter 6 that this prefactorization algebra is actually a prefactorization algebra.)

In the case of classical observables of the free scalar field theory, we can think of $\mathrm{Obs}^{cl}(U)$ as the space of polyvector fields on $C^\infty(U)$ along the foliation of $C^\infty(U)$ given by the subspace $C_c^\infty(U) \subset C^\infty(U)$. The Poisson bracket we have just defined is the Schouten bracket on polyvector fields.

4.2.6 The Quantum Observables of a Free Field Theory

In Chapter 2, we constructed a prefactorization algebra that we called $H^0(\mathrm{Obs}^q)$, the quantum observables of a free scalar field theory on a manifold. This space is defined as a space of functions on the space of fields, modulo the image of a certain divergence operator. The aim of this section is to lift this vector space $H^0(\mathrm{Obs}^q)$ to a cochain complex Obs^q. As we explained in Section 4.1, this cochain complex will be the analog of the divergence complex of a measure in finite dimensions.

In Section 4.1, we explained that for a quadratic function q on a vector space V, the divergence complex for the measure $e^{q/\hbar}\omega_0$ on V (where ω_0 is the Lebesgue measure) can be realized as the Chevalley–Eilenberg chain complex of a certain Heisenberg Lie algebra. We follow this procedure in defining Obs^q.

We construct the prefactorization algebra $\mathrm{Obs}^q(U)$ as a twisted factorization envelope of a sheaf of Lie algebras. Let

$$\widehat{\mathscr{E}_c}(U) = \mathscr{E}_c(U) \oplus \mathbb{R} \cdot \hbar$$

where $\mathbb{R}\hbar$ denotes the one-dimensional real vector space situated in degree 1 and spanned by $\hbar$. We give $\widehat{\mathscr{E}_c}(U)$ a Lie bracket by saying that, for $\alpha, \beta \in \mathscr{E}_c(U)$,

$$[\alpha, \beta] = \hbar \langle \alpha, \beta \rangle .$$

Thus, $\widehat{\mathscr{E}_c}(U)$ is a graded version of a Heisenberg Lie algebra, centrally extending the Abelian dg Lie algebra $\mathscr{E}_c(U)$.

Let

$$\mathrm{Obs}^q(U) = C_*(\widehat{\mathscr{E}_c}(U)),$$

where C_* denotes the Chevalley–Eilenberg complex for the Lie algebra homology of $\widehat{\mathscr{E}_c}(U)$, defined using the tensor product $\widehat{\otimes}_\beta$ on the category of convenient vector spaces, as discussed in Section 3.5 in Chapter 3. Thus,

$$\mathrm{Obs}^q(U) = \left(\mathrm{Sym}\left(\widehat{\mathscr{E}_c}(U)[1]\right), \mathrm{d}\right)$$
$$= \left(\mathrm{Obs}^{cl}(U)[\hbar], \mathrm{d}\right)$$

where the differential arises from the Lie bracket and differential on $\widehat{\mathscr{E}_c}(U)$. The symbol [1] indicates a shift of degree down by one. (Recall that we always work with cochain complexes, so our grading convention of C_* is the negative of one common convention.)

Remark: Those readers who are operadically inclined might notice that the Lie algebra chain complex of a Lie algebra $\mathfrak{g}$ is the E_0 version of the universal enveloping algebra of a Lie algebra. Thus, our construction is an E_0 version of the familiar construction of the Weyl algebra as a universal enveloping algebra of a Heisenberg algebra.   ◊

Since this is an example of the general construction we discussed in Section 3.6 in Chapter 3, we see that $\text{Obs}^q(U)$ has the structure of a prefactorization algebra.

As we discussed in Section 4.2.2, we can view classical observables on an open set U for the free scalar theory as a certain complex of polyvector fields on the space $C^\infty(U)$:

$$\text{Obs}^{cl}(U) = \left(\text{PV}_c(C^\infty(U)), \vee dS \right).$$

By $\text{PV}_c(C^\infty(U))$ we mean polyvector fields along the foliation $T_c C^\infty(U) \subset TC^\infty(U)$ of compactly supported variants of a field. Concretely, the cochain complex $\text{Obs}^{cl}(U)$ is

$$\cdots \xrightarrow{\vee dS} \wedge^2 C_c^\infty(U) \otimes \text{Sym}\, C_c^\infty(U) \xrightarrow{\vee dS} C_c^\infty(U)$$
$$\otimes \text{Sym}\, C_c^\infty(U) \xrightarrow{\vee dS} \text{Sym}\, C_c^\infty(U),$$

where all tensor products appearing in this expression are $\widehat{\otimes}_\beta$.

In a similar way, the cochain complex of quantum observables $\text{Obs}^q(U)$ involves just a modification of the differential. It is

$$\cdots \xrightarrow{\vee dS + \hbar\, \text{Div}} C_c^\infty(U) \otimes \text{Sym}\, C_c^\infty(U) \xrightarrow{\vee dS + \hbar\, \text{Div}} \text{Sym}\, C_c^\infty(U).$$

The operator Div is the extension to all polyvector fields of the operator (see Definition 2.2.1) defined in Chapter 2 as a map from polynomial vector fields on $C^\infty(U)$ to polynomial functions. Thus, $H^0(\text{Obs}^q(U))$ is the same vector space (and same prefactorization algebra) that we defined in Chapter 2.

As a graded vector space, there is an isomorphism

$$\text{Obs}^q(U) = \text{Obs}^{cl}(U)[\hbar],$$

but it does not respect the differentials. In particular, the differential d on the quantum observables satisfies

(i) Modulo $\hbar$, d coincides with the differential on $\text{Obs}^{cl}(U)$, and
(ii) the equation

$$d(a \cdot b) = (da) \cdot b + (-1)^{|a|} a \cdot b + (-1)^{|a|} \hbar \{a, b\}. \qquad (4.2.6.1)$$

Here, · indicates the commutative product on $\text{Obs}^{cl}(U)$. As we explain in Volume 2, these properties imply that Obs^q defines a prefactorization algebra valued in Beilinson–Drinfeld algebras and that Obs^q quantizes the prefactorization algebra Obs^{cl} valued in P_0 algebras. (See Section A.3.2 in Appendix A for the definition of these types of algebras.) It is a characterizing feature of the Batalin–Vilkovisky formalism that the Poisson bracket measures the failure of the differential d to be a derivation, and the language of P_0 and BD algebras is an operadic formalization of this concept. We prove in Section 6.5 in Chapter 6 that Obs^{cl} is a factorization algebra, which implies Obs^q is a factorization algebra over $\mathbb{R}[\hbar]$.

4.3 Quantum Mechanics and the Weyl Algebra

We will now show that our construction of the free scalar field on the real line $\mathbb{R}$ recovers the Weyl algebra, which is the associative algebra of observables in quantum mechanics . We already showed in Section 4.2.4 that the classical observables recovers functions on the phase space.

First, we must check that this prefactorization algebra is locally constant and so gives us an associative algebra.

Lemma 4.3.1 *The prefactorization algebra* Obs^q *on* $\mathbb{R}$ *constructed from the free scalar field theory with mass m is locally constant.*

Proof Recall that $\text{Obs}^q(U)$ is the Chevalley–Eilenberg chains on the Heisenberg Lie algebra $\mathcal{H}(U)$ built as a central extension of $C_c^\infty(U) \xrightarrow{\Delta+m^2} C_c^\infty(U)[-1]$. Let us filter $\text{Obs}^q(U)$ by saying that

$$F^{\leq k} \text{Obs}^q(U) = \text{Sym}^{\leq k}(\mathcal{H}(U)[1]).$$

The associated graded for this filtration is $\text{Obs}^{cl}(U)[\hbar]$. Thus, to show that $H^* \text{Obs}^q$ is locally constant, it suffices to show that $H^* \text{Obs}^{cl}$ is locally constant, by considering the spectral sequence associated to this filtration. We have already seen that $H^*(\text{Obs}^{cl}(a, b)) = \mathbb{R}[p, q]$ for any interval (a, b), and that the inclusion maps $(a, b) \to (a', b')$ induces isomorphisms. Thus, the cohomology of Obs^{cl} is locally constant, as desired. □

It follows that the cohomology of this prefactorization algebra is an associative algebra, which we call A_m. We will show that A_m is the Weyl algebra, no matter the mass m.

In fact, we will show more, showing that the prefactorization algebra does know the mass m, encoded in a *Hamiltonian* operator. The prefactorization algebra for the free scalar field theory on $\mathbb{R}$ is built as Chevalley–Eilenberg chains of the Heisenberg algebra based on $C_c^\infty(U) \xrightarrow{\Delta+m^2} C_c^\infty(U)[-1]$. The operator $\frac{\partial}{\partial x}$ of infinitesimal translation acts on $C_c^\infty(U)$, commutes with the operator $\Delta + m^2$, and preserves the cocycle defining the central extension. Therefore, it acts naturally on the Chevalley–Eilenberg chains of the Heisenberg algebra. One can check that this operator is a derivation for the factorization product. That is, the operator $\frac{\partial}{\partial x}$ commutes with inclusions of one open subset into another, and if U, V are disjoint and $\alpha \in \mathrm{Obs}^q(U)$, $\beta \in \mathrm{Obs}^q(V)$, we have

$$\frac{\partial}{\partial x}(\alpha \cdot \beta) = \frac{\partial}{\partial x}(\alpha) \cdot \beta + \alpha \cdot \frac{\partial}{\partial x}(\beta) \in \mathrm{Obs}^q(U \sqcup V).$$

In other words, the prefactorization algebra is, in a sense, translation-invariant! (Derivations, and this interpretation, are discussed in more detail in Section 4.8.)

It follows immediately that $\frac{\partial}{\partial x}$ defines a derivation of the associative algebra A_m coming from the cohomology of $\mathrm{Obs}^q((0, 1))$.

Remark: These observations about $\frac{\partial}{\partial x}$ apply to the classical observables, too. Indeed, one discovers that infinitesimal translation induces a derivation of the Poisson algebra $\mathbb{R}[p, q]$. As this derivation preserves the Poisson bracket, we know there is an element H of $\mathbb{R}[p, q]$ such that $\{H, -\}$ is the derivation. Here, $H = p^2 - m^2 q^2$. (On a symplectic vector space, every symplectic vector field is Hamiltonian.) $\Diamond$

Definition 4.3.2 The *Hamiltonian H* is the derivation of the associative algebra A_m arising from the derivation $-\frac{\partial}{\partial x}$ of the prefactorization algebra Obs^q of observables of the free scalar field theory with mass m.

Proposition 4.3.3 *The associative algebra A_m coming from the free scalar field theory with mass m is the Weyl algebra, generated by $p, q, \hbar$ with the relation $[p, q] = \hbar$ and all other commutators being zero.*

The Hamiltonian H is the derivation

$$H(a) = \tfrac{1}{2\hbar}[p^2 - m^2 q^2, a].$$

Note that it is an inner, or principal, derivation.

Remark: It might seem a priori that different action functionals, which encode different theories, should have different algebras of observables. In this case, we see that all the theories lead to the same associative algebra. The Hamiltonian is what distinguishes these algebras of observables; that is, different actions differ by "how to relate the same observable applied at different moments in time."

There is an important feature of the Weyl algebra that illuminates the role of the Hamiltonian. The Weyl algebra is rigid: the Hochschild cohomology of the Weyl algebra $A^{(n)}$ for a symplectic vector space of dimension $2n$ vanishes, except in degree 0, where the cohomology is one-dimensional. Hence, in particular, $HH^2(A^{(n)}) = 0$, so there are no nontrivial deformations of the Weyl algebra. In consequence, any action functional whose underlying free theory is the free scalar field (when the coupling constants are formal parameters) will have the same algebra of observables, as an associative algebra. Since $HH^1(A^{(n)}) = 0$, we know that every derivation is inner and hence is represented by some element of $A^{(n)}$. Thus, the derivation arising from translation must have a Hamiltonian operator.   ◇

Proof We start by writing down elements of Obs^q corresponding to the position and momentum observables. Recall that

$$\mathrm{Obs}^q((a,b)) = \mathrm{Sym}\left(C_c^\infty((a,b))^{-1} \oplus C_c^\infty((a,b))^0\right)[\hbar]$$

with a certain differential.

We let $\phi_q, \phi_p \in C^\infty(\mathbb{R})$ be the functions introduced in the proof of Lemma 4.2.4. Explicitly, if $m = 0$, then $\phi_q(x) = 1$ and $\phi_p(x) = x$, whereas if $m \neq 0$ we have

$$\phi_q(x) = \tfrac{1}{2}\left(e^{mx} + e^{-mx}\right),$$
$$\phi_p(x) = \tfrac{1}{2m}\left(e^{mx} - e^{-mx}\right).$$

Thus, ϕ_q, ϕ_p are both in the null space of the operator $-\partial_x^2 + m^2$, with the properties that $\partial_x\phi_p = \phi_q$ and that ϕ_q is symmetric under $x \mapsto -x$, whereas ϕ_p is antisymmetric.

As in the proof of Lemma 4.2.4, choose a function $f \in C_c^\infty((-\tfrac{1}{2}, \tfrac{1}{2}))$ that is symmetric under $x \mapsto -x$ and has the property that

$$\int_{-\infty}^{\infty} f(x)\phi_q(x)\,dx = 1.$$

The symmetry of f implies that the integral of f against ϕ_p is zero.

Let $f_t \in C_c^\infty((t - \tfrac{1}{2}, t + \tfrac{1}{2}))$ be $f_t(x) = f(x - t)$. We define observables P_t, Q_t by

$$Q_t = f_t,$$
$$P_t = -f_t'.$$

The observables Q_t, P_t are in the space $C_c^\infty(I_t)[1] \oplus C_c^\infty(I)$ of linear observables for the interval $I_t = ((t - \frac{1}{2}, t + \frac{1}{2})$. They are also of cohomological degree 0. If we think of observables as functionals of a field in $C^\infty(I_t) \oplus C^\infty(I_t)[-1]$, then these are linear observables given by integrating f_t or $-f_t'$ against the field ϕ.

Thus, Q_t and P_t represent average measurements of positions and momenta of the field ϕ in a neighborhood of t.

Because the cohomology classes $[P_0], [Q_0]$ generate the commutative algebra $H^*(\mathrm{Obs}^{cl}(\mathbb{R}))$, it is automatic that they still generate the associative algebra $H^0\,\mathrm{Obs}^q(\mathbb{R})$. We thus need to show that they satisfy the Heisenberg commutation relation

$$[[P_0], [Q_0]] = \hbar$$

for the associative product on $H^0\,\mathrm{Obs}^q(\mathbb{R})$, which is an associative algebra by virtue of the fact that $H^*\,\mathrm{Obs}^q$ is locally constant. (The other commutators among $[P_0]$ and $[Q_0]$ vanish automatically.)

We also need to calculate the Hamiltonian. We see that

$$-\frac{\partial}{\partial x} Q_t = -f_t' = P_t = \tfrac{\mathrm{d}}{\mathrm{d}t} f(x - t) = \tfrac{\mathrm{d}}{\mathrm{d}t} Q_t,$$

$$-\frac{\partial}{\partial x} P_t = f_t''.$$

Hence, the Hamiltonian acting on $[P_0]$ gives the t-derivative of $[P_t]$ at $t = 0$ and similarly for Q_0. In other words, the parameter t encodes translation of the real line, and the infinitesimal action of translation acts correctly on the observables.

In cohomology, the image of $-\partial_x^2 + m^2$ is zero, we see that

$$\tfrac{\mathrm{d}}{\mathrm{d}t}[P_t] = 0 + m^2[f(x - t)] = m^2[Q_t],$$

as we expect from usual classical mechanics. In particular, when $m = 0$, $[P_t]$ is independent of t: in other words, momentum is conserved.

Thus, the Hamiltonian H satisfies

$$H([P_0]) = m^2[Q_0]$$
$$H([Q_0]) = [P_0].$$

If we assume that the commutation relation $[[P_0], [Q_0]] = \hbar$ holds, then

$$H(a) = \tfrac{1}{2\hbar}\left[[P_0]^2 - m^2[Q_0]^2, a\right],$$

as desired.

Thus, to complete the proof, it only remains to verify the Heisenberg commutation relation.

If $a(t)$ is a function t, let $\dot{a}(t)$ denote the t-derivative. It follows from the equations above for the t-derivatives of $[P_t]$ and $[Q_t]$ that for any function $a(t)$ satisfying $\ddot{a}(t) = m^2 a(t)$, the observable

$$a(t)P_t - \dot{a}(t)Q_t$$

is independent of t at the level of cohomology.

More precisely, if $\widetilde{Q}_t$ is the linear observable of cohomological degree -1 given by the function f_t in $C_c^\infty((t - \frac{1}{2}, t + \frac{1}{2})[1]$, we have

$$\frac{\partial}{\partial t}(a(t)P_t - \dot{a}(t)Q_t) = -\mathrm{d}(a(t)\widetilde{Q}_t)$$

where d is the differential on observables. Explicitly, we compute

$$\frac{\partial}{\partial t}(a(t)P_t - \dot{a}(t)Q_t) = -\dot{a}(t)\frac{\partial}{\partial x}f_t(x) - a(t)\frac{\partial}{\partial t}\frac{\partial}{\partial x}f_t(x) - \ddot{a}(t)f_t(x) - \dot{a}(t)\frac{\partial}{\partial t}f_t(x)$$

$$= a(t)\frac{\partial^2}{\partial x^2}f_t(x) - \ddot{a}(t)f_t(x)$$

$$= a(t)\frac{\partial^2}{\partial x^2}f_t(x) - m^2 a(t)f_t(x)$$

$$= -\mathrm{d}(a(t)\widetilde{Q}_t).$$

We will define modified observables $\mathscr{P}_t$, $\mathscr{Q}_t$ that are independent of t at the cohomological level. We let

$$\mathscr{P}_t = \phi_q(t)P_t - \phi_q'(t)Q_t$$
$$\mathscr{Q}_t = \phi_p(t)P_t - \phi_p'(t)Q_t.$$

Since ϕ_p and ϕ_q are in the null space of the operator $-\partial_x^2 + m^2$, the observables $\mathscr{P}_t$, $\mathscr{Q}_t$ are independent of t at the level of cohomology. In the case $m = 0$, then $\phi_q(t) = 1$ so that $\mathscr{P}_t = P_t$. The statement that $\mathscr{P}_t$ is independent of t corresponds, in this case, to conservation of momentum.

In general, $\mathscr{P}_0 = P_0$ and $\mathscr{Q}_0 = Q_0$. We also have

$$\frac{\partial}{\partial t}\mathscr{P}_t = -\mathrm{d}(\phi_q(t)\widetilde{Q}_t),$$

and similarly for $\mathscr{Q}_t$.

It follows that if we define a linear degree -1 observable $h_{s,t}$ by

$$h_{s,t} = \int_{u=s}^t \phi_q(u)\widetilde{Q}_u(x)\,\mathrm{d}u,$$

then

$$\mathrm{d}h_{s,t} = \mathscr{P}_s - \mathscr{P}_t.$$

Note that if $|t| > 1$, the observables P_t and Q_t have disjoint support from the observables P_0 and Q_0 . Thus, we can use the prefactorization structure map

$$\text{Obs}^q((-\tfrac{1}{2}, \tfrac{1}{2})) \otimes \text{Obs}^q((t - \tfrac{1}{2}, t + \tfrac{1}{2})) \to \text{Obs}^q(\mathbb{R})$$

to define a product observable

$$Q_0 \cdot \mathscr{P}_t \in \text{Obs}^q(\mathbb{R}).$$

We will let $\star$ denote the associative multiplication on $H^0 \text{Obs}^q(\mathbb{R})$. We defined this multiplication by

$$[Q_0] \star [P_0] = [Q_0 \cdot \mathscr{P}_t] \text{ for } t > 1$$
$$[P_0] \star [Q_0] = [Q_0 \cdot \mathscr{P}_t] \text{ for } t < -1.$$

Thus, it remains to show that, for $t > 1$,

$$[Q_0 \cdot \mathscr{P}_t] - [Q_0 \cdot \mathscr{P}_{-t}] = \hbar.$$

We will construct an observable whose differential is the difference between the left- and right-hand sides. Consider the observable

$$S = f(x)h_{-t,t}(y) \in C_c^\infty(\mathbb{R}) \otimes C_c^\infty(\mathbb{R})[1],$$

where the functions f and $h_{-t,t}$ were defined earlier. We view $h_{-t,t}$ as being of cohomological degree -1, and f as being of cohomological degree 1.

Recall that the differential on $\text{Obs}^q(\mathbb{R})$ has two terms: one coming from the Laplacian $\triangle + m^2$ mapping $C_c^\infty(\mathbb{R})^{-1}$ to $C_c^\infty(\mathbb{R})^0$, and one arising from the bracket of the Heisenberg Lie algebra. The second term maps

$$\text{Sym}^2 \left(C_c^\infty(\mathbb{R})^{-1} \oplus C_c^\infty(\mathbb{R})^0 \right) \to \mathbb{R}\,\hbar.$$

Applying this differential to the observable S, we find that

$$(dS) = f(x)(-\partial^2 + m^2)h_{-t,t}(y) + \hbar \int_{\mathbb{R}} h_{-t,t}(x)f(x)\, dx$$

$$= Q_0 \cdot (\mathscr{P}_{-t} - \mathscr{P}_t) + \hbar \int f(x)h_{-t,t}(x).$$

Therefore

$$[Q_0\mathscr{P}_t] - [Q_0\mathscr{P}_{-t}] = \hbar \int f(x)h_{-t,t}(x).$$

It remains to compute the integral. This integral can be rewritten as

$$\int_{u=-t}^{t} \int_{x=-\infty}^{\infty} f(x)f(x-u)\phi_q(u)\, du.$$

Note that the answer is automatically independent of t for t sufficiently large, because $f(x)$ is supported near the origin so that $f(x)f(x - u) = 0$ for u sufficiently large. Thus, we can sent $t \to \infty$.

Since f is also symmetric under $x \to -x$, we can replace $f(x - u)$ by $f(u - x)$. We can perform the u-integral by changing coordinates $u \to u - x$, leaving the integrand as $f(x)f(u)\phi_q(u + x)$. Note that

$$\phi_q(u + x) = \tfrac{1}{2}\left(e^{m(x+u)} + e^{-m(x+u)}\right).$$

Now, by assumption on f,

$$\int f(x)e^{mx}\,\mathrm{d}x = \int f(x)e^{-mx}\,\mathrm{d}x = 1.$$

It follows that

$$\int_{u=-\infty}^{\infty} \phi_q(u + x)f(u)\,\mathrm{d}u = \phi_q(x).$$

We can then perform the remaining x integral $\int \phi_q(x)f(x)\,\mathrm{d}x$, which gives 1, as desired.

Thus, we have proven

$$[[Q_0],[P_0]] = \hbar,$$

as desired.   □

4.4 Pushforward and Canonical Quantization

Consider the free scalar field theory on a manifold of the form $N \times \mathbb{R}$, equipped with the product metric. We assume for simplicity that N is compact. Let Obs^q denote the prefactorization algebra of observables of the free scalar field theory with mass m on $N \times \mathbb{R}$. Let $\pi : N \times \mathbb{R} \to \mathbb{R}$ be the projection map. There is a *pushforward prefactorization algebra* $\pi_* \mathrm{Obs}^q$ living on $\mathbb{R}$ and defined by

$$(\pi_* \mathrm{Obs}^q)(U) = \mathrm{Obs}^q(\pi^{-1}(U)).$$

(This pushforward construction is a version of compactification in physics.) In this section, we explain how to relate this prefactorization algebra on $\mathbb{R}$ to an infinite tensor product of the prefactorization algebras associated to quantum mechanics on $\mathbb{R}$. See Figure 4.1.

Let $\{e_i\}_{i \in I}$ be an orthonormal basis of eigenvectors of the operator $\triangle + m^2$ on $C^\infty(N)$, where e_i has eigenvalue λ_i. The direct sum $\bigoplus_{i \in I} \mathbb{R}e_i$ is a dense subspace of $C^\infty(N)$.

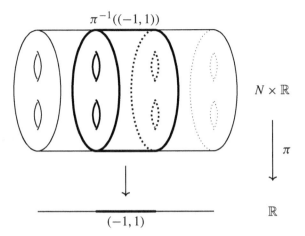

Figure 4.1. The value of $\pi_*\mathrm{Obs}^q$ on $(-1, 1)$ is the value of Obs^q on the preimage $\pi^{-1}((-1, 1))$.

For $m \in \mathbb{R}$, let A_m denote the cohomology of the prefactorization algebra associated to the free one-dimensional scalar field theory with mass m. Thus, A_m is the Weyl algebra generated by p, q, and $\hbar$ with commutator $[p, q] = \hbar$. The dependence on m appears only through the Hamiltonian.

Consider the tensor product of algebras

$$A_N = A_{\sqrt{\lambda_1}} \otimes_{\mathbb{R}[\hbar]} A_{\sqrt{\lambda_2}} \otimes_{\mathbb{R}[\hbar]} \cdots$$

over the entire spectrum $\{\lambda_i\}_{i \in I}$. (The infinite tensor product is defined to be the colimit of the finite tensor products, where the maps in the colimit use the unit in each algebra. See the example in Section 3.1.1 for more discussion.)

Proposition 4.4.1 *There is a dense sub-prefactorization algebra of $\pi_* \mathrm{Obs}^q$ that is locally constant and whose cohomology prefactorization algebra corresponds to the associative algebra A_N.*

Remark: This proposition encodes, in essence, the procedure usually known as canonical quantization. (In a physics textbook, N is typically a torus $S^1 \times \cdots \times S^1$ whose radii are eventually "sent to infinity.") The Weyl algebra $A_{\sqrt{\lambda_i}}$ for eigenvalue λ_i corresponds to "a quantum mechanical particle evolving in N with mass m and energy λ_i." (An important difference is that a quantum field theory textbook often works with Lorentzian signature, rather than Euclidean signature, as we do.) The time-ordered observables of the whole free scalar field $\pi_* \mathrm{Obs}^q$ are essentially described by the combination of the algebras for all these energy levels. The different energy levels do not affect one another, as it is a free theory. When an interaction term is added to the action functional, these energy levels do interact.

The prefactorization algebra $\pi_* \text{Obs}^q$ has a derivation, the Hamiltonian, coming from infinitesimal translation in $\mathbb{R}$. The prefactorization algebra $A_{\sqrt{\lambda_i}}$ also has a Hamiltonian, given by bracketing with $\frac{1}{2\hbar}[p^2 - \lambda_i q^2, -]$. The map from $\bigotimes_i A_{\sqrt{\lambda_i}}$ to $H^*(\pi_* \text{Obs}^q)$ intertwines these derivations. $\Diamond$

Proof The prefactorization algebra $\pi_* \text{Obs}^q$ on $\mathbb{R}$ assigns to an open subset $U \subset \mathbb{R}$ the Chevalley–Eilenberg chains of a Heisenberg Lie algebra given by a central extension of

$$C_c^\infty(U \times N) \xrightarrow{\Delta + m^2} C^\infty(U \times N)[-1].$$

A dense subcomplex of these linear observables is

$$\bigoplus_{i \in I} \left(C_c^\infty(U)e_i \xrightarrow{\Delta_{\mathbb{R}} + \lambda_i} C_c^\infty(U)e_i[-1] \right), \qquad (\dagger)$$

because the topological vector space of smooth functions is a certain completion of this direct sum of eigenspaces.

Let $\mathcal{F}_i$ be the prefactorization algebra on $\mathbb{R}$ associated to quantum mechanics with mass $\sqrt{\lambda_i}$. This prefactorization algebra is the envelope of the Heisenberg central extension of

$$C_c^\infty(U)e_i \xrightarrow{\Delta_{\mathbb{R}} + \lambda_i} C_c^\infty(U)e_i[-1].$$

Note that $\mathcal{F}_i$ is a prefactorization algebra in modules over $\mathbb{R}[\hbar]$. We can define then the tensor product prefactorization algebra

$$\mathcal{F} = \mathcal{F}_1 \otimes_{\mathbb{R}[\hbar]} \mathcal{F}_2 \otimes_{\mathbb{R}[\hbar]} \cdots$$

to be the colimit of the finite tensor products of the $\mathcal{F}_i$ under the inclusion maps coming from the unit $1 \in \mathcal{F}_i(U)$ for any open subset. Equivalently, we can view this tensor product as being associated to the Heisenberg central extension of the complex ($\dagger$) above.

Because the complex ($\dagger$) is a dense subspace of the complex whose Heisenberg extension defines $\pi_* \text{Obs}^q$, we see that there is a map of prefactorization algebras with dense image

$$\mathcal{F} \to \pi_* \text{Obs}^q.$$

Passing to cohomology, we have a map

$$H^* \mathcal{F} \to H^*(\pi_* \text{Obs}^q).$$

As the prefactorization algebra $H^*(\mathcal{F}_i)$ corresponds to the Weyl algebra $A_{\sqrt{\lambda_i}}$, we see that $H^*\mathcal{F}$ corresponds to the algebra A_N. $\qquad\square$

4.5 Abelian Chern–Simons Theory

We have discussed the free scalar field in some detail, and we have recovered several aspects of the "usual" story. (Further aspects of the free scalar field occupy much of the rest of this chapter.) Another important example of a free field theory is Abelian Chern–Simons theory, which is a particularly simple example of both a topological field theory and a gauge theory. The theory examined here is the perturbative facet of Chern–Simons theory with gauge group $U(1)$.

Let M be an orientable manifold of dimension 3. The space of fields is $\mathscr{E} = \Omega_M^*[1]$, equipped with the exterior derivative d as its differential. Integration provides a skew-symmetric pairing of degree -1:

$$\langle \alpha, \beta \rangle = (-1)^{|\beta|} \int_M \alpha \wedge \beta,$$

where α, β live in $\mathscr{E}_c$ and each has degree 1 less than its usual degree, due to the shift. (Inside the integral, we simply view them as ordinary differential forms, without the shift.) The action functional of Abelian Chern–Simons theory is then

$$S(\alpha) = \langle \alpha, \mathrm{d}\alpha \rangle,$$

and the associated equation of motion is that $\mathrm{d}\alpha = 0$. In other words, it picks out flat connections.

To be more precise, this complex describes the derived space of solutions to the problem of finding flat connections on the trivial complex line bundle on M, up to gauge equivalence. If we focus on the fields of degree 0 – namely, 1-forms – then a solution is precisely a flat connection. Two solutions α and α' such that $\alpha - \alpha' = \mathrm{d}f$, for $f \in \Omega^0[1]$, are viewed as equivalent solutions. The function $s = e^f$ is a nowhere-vanishing section of the complex line bundle, and we can use it to relate the two connections, as follows. If g is a function, then

$$
\begin{aligned}
s^{-1}(\mathrm{d} + \alpha')(sg) &= s^{-1}\left((\mathrm{d}s)g + s\mathrm{d}g + \alpha'sg\right) \\
&= (\mathrm{d}g + \alpha'g) + (\mathrm{d}f)g \\
&= (\mathrm{d} + \alpha)g,
\end{aligned}
$$

since $\mathrm{d}s = s\mathrm{d}f$. In other words, s provides an automorphism, or gauge transformation, intertwining the two flat connections.

We have explained why Ω^1 appears in degree 0 (the connection 1-forms are the main actors) and why Ω^0 appears in degree 1 (functions are the Lie algebra of gauge automorphisms). We now explain why Ω^2 and Ω^3 appear.

The action functional S is naturally a function on this linear derived space

$$\Omega^0[1] \overset{\text{d}}{\to} \Omega^1,$$

so that its exterior derivative $\text{d}S$ is a section of the cotangent bundle of this linear derived space. As S is quadratic, $\text{d}S$ is linear, so the section is linear. From the physical perspective, we want to study the derived intersection of the zero section and $\text{d}S$ in this cotangent bundle: this amounts to studying the derived critical locus of S (i.e., the points in the linear derived stack where $\text{d}S$ vanishes). This derived intersection is itself a linear derived stack and is described by the full de Rham complex.

4.5.1 Observables

The classical observables for Abelian Chern–Simons theory are easy to describe and to compute. For U an open subset of M, we have

$$\text{Obs}^{cl}(U) = \text{Sym}^* \left(\Omega_c^0[2] \overset{\text{d}}{\to} \Omega_c^1[1] \overset{\text{d}}{\to} \Omega_c^2 \overset{\text{d}}{\to} \Omega_c^3[-1] \right).$$

Thus, we see that

$$H^* \text{Obs}^{cl}(U) \cong \text{Sym}^* \left(H_c^*(U)[2] \right).$$

In contrast to the free scalar field, where the cohomology of the linear observables was often infinite-dimensional, the cohomology of the linear observables is finite-dimensional, for any open U. Thus, the cohomology prefactorization algebra $H^* \text{Obs}^{cl}$ is relatively easy to understand.

This pleasant situation continues with the quantum observables. Consider the natural filtration on $\text{Obs}^q(U)$ by

$$F^k \text{Obs}^q(U) = \text{Sym}^{\leq k} \left(\Omega_c^*(U)[2] \right) [\hbar].$$

This filtration induces a spectral sequence, and the first page is $H^* \text{Obs}^{cl}(U)[\hbar]$, which we have already computed. The differential on the first page is zero, but the differential on the second page arises from the pairing on linear observables (recall that the pairing provides the Lie bracket of the Heisenberg Lie algebra, and this bracket provides the differential in the Chevalley–Eilenberg chain complex). Note that the pairing on linear observables descends to the cohomology of the linear observables by

$$\langle [\alpha], [\beta] \rangle_{H^*} = \langle \alpha, \beta \rangle.$$

Hence, the differential d_2 on the second page of the spectral sequence vanishes on constant and linear terms (i.e., from Sym^0 and Sym^1) and satisfies

$$d_2([\alpha][\beta]) = \hbar \langle [\alpha], [\beta] \rangle_{H^*}$$

for a pure quadratic term. The BD algebra axiom (recall Eq. (4.2.6.1)) then determines d_2 on terms that are cubic and higher.

Thus, the cohomology of the quantum observables reduces to understanding the integration pairing between cohomology classes of compactly supported de Rham forms. In particular, Poincaré duality implies the following.

Lemma 4.5.1 *For M a connected, closed 3-manifold,*

$$H^* \text{Obs}^q(M) \cong \mathbb{R}[\hbar][1 - \dim H^2(M)].$$

Proof Let $b_2 = \dim H^2(M)$. Pick a basis $\alpha_1, \ldots, \alpha_{b_2}$ for $H^2(M)$ and a dual basis $\beta_1, \ldots, \beta_{b_2}$ for $H^1(M)$:

$$\langle \alpha_j, \beta_k \rangle_{H^*} = \delta_{jk}.$$

Let ν denote the generator for $H^0(M)$ and $[M]$ the dual generator for $H^3(M)$. Thus, $d_2(\alpha_j \beta_k) = \hbar \delta_{jk}$ and $d_2(\nu[M]) = \hbar$, and d_2 vanishes on all other quadratic monomials. Explicit computation then shows that the pure odd term $\nu \alpha_1 \cdots \alpha_{b_2}$ is closed but not exact, and it generates the cohomology.   □

4.5.2 Compactifying Along a Closed Surface

The next case to consider is a 3-manifold of the form $M = \mathbb{R} \times \Sigma$, where Σ is a closed, orientable surface. Let $\pi : M \to \mathbb{R}$ denote the obvious projection map. We will show that the pushforward prefactorization algebra $\pi_* \text{Obs}^q$ is locally constant, so that it corresponds to an associative algebra. This associative algebra is the Weyl algebra for the graded symplectic vector space $H^*(\Sigma)[1]$. (Recall Figure 4.1.)

Some notation will make it easier to be precise. Let g denote the genus of Σ. Pick a symplectic basis $\{\alpha_1, \ldots, \alpha_g, \beta_1, \ldots, \beta_g\}$ for $H^1(\Sigma)$ in the sense that

$$\int_\Sigma \alpha_j \wedge \beta_k = \delta_{jk}.$$

Let ν denote the basis for $H^0(\Sigma)$ given by the constant function 1. Let μ denote a basis for $H^3(M)$ such that $\int_\Sigma \mu = 1$. Thus, the graded vector space $H^*(\Sigma)[1]$ has a natural symplectic structure given by the integration pairing. The odd elements ν and μ are dual under this pairing.

Proposition 4.5.2 *The prefactorization algebra $H^*\pi_*\operatorname{Obs}^q$ is locally constant, and it corresponds to the Weyl algebra A_Σ for $H^*(\Sigma)[1]$. In terms of our chosen generators, this Weyl algebra has commutators $[\nu, \mu] = \hbar$, $[\alpha_j, \beta_k] = \hbar\delta_{jk}$ and all others vanish.*

The Hamiltonian in this Weyl algebra is zero.

The fact that the Hamiltonian is zero is what makes Abelian Chern–Simons a topological field theory . Note as well the appearance of the Weyl algebra.

Proof This argument is much simpler than the argument for the free scalar field in one dimension. We will use the Hodge theorem to replace the big complex $\Omega_c^*(\mathbb{R}) \otimes \Omega^*(\Sigma)$ by the smaller complex $\Omega_c^*(\mathbb{R}) \otimes H^*(\Sigma)$, and then exploit our construction for the universal enveloping algebra (recall Proposition 3.4.1) to obtain the result.

Let $\mathcal{L} = \pi_*\Omega_M^*[1]$. This is the pushforward of the local Lie algebra of linear observables (it is an Abelian Lie algebra, as we are working with a free theory). Then $\pi_*\operatorname{Obs}^{cl} \cong C_*\mathcal{L}_c$. Let $\widehat{\mathcal{L}}$ denote the pushforward of the Heisenberg central extension. Then $\pi_*\operatorname{Obs}^q \cong C_*\widehat{\mathcal{L}}_c$, the factorization envelope.

By the Hodge theorem, we can obtain a homotopy equivalence between $\Omega^*(\Sigma)$ and its cohomology $H^*(\Sigma)$. Tensoring with Ω_c^* on $\mathbb{R}$, we obtain a homotopy equivalence

$$Ł_c \simeq \Omega_c^* \otimes H^*(\Sigma)[1] = \mathfrak{g}$$

of Abelian dg Lie algebras.Thus, we see that $C_*\mathfrak{g}$ is homotopy equivalent to $\pi_*\operatorname{Obs}^{cl}$. Similarly, $C_*\widehat{\mathfrak{g}}$ is homotopy equivalent to $\pi_*\operatorname{Obs}^q$, where $\widehat{\mathfrak{g}}$ denotes the Heisenberg central extension (i.e., we use the Poincaré pairing on $H^*(\Sigma)$ tensored with the integration pairing on $\mathbb{R}$).

The cohomology prefactorization algebra $H^*\pi_*\operatorname{Obs}^q$ is isomorphic to the cohomology prefactorization algebra $H^*(C_*\widehat{\mathfrak{g}})$. By Proposition 3.4.1, we know this corresponds to the associative algebra $UH^*(\Sigma)[1]$, where the central element is denoted $\hbar$.

Now we compute the Hamiltonian. Recall that the Hamiltonian arose in the free scalar field via the translation action on $\mathbb{R}$. In the text that follows, we will explicitly compute the Hamiltonian in a manner parallel to Proposition 4.3.3. But there is a conceptual reason for the vanishing of the Hamiltonian: the theory, because it is simply the de Rham complex, is homotopy-invariant under diffeomorphism. Not only are the observables preserved by translation along $\mathbb{R}$; they are preserved under any diffeomorphism of $\mathbb{R}$.

Let x denote a coordinate on $\mathbb{R}$. Infinitesimal translation $\frac{\partial}{\partial x}$ preserves everything in the prefactorization algebra $\pi_*\operatorname{Obs}^q$, just as in Proposition 4.3.3. In this case, however, the solutions to the equations of motion – namely, de Rham cohomology classes of Σ – are translation-invariant, so the Hamiltonian is trivial, as follows. Fix a function $f \in C_c^\infty((-\frac{1}{2}, \frac{1}{2}))$ such that $\int_{\mathbb{R}} f(x)dx = 1$. Hence the 1-form $f(x)dx$ generates $H_c^1(\mathbb{R})$. Let $f_t(x) = f(x - t)$. Then we get linear observables in $C_*\widehat{\mathfrak{g}}((t - \frac{1}{2}, t + \frac{1}{2}))$ by

$$A_{t,j} = f_t(x)\,dx \otimes \alpha_j,$$
$$B_{t,k} = f_t(x)\,dx \otimes \beta_k,$$
$$\nu_t = f_t(x)\,dx \otimes \nu,$$
$$\mu_t = f_t(x)\,dx \otimes \mu.$$

These elements are all cohomologically nontrivial, and their cohomology classes generate the Weyl algebra. Observe that, for instance,

$$\frac{\partial}{\partial x}A_{t,j} = f_t'(x)\,dx \otimes \alpha_j = df_t \otimes \alpha_j,$$

so that the infinitesimal translation is cohomologically trivial. This argument applies to all the generators, so we see that the derivation on the Weyl algebra induced from $\frac{\partial}{\partial x}$ must be trivial. $\qquad\qquad\square$

Remark: There is a closely related free field theory on $M = \mathbb{R} \times \Sigma$. Equip Σ with a complex structure, and consider the free theory on M whose fields are

$$\Omega^*(\mathbb{R}) \otimes \Omega^{0,*}(\Sigma)[1] \oplus \Omega^*(\mathbb{R}) \otimes \Omega^{1,*}(\Sigma),$$

concentrated from degrees -1 to 2. (Just to pin down our conventions: $\Omega^0(\mathbb{R})\otimes \Omega^{0,0}(\Sigma)$ is in degree -1, and $\Omega^0(\mathbb{R}) \otimes \Omega^{1,0}(\Sigma)$ is in degree 0.) This theory is topological along $\mathbb{R}$ and holomorphic along Σ (see Chapter 5 for more discussion of this notion). Notice that if we added the differential ∂ for Σ to this complex, we would recover Abelian Chern–Simons on $\mathbb{R} \times \Sigma$.

Let us denote the observables for this theory by $\mathrm{Obs}_{\overline{\partial}}^q$. It is straightforward to mimic the proof above to show that the cohomology of $\pi_* \mathrm{Obs}_{\overline{\partial}}^q$ corresponds to the Weyl algebra on the graded symplectic vector space $H^{*,*}(\Sigma)[1]$, the Dolbeault cohomology of Σ. As this cohomology is isomorphic to the de Rham cohomology of Σ, we see that the cohomology prefactorization algebras $H^*\pi_* \mathrm{Obs}^q$ and $H^*\pi_* \mathrm{Obs}_{\overline{\partial}}^q$ are isomorphic. $\qquad\qquad\Diamond$

4.5.3 3-Manifolds with Boundary

We now examine the observables on a compact, connected, orientable 3-manifold $\overline{M}$ with connected boundary $\partial\overline{M} = \Sigma$ of genus g. Let M denote the open interior, so $M = \overline{M} - \partial\overline{M}$. Speaking loosely, we can view M as a

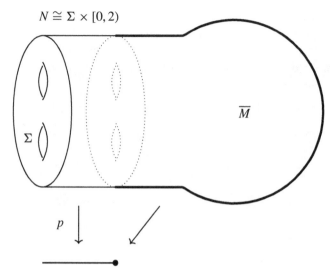

$$N \cong \Sigma \times [0, 2)$$

Figure 4.2. Everything right of the dotted copy of Σ maps to the right endpoint.

"module" over the "algebra" $\Sigma \times \mathbb{R}$, because we can keep stretching the end of M to include a copy of $\Sigma \times \mathbb{R}$. We will show that this picture leads to a precise statement about the observables: the cohomology $H^* \operatorname{Obs}^q(M)$ is a module for the Weyl algebra A_Σ associated to $H^*(\Sigma)[1]$.

Pick a parametrization of a collar neighborhood N of the boundary $\partial \overline{M}$ such that we have a diffeomorphism $\rho : N \xrightarrow{\cong} \Sigma \times [0, 2)$. Let $\pi : N \to [0, 2)$ denote composition with the projection map. Let $N' = \pi^{-1}((0, 1))$ denote a neighborhood of the end of M.

Consider the map $p : M \to (0, 1]$ where

$$p(x) = \begin{cases} \pi(x), & x \in \pi^{-1}((0, 1)) \\ 1, & \text{else} \end{cases},$$

which collapses everything in M outside of N' down to a point. By construction, $p^{-1}((a, 1])$ is diffeomorphic to M for any $0 < a$. Likewise, $p^{-1}((a, b))$ is diffeomorphic to $\Sigma \times (a, b)$ for any $0 < a < b < 1$. See Figure 4.2.

It is clear that $p_* \operatorname{Obs}^q$ on the interval $(0, 1)$ is simply the observables compactified along a surface, as in Proposition 4.5.2. To any open set of the form $(a, 1]$ in $[0, 1]$, we see that $p_* \operatorname{Obs}^q((a, 1])$ is quasi-isomorphic to $\operatorname{Obs}^q(M)$.

Proposition 4.5.3 *The cohomology prefactorization algebra $H^* p_* \operatorname{Obs}^q$ is constructible with respect to the stratification $(0, 1] = (0, 1) \cup \{1\}$. In particular,*

(1) *On the stratum* $(0, 1)$, $H^* p_* \operatorname{Obs}^q$ *corresponds to the Weyl algebra* A_Σ.

(2) *For any open* $(a, 1]$, $H^* p_* \operatorname{Obs}^q((a, 1]) \cong H^*(\operatorname{Obs}^q(M))$.

(3) *The structure maps equip* $H^*(\operatorname{Obs}^q(M))$ *with the structure of a left* A_Σ *module, whose annihilator is the ideal generated by the Lagrangian subspace* $\operatorname{Im} i^* \subset H^*(\Sigma)[1]$ *for the map* $i^* : H^*(\overline{M}; \Sigma)[1] \to H^*(\Sigma)[1]$ *arising from* $i : \Sigma \hookrightarrow \overline{M}$.

Remark: This proposition – more accurately, its proof – indicates how to construct a topological field theory, in the functorial sense, from the bordism category of oriented 2-manifolds with oriented cobordisms as morphisms to the category of graded algebras with bimodules as morphisms. To a surface Σ, we assign the Weyl algebra A_Σ. To a 3-manifold, we associate to the bimodule over Weyl algebras (extending the proposition above to more general 3-manifolds with boundary). The final step in constructing the functor – verifying composition – relies on the gluing properties of prefactorization algebras, which we discuss in Chapter 6, although one could verify it directly by computation in this case. ◇

Proof We have already explained claims (1) and (2), so we tackle (3). It is manifest that $H^*(\operatorname{Obs}^q(M))$ is a left A_Σ module, because $p_* \operatorname{Obs}^q$ is a prefactorization algebra. Thus, it remains to describe the annihilator of this module.

Recall the "half lives, half dies" principle for oriented 3-manifolds: for $\overline{M}$ a compact, orientable 3-manifold with boundary $i : \partial \overline{M} \hookrightarrow \overline{M}$, the image of the map $i^* : H^1(\overline{M}) \to H^1(\partial \overline{M})$ has dimension half that of $H^1(\partial \overline{M})$. (For a proof, see lemma 3.5 of Hatcher (2007).) The image $L = \operatorname{Im} i^*$ is, in fact, a Lagrangian subspace of $H^1(\partial \overline{M})[1]$.

Now consider our situation, where $M = \overline{M} - \partial \overline{M}$ and $\partial \overline{M} = \Sigma$. We know that the neighborhood $N' = p^{-1}((0, 1))$ of the end of M is diffeomorphic to $\Sigma \times (0, 1)$. Let $j : N' \hookrightarrow M$ denote the inclusion. By Poincaré duality for compactly supported de Rham cohomology, we know $H_c^2(N') \cong H^1(\Sigma)$ and $H_c^2(M) \cong H^1(M)$. Hence, the "half lives, half dies" principle shows that for the extension-by-zero map $j_! : H_c^2(N') \to H_c^2(M)$, the kernel $\operatorname{Ker} j_!$ is a Lagrangian subspace. We also see that $j_! : H_c^3(N') \cong H_c^3(M)$ and $j_! : H_c^1(N') \to H_c^1(M)$ is the zero map.

Thus, the graded subspace $\operatorname{Ker} j_! \subset H_c^*(N')$ is Lagrangian.

For the observables, the map $j_!$ describes how the linear observables on N' behave as linear observables on M. The observables in $\operatorname{Ker} j_!$ thus vanish on M, and so the ideal they generate in A_Σ vanishes on the module $H^* \operatorname{Obs}^q(M)$. We need to show this ideal is the entire annihilator. Note that the action of A_Σ is governed by the integration pairing in $H_c^*(M)$, and the extension-by-zero map

$H_c^*(N') \to H_c^*(M)$ respects the integration pairing, so there are no nontrivial relations involving nonlinear observables. □

4.5.4 Knots and Links

Chern–Simons theory is famous for its relationship with knot theory, notably the Vassiliev invariants. Here we will explain a classic example: how the Gauss linking number appears in Abelian Chern–Simons theory. For simplicity, we work in $\mathbb{R}^3$.

Let $K : S^1 \hookrightarrow \mathbb{R}^3$ and $K' : S^1 \hookrightarrow \mathbb{R}^3$ be two disjoint knots. The *linking number* $\ell(K, K')$ of K and K' is the degree of the Gauss map $G_{K,K'} : S^1 \times S^1 \to S^2$ that sends a pair of points $(x, y) \in K \times K'$ to the unit vector $(x - y)/|x - y|$. One way to compute it – indeed, the way Gauss first introduced it – is to pull back the standard volume form on S^2 and integrate over $K \times K'$. Another way is to pick an oriented disc whose boundary is K' and to count, with signs, the number of times K intersects D (we may wiggle D gently to make it transverse to K).

To relate this notion to Chern–Simons theory, we need to find an observable associated to a knot. The central object of Chern–Simons theory is a 1-form A, which we view as describing a connection $d + A$ for the trivial line bundle on $\mathbb{R}^3$. Thus, the natural observable for the knot $K : S^1 \to \mathbb{R}^3$ is

$$O_K(A) = \int_K A,$$

where we have fixed an orientation on S^1. Note that this operator is linear in A.

Remark: The *Wilson loop observable* W_K is $\exp(O_K)$, which is the holonomy of the connection $d + A$ around K. We do not work with it here because we want to work with the symmetric algebra on linear observables, rather than the completed symmetric algebra. In the non-Abelian setting, we do work with the completed symmetric algebra, so the analogous Wilson loop observable lives in our quantum observables for non-Abelian Chern–Simons theory. ◊

There is a minor issue with this observable, however: it is distributional in nature, as its support sits on the knot K. It is easy to find a smooth (or "smeared") observable that captures the same information. (We know such a smooth observable exists abstractly by Lemma 4.2.1, but here we will give an explicit representative.) The key is again to use the pushforward.

Consider the 3-manifold $T = S^1 \times \mathbb{R}^2$. We view the core $C = S^1 \times (0, 0) \subset T$ as the "distinguished" knot in T. Any embedding $\kappa : T \hookrightarrow \mathbb{R}^3$ induces a pushforward of observables so that we can understand the observables supported in tubular neighborhood $\kappa(T)$ of the knot $\kappa(C)$ by understanding the observables on T.

We now explain how to find a smeared analog of O_C, where C is the distinguished knot, in $\mathrm{Obs}^q(T)$. Fix an orientation of T and an orientation of the embedded circle $C \subset T$. Let μ be a compactly supported volume form on $\mathbb{R}^2$ whose support contains the origin and whose total integral is 1. Consider its pullback $\mu_T = \pi^*\mu$ along the projection $\pi : T \to \mathbb{R}^2$. Using Cartesian coordinates r, z on $\mathbb{R}^2$ and coordinates θ, r, z on T, we have $\mu_T = \phi(r, z)\,\mathrm{d}r \wedge \mathrm{d}z$, where ϕ is a nonnegative bump function with support at the origin $(0, 0)$. Note that μ_T is independent of θ. We thus have a linear function, or observable,

$$O_\mu : A \mapsto \int_T A \wedge \mu_T$$

on any 1-form A. Using the coordinates from the preceding, we see

$$O_\mu(A) = \int_{\mathbb{R}^2} \phi(x, y) O_{S^1 \times (r,z)}(A)\,\mathrm{d}r\,\mathrm{d}z,$$

so that O_μ is just a smoothed version of O_C.

Let's return to the linking number. Given two knots K and K', we can find thickened embeddings $\kappa : T \to \mathbb{R}^3$ and $\kappa' : T \to \mathbb{R}^3$ whose images are disjoint and where $\kappa(C) = K$ and $\kappa'(C) = K'$. Let μ_κ denote the image of μ_T in $\mathrm{Obs}^q(\kappa(T))$ under the pushforward by κ, and likewise for $\mu_{\kappa'}$. This observable μ_κ is a smeared version of O_K.

Lemma 4.5.4 *In $H^0(\mathrm{Obs}^q(\mathbb{R}^3)) \cong \mathbb{R}[\hbar]$, the cohomology class $[\mu_\kappa \mu_{\kappa'}]$ of the product observable $\mu_\kappa \mu_{\kappa'}$ is equal to $\hbar\ell(K, K')$.*

Proof We outline here one approach, in the spirit of our preceding work. Below, in the example in Section 4.6.3 we explain the standard proof.

The Poincaré lemma for compactly supported forms tells us that $H_c^2(\mathbb{R}^3) = 0$. As $\mu_{\kappa'}$ is closed, we know there is some 1-form ρ such that $\mathrm{d}\rho = \mu_{\kappa'}$. The product observable $\mu_\kappa \rho$ then has cohomological degree -1, and by construction we have

$$(\mathrm{d} + \hbar\triangle)\mu_\kappa \rho = \mu_\kappa \mu_{\kappa'} + \hbar \int_{\mathbb{R}^3} \mu_\kappa \wedge \rho.$$

Thus we need to verify that, up to a sign, the integral agrees with the linking number.

We do this by making a good choice of ρ. There is a smooth embedding of a closed 2-disc $D \hookrightarrow \mathbb{R}^3$ whose oriented boundary is K'. Extend this embedding to a thickened embedding $i : \mathbb{R}^3 \hookrightarrow \mathbb{R}^3$, where D sits inside the input copy of $\mathbb{R}^3$ as $D \times 0$ and where this thickened embedding contains the image of κ'. Then $i^*\mu_{\kappa'}$ is cohomologically trivial, by the Poincaré lemma. Pick a 1-form

ρ such that $d\rho = i^*\mu_{\kappa'}$. Note that ρ has support in $i(\mathbb{R}^3)$. The product $\mu_\kappa \wedge \rho$ is smeared version, under Poincaré duality, of the intersection of $i(D)$ with K. Thus its integral counts the signed intersection. $\qquad\square$

4.6 Another Take on Quantizing Classical Observables

We have given an abstract definition of the prefactorization algebra of quantum observables of a free field theory as the factorization envelope of a certain Heisenberg dg Lie algebra. Our goal in this section is to provide another useful description.

The prefactorization algebra of quantum observables Obs^q, viewed as a graded prefactorization algebra with no differential, coincides with $\mathrm{Obs}^{cl}[\hbar]$. In particular, the structure maps are the same. The only difference between Obs^q and $\mathrm{Obs}^{cl}[\hbar]$ is in the differential.

In deformation quantization, we deform, however, the product of observables. We saw in Section 4.3 that taking cohomology, the structure maps for the quantum observables *do* change and correspond to the Weyl algebra. The change in differential modifies the structure maps at the level of cohomology. However, it is possible to approach quantization differently and deform the structure maps at the cochain level, instead of the differential.

Hence, in this section we will give an alternative, but equivalent, description of Obs^q. We will construct an isomorphism of precosheaves $\mathrm{Obs}^q \cong \mathrm{Obs}^{cl}[\hbar]$ that is compatible with differentials. This isomorphism is not compatible, however, with the factorization product. Thus, this isomorphism induces a deformed factorization product on $\mathrm{Obs}^{cl}[\hbar]$ corresponding to the factorization product on Obs^q.

In other words, instead of viewing Obs^q as being obtained from Obs^{cl} by keeping the factorization product fixed but deforming the differential, we will show that it can be obtained from Obs^{cl} by keeping the differential fixed but deforming the product.

One advantage of this alternative description is that it is easier to construct correlation functions and vacua in this language. We will now develop these ideas in the case of the free scalar field.

4.6.1 Green's Functions

The isomorphism we will construct between the cochain complexes of quantum and classical observables relies on a Green's function for the Laplacian.

Definition 4.6.1 A *Green's function* is a distribution G on $M \times M$, preserved under reflection across the diagonal (i.e., symmetric) and satisfying

$$(\triangle \otimes 1)G = \delta_\triangle,$$

where $\delta_\triangle$ is the δ-distribution on the diagonal $M \hookrightarrow M \times M$.

A Green's function for the Laplacian with mass satisfies the equation

$$(\triangle \otimes 1)G + m^2 G = \delta_\triangle.$$

(The convention is that $\triangle$ has positive eigenvalues, so that on $\mathbb{R}^n$, $\triangle = -\sum \frac{\partial}{\partial x_i}^2$.)

If M is compact, then there is no Green's function for the Laplacian. Instead, there is a unique function $\widetilde{G}$ satisfying

$$(\triangle \otimes 1)\widetilde{G} = \delta_\triangle - 1.$$

However, if we introduce a nonzero mass term, then the operator $\triangle + m^2$ on $C^\infty(M)$ is an isomorphism, so that there is a unique Green's function.

If M is noncompact, then there can be a Green's function for the Laplacian without mass term. For example, if M is $\mathbb{R}^n$, then a choice of Green's function is

$$G(x, y) = \begin{cases} \frac{1}{4\pi^{d/2}} \Gamma(d/2 - 1) \, |x - y|^{2-n} & \text{if } n \neq 2 \\ -\frac{1}{2\pi} \log |x - y| & \text{if } n = 2. \end{cases}$$

As the reader will see, the role of the Green's function is to produce an isomorphism, so different choices lead to different isomorphisms. These choices do *not* change the quantum observables themselves, merely how one "promotes" a classical observable to a quantum one.

4.6.2 The Isomorphism of Graded Vector Spaces

Let us now turn to the construction of the isomorphism of graded vector spaces between $\mathrm{Obs}^{cl}(U)$ and $\mathrm{Obs}^q(U)$ in the presence of a Green's function.

The underlying graded vector space of $\mathrm{Obs}^q(U)$ is

$$\mathrm{Sym}(C_c^\infty(U)[1] \oplus C_c^\infty(U))[\hbar].$$

In general, for any vector space V, each element $P \in (V^\vee)^{\otimes 2}$ defines a differential operator ∂_P of order two on $\mathrm{Sym}\, V$, viewed as a commutative algebra, where ∂_P is uniquely characterized by the conditions that it is zero on $\mathrm{Sym}^{\leq 1} V$ and that on $\mathrm{Sym}^2 V$ it is given by contraction with P. The same holds when we define the symmetric algebra using the completed tensor product.

In the same way, for every distribution P on $U \times U$, we can define a continuous, second-order differential operator on $\mathrm{Sym}(C_c^\infty(U)[1] \oplus C_c^\infty(U))$ that is uniquely characterized by the properties that it vanishes on $\mathrm{Sym}^{\leq 1}$, that it vanishes elements of negative cohomological degree in $\mathrm{Sym}^2(C_c^\infty(U)[1] \oplus C_c^\infty(U))$, and that

$$\partial_P(\phi\psi) = \int_{U \times U} P(x,y)\phi(x)\psi(y)$$

for any $\phi, \psi \in C_c^\infty(U)$ of cohomological degree zero. (On the left-hand side, we view $\phi\psi$ as an element of Sym^2.)

Choose a Green's function G for the Laplacian on M. Thus, G restricts to a Green's function for the Laplacian on any open subset U of M.

Therefore, we can define a second-order differential operator ∂_G on the algebra $\mathrm{Sym}(C_c^\infty(U)[1] \oplus C_c^\infty(U))$. We extend this operator by $\mathbb{R}[\hbar]$-linearity to an operator on the graded vector space $\mathrm{Obs}^q(U)$.

Now, the differential on $\mathrm{Obs}^q(U)$ can be written as $\mathrm{d} = \mathrm{d}_1 + \mathrm{d}_2$, where d_1 is a first-order differential operator and d_2 is a second-order operator. The operator d_1 is the derivation arising from the differential on the complex $C_c^\infty(U) \xrightarrow{\Delta} C_c^\infty(U)$. The operator d_2 arises from the Lie bracket on the Heisenberg dg Lie algebra; it is a continuous, $\hbar$-linear, second-order differential operator uniquely characterized by the property that

$$\mathrm{d}_2(\phi^{-1}\phi^0) = \hbar \int_M \phi^{-1}(x)\phi^0(x)$$

for ϕ^k in the copy of $C_c^\infty(U)$ in degree k, sitting inside of $\mathrm{Obs}^q(U)$.

The Green's function G satisfies

$$((\Delta + m^2) \otimes 1)G = \delta_\Delta$$

where δ_Δ is the Green's function on the diagonal. It follows that

$$[\hbar\partial_G, \mathrm{d}_1] = \mathrm{d}_2.$$

Indeed, both sides of this equation are second-order differential operators, so to check the equation, it suffices to calculate how the act on an element of $\mathrm{Sym}^2(C_c^\infty(U)[1] \oplus C_c^\infty(U))$. If ϕ^k denotes an element of the copy of $C_c^\infty(U)$

in cohomological degree k, we have

$$
\begin{aligned}
\hbar \partial_G d_1(\phi^0 \phi^{-1}) &= \hbar \partial_G(\phi^0(\triangle + m^2)\phi^{-1}) \\
&= \int_{U \times U} G(x,y)\phi^0(x)((\triangle + m^2)\phi^{-1})(y) \\
&= \int_{U \times U} ((\triangle_y + m^2)G(x,y))\phi^0(x)\phi^{-1}(y) \\
&= \int_U \phi^0(x)\phi^{-1}(x) \\
&= d_2(\phi^0\phi^{-1}).
\end{aligned}
$$

On the first line, the element $\phi^0\phi^{-1}$ lives in Sym^2, as does its image under d_1. On the fourth line, we have used the fact that $\triangle_y + m^2$ applied to $G(x,y)$ is the delta-distribution on the diagonal.

It is also immediate that ∂_G commutes with d_2. Thus, if we define $W(\alpha) = e^{\hbar \partial_G}(\alpha)$, we have

$$(d_1 + d_2)W(\alpha) = W(d_1\alpha).$$

In other words, there is an $\mathbb{R}[\hbar]$-linear cochain isomorphism

$$W_U : \mathrm{Obs}^{cl}(U)[\hbar] \to \mathrm{Obs}^q(U),$$

where the domain is the complex $\mathrm{Obs}^{cl}(U)[\hbar]$ with differential d_1 and the codomain is the same graded vector space with differential $d_1 + d_2$. This isomorphism holds for every open U.

Note that W is *not* a map of prefactorization algebras. Thus, W induces a factorization product (i.e., structure maps) on classical observables that quantizes the original factorization product. Let us denote this quantum product by $\star_\hbar$, whereas the original product on classical observables will be denoted by $\cdot$. If U, V are disjoint open subsets of M with $\alpha \in \mathrm{Obs}^{cl}(U)$, $\beta \in \mathrm{Obs}^{cl}(V)$, we have

$$\alpha \star_\hbar \beta = e^{-\hbar \partial_G}\left(\left(e^{\hbar \partial_G}\alpha\right) \cdot \left(e^{\hbar \partial_G}\beta\right)\right).$$

This product is an analog of the Moyal formula for the product on the Weyl algebra.

4.6.3 Beyond the Free Scalar Field

We have described this construction for the case of a free scalar field theory. This construction can be readily generalized to the case of an arbitrary free theory. Suppose we have such a theory on a manifold M, with space of fields

$\mathscr{E}(M)$ and differential d. Instead of a Green's function, we require a symmetric and continuous linear operator $G : (\mathscr{E}(M)[1])^{\otimes 2} \to \mathbb{R}$ such that

$$G\,\mathrm{d}(e_1 \otimes e_2) = \langle e_1, e_2 \rangle$$

where $\langle -, - \rangle$ is the pairing on $\mathscr{E}(M)$, which is part of the data of a free field theory. (When M is compact and $H^*(\mathscr{E}(M)) = 0$, the propagator of the theory satisfies this property. For $M = \mathbb{R}^n$, we can generally construct such a G from the Green's function for the Laplacian.) In this context, the operator $e^{\hbar \partial_G}$ is, in the terminology of Costello (2011b), the renormalization group flow operator from scale zero to scale ∞.

This construction often makes it easier to analyze the structure of the quantum observables. We will use it, for instance, to understand the correlation functions of some chiral conformal field theories in the next chapter.

Example: We can also use it to provide another proof of Lemma 4.5.4, which asserts that Abelian Chern–Simons knows about linking number. Recall that for two disjoint knots K and K', we introduced smoothed versions, μ_K and $\mu_{K'}$, of the currents with support along the knots.

It is easy to check that, up to a constant, the 2-form

$$\omega = \frac{1}{|x-y|^3} \sum_{i=1}^{3} (-1)^i (x_i - y_i)\, \mathrm{d}(x_1 - y_1) \wedge \mathrm{d}\widehat{(x_i - y_i)} \wedge \mathrm{d}(x_3 - y_3)$$

on $\mathbb{R}^3 \times \mathbb{R}^3$ acts as a Green's function for the de Rham complex on $\mathbb{R}^3$. It is the pullback of the standard volume form on S^2 along the projection

$$(x, y) \mapsto (x - y)/|x - y|$$

from the configuration space of two distinct points in $\mathbb{R}^3$ to S^2. (Indeed, ω is the standard propagator, as discussed in Costello (2007a) or Kontsevich (1994).) Using the $\star_\hbar$ formula from given earlier, one can see that the cohomology class of a product observable $\mu_K \mu_{K'}$ is precisely the integral formula for the Gauss linking number.

To be explicit, we see that

$$\mu_K \star_\hbar \mu_{K'} = \mu_K \mu_{K'} + \hbar \int_{x,y \in \mathbb{R}^3} \mu_K(x)\mu_{K'}(y)\omega(x,y).$$

Our tubular embeddings, $\kappa : T \to \mathbb{R}^3$ and $\kappa' : T \to \mathbb{R}^3$, allow us to rewrite the integral as an integral over $T \times T \cong C \times C \times \mathbb{R}^4$. The product $C \times C$ of two circles parametrize $K \times K'$, and the last four coordinates parametrize small shifts of $K \times K'$ within their tubular neighborhoods. Fixing such a shift, we are integrating ω over $K \times K'$, and hence obtain the linking number $\ell(K, K')$

because ω is a pullback of the volume form on S^2. When we integrate over the shifts, we are integrating $\mu_T(x) \wedge \mu_T(y)$, which integrates to 1 by construction. (We have not checked the overall sign, of course.) ◇

4.7 Correlation Functions

Suppose we have a free field theory on a compact manifold M, with the property that $H^*(\mathscr{E}(M)) = 0$. As an example, consider the massive scalar field theory on M where

$$\mathscr{E} = C_M^\infty \xrightarrow{\Delta + m^2} C_M^\infty.$$

Our conventions are such that the eigenvalues of Δ are nonnegative. Adding a nonzero mass term m^2 gives an operator with no zero eigenvalues, so that this complex has no cohomology, by Hodge theory.

In this situation, the dg Lie algebra $\widehat{\mathscr{E}}(M)$ we constructed above has cohomology spanned by the central element $\hbar$ in degree 1. It follows that there is an isomorphism of $\mathbb{R}[\hbar]$-modules

$$H^*(\text{Obs}^q(M)) \cong \mathbb{R}[\hbar].$$

Let us normalize this isomorphism by asking that the element 1 in

$$\text{Sym}^0(\widehat{\mathscr{E}}[1]) = \mathbb{R} \subset C_*(\widehat{\mathscr{E}}[1])$$

is sent to $1 \in \mathbb{R}[\hbar]$.

Definition 4.7.1 In this situation, we define the *correlation functions* of the free theory as follows. For a finite collection $U_1, \ldots, U_n \subset M$ of disjoint opens and a closed element $O_i \in \text{Obs}^q(U_i)$ for each open, we define

$$\langle O_1 \cdots O_n \rangle = [O_1 \cdots O_n] \in H^*(\text{Obs}^q(M)) = \mathbb{R}[\hbar].$$

Here $O_1 \cdots O_n \in \text{Obs}^q(M)$ denotes the image under the structure map

$$\bigotimes_i \text{Obs}^q(U_i) \to \text{Obs}^q(M)$$

of the tensor product $O_1 \otimes \cdots \otimes O_n$.

In other words, our normalization gives a value in $\mathbb{R}[\hbar]$ for each cohomology class, and we use this to read off the "expected value" of the product observable.

The map W constructed in the previous section allows us to calculate correlation functions. Since we have a nonzero mass term and M is compact, there is a unique Green's function for the operator $\triangle + m^2$.

Lemma 4.7.2 (Wick's lemma) *Let* $U_1, \ldots, U_n$ *be pairwise disjoint open sets. Let*

$$\alpha_i \in \mathrm{Obs}^{cl}(U_i) = \mathrm{Sym}(C_c^\infty(U_i)[1] \oplus C_c^\infty(U_i))$$

be classical observables, and let

$$W(\alpha_i) = e^{\hbar\partial_G}\alpha_i \in \mathrm{Obs}^q(U_i)$$

be the corresponding quantum observables under the isomorphism W of cochain complexes. Then

$$\langle W(\alpha_1)\cdots W(\alpha_n)\rangle = W^{-1}\left(W(\alpha_1)\cdots W(\alpha_n)\right)(0) \in \mathbb{R}[\hbar].$$

On the right-hand side, $W(\alpha_1) \cdot W(\alpha_2)$ *indicates the product in the algebra* $\mathrm{Sym}(C^\infty(M)[1] \oplus C^\infty(M))[\hbar]$. *The symbol* (0) *indicates evaluating a function on* $C^\infty(M)[1] \oplus C^\infty(M)[\hbar]$ *at zero, i.e., taking its component in* Sym^0.

Proof The map W is an isomorphism of cochain complexes between $\mathrm{Obs}^q(U)$ and $\mathrm{Obs}^{cl}(U)[\hbar]$ for every open subset U of M. As above, let $\star_\hbar$ denote the factorization product on $\mathrm{Obs}^{cl}[\hbar]$ corresponding, under the isomorphism W, to the factorization product on Obs^q. That is, we apply W to each input, multiply in Obs^q, and then apply W^{-1} to the output: explicitly, we have

$$\alpha_1 \star_\hbar \cdots \star_\hbar \alpha_n = e^{-\hbar\partial_G}\left((e^{\hbar\partial_G}\alpha_1)\cdots(e^{\hbar\partial_G}\alpha_n)\right).$$

Since W gives an isomorphism of prefactorization algebras between Obs^q and $(\mathrm{Obs}^{cl}[\hbar], \star_\hbar)$, the correlation function for the observables $W(\alpha_i)$ is the cohomology class of $\alpha_1 \star_\hbar \cdots \star_\hbar \alpha_n$ in $\mathrm{Obs}^{cl}(M)[\hbar]$. The map $\mathrm{Obs}^{cl}(M)[\hbar] \to \mathbb{R}[\hbar]$ sending an observable α to $\alpha(0)$ is an $\mathbb{R}[\hbar]$-linear cochain map inducing an isomorphism on cohomology, and it sends 1 to 1. There is a unique such map (unique up to cochain homotopy). Therefore,

$$\langle W(\alpha_1)\cdots W(\alpha_n)\rangle = \left(\alpha_1 \star_\hbar \cdots \star_\hbar \alpha_n\right)(0),$$

as claimed.   □

This statement does not resemble the usual statement in a physics text, so we unpack it in an example to clarify the relationship with other versions of Wick's lemma.

As an example, let us suppose that we have two linear observables α_1, α_2 of cohomological degree 0, defined on open sets U, V. Thus, $\alpha_1 \in C_c^\infty(U)$ and

$\alpha_2 \in C_c^{\infty}(V)$. As these are linear observables, the second-order operator ∂_G does nothing, so $W(\alpha_k) = \alpha_k$. But the product is quadratic so we have

$$W^{-1}(\alpha_1 \alpha_2) = \alpha_1 \alpha_2 - \hbar \partial_G(\alpha_1 \alpha_2).$$

As

$$\partial_G(\alpha_1 \alpha_2) = \int_{M \times M} G(x, y) \alpha_1(x) \alpha_2(y),$$

it follows that

$$\langle W(\alpha_1) W(\alpha_2) \rangle = -\hbar \int_{M \times M} G(x, y) \alpha_1(x) \alpha_2(y),$$

which is what a physicist would write down as the expected value of two linear observables.

Remark: We have set things up so that we are computing the functional integral against the measure $e^{S/\hbar} \, d\mu$,where S is the action functional and $d\mu$ is the (non-existent) "Lebesgue measure" on the space of fields. Physicists often use the measures $e^{-S/\hbar}$ or $e^{iS/\hbar}$. By changing $\hbar$ (e.g., $\hbar \mapsto -\hbar$), one can move between these different conventions. ◇.

4.8 Translation-Invariant Prefactorization Algebras

In this section we will analyze in detail the notion of *translation-invariant* prefactorization algebras on $\mathbb{R}^n$. Most theories from physics possess this property. For one-dimensional field theories, one often uses the phrase "time-independent Hamiltonian" to indicate this property, and in this section we will explain, in examples, how to relate the Hamiltonian formalism of quantum mechanics to our approach.

4.8.1 The Definition

We now turn to the definition of a translation-invariant prefactorization algebra. It is a special case of the definition of an equivariant prefactorization algebra in Section 3.7 in Chapter 3, but we go through the definition here in some detail.

Remark: We use the term "translation-invariance" here in place of translation-equivariance. The motivation is that a field theory on $\mathbb{R}^n$ whose action functional is translation-invariant will produce a translation-equivariant prefactorization algebra. We hope this slight inconsistency of terminology causes no confusion. ◇

If $U \subset \mathbb{R}^n$ and $x \in \mathbb{R}^n$, let

$$\tau_x U = \{y \; : \; y - x \in U\}$$

denote the translate of U by x.

Definition 4.8.1 A prefactorization algebra $\mathcal{F}$ on $\mathbb{R}^n$ is *discretely translation-invariant* if we have isomorphisms

$$\tau_x : \mathcal{F}(U) \cong \mathcal{F}(\tau_x U)$$

for all $x \in \mathbb{R}^n$ and all open subsets $U \subset \mathbb{R}^n$. These isomorphisms must satisfy a few conditions. First, we require that $\tau_x \circ \tau_y = \tau_{x+y}$ for every $x, y \in \mathbb{R}^n$. Second, for all disjoint open subsets $U_1, \ldots, U_k$ in V, the diagram

$$
\begin{array}{ccc}
\mathcal{F}(U_1) \otimes \cdots \otimes \mathcal{F}(U_k) & \xrightarrow{\tau_x} & \mathcal{F}(\tau_x U_1) \otimes \cdots \otimes \mathcal{F}(\tau_x U_k) \\
\downarrow & & \downarrow \\
\mathcal{F}(V) & \xrightarrow{\quad\tau_x\quad} & \mathcal{F}(\tau_x V)
\end{array}
$$

commutes. (Here the vertical arrows are the structure maps of the prefactorization algebra.)

Example: Consider the prefactorization algebra of quantum observables of the free scalar field theory on $\mathbb{R}^n$, as defined in Section 4.2.6. This theory has as its complex of fields

$$\mathscr{E} = \left\{ C^\infty \xrightarrow{\triangle} C^\infty[-1] \right\}.$$

By definition, $\mathrm{Obs}^q(U)$ is the Chevalley–Eilenberg chains of a -1-shifted central extension $\widehat{\mathscr{E}_c}(U)$ of $\mathscr{E}_c(U)$, with cocycle defined by $\int \phi^0 \phi^1$ where ϕ^k denotes the field in cohomological degree k.

This Heisenberg algebra is defined using only the Riemannian structure on $\mathbb{R}^n$, and it is therefore automatically invariant under all isometries of $\mathbb{R}^n$. In particular, the resulting prefactorization algebra is discretely translation-invariant. $\Diamond$

We are interested in a refined version of this notion, where the structure maps of the prefactorization algebra depend smoothly on the position of the open sets. It is a bit subtle to talk about "smoothly varying an open set," and to do this, we introduce some notation.

Recall the notion of a *derivation* of a prefactorization algebra on a manifold M.

Definition 4.8.2 A *degree k derivation* of a prefactorization algebra $\mathcal{F}$ is a collection of maps $D_U : \mathcal{F}(U) \to \mathcal{F}(U)$ of cohomological degree k for each open subset $U \subset M$, with the property that for any finite collection $U_1, \ldots, U_n \subset V$ of disjoint opens and an element $\alpha_i \in \mathcal{F}(U_i)$ for each open, the derivation acts by a Leibniz rule on the structure maps:

$$D_V m_V^{U_1, \ldots, U_n}(\alpha_1, \ldots, \alpha_n) = \sum_i (-1)^{k(|\alpha_1| + \cdots + |\alpha_{i-1}|)}$$
$$m_V^{U_1, \ldots, U_n}(\alpha_1, \ldots, D_{U_i}\alpha_i, \ldots, \alpha_n),$$

where the sign arises from the usual Koszul sign rule.

Example: Let us consider, again, the prefactorization algebra of the free scalar field on $\mathbb{R}^n$. Observables on U are Chevalley–Eilenberg chains of the Heisenberg algebra $\widehat{\mathscr{E}_c}(U)$.

Let X denote a Killing vector field on $\mathbb{R}^n$ (i.e., X is an infinitesimal isometry). For example, we could take X to be the translation vector field $\frac{\partial}{\partial x^j}$. Then the Heisenberg dg Lie algebra has a derivation where $\phi^k \mapsto X\phi^k$, for $k = 0, 1$, and the central element $\hbar$ is annihilated. (Recall that ϕ^k is notation for an element of the copy of $C_c^\infty(U)$ situated in degree k.) Because X is a Killing vector field on $\mathbb{R}^n$, it commutes with the differential on the Heisenberg algebra and satisfies the equality

$$\int (X\phi^0)\phi^1 + \int \phi^0(X\phi^1) = 0.$$

Hence, X is a derivation of dg Lie algebras.

By naturality, X extends to an endomorphism of $\mathrm{Obs}^q(U) = C_*(\widehat{\mathscr{E}_c}(U))$. This endomorphism defines a derivation of the prefactorization algebra Obs^q of observables of the free scalar field theory. $\Diamond$

All the derivations together $\mathrm{Der}^*(\mathcal{F})$ form a differential graded Lie algebra. The differential is defined by $(dD)_U = [d_U, D_U]$, where d_U is the differential on $\mathcal{F}(U)$. The Lie bracket is defined by $[D, D']_U = [D_U, D'_U]$.

The concept of derivation allows us to talk about the action of a dg Lie algebra $\mathfrak{g}$ on a prefactorization algebra $\mathcal{F}$. Such an action is simply a homomorphism of differential graded Lie algebras $\mathfrak{g} \to \mathrm{Der}^*(\mathcal{F})$.

Next, we introduce some notation that will help us describe the smoothness conditions for a discretely translation-invariant prefactorization algebra.

Let $U_1, \ldots, U_k \subset V$ be disjoint open subsets. Let $\mathrm{Disj}(U_1, \ldots, U_k \mid V) \subset (\mathbb{R}^n)^k$ be the set of k-tuples $(x_1, \ldots, x_k)$ such that the translates $\tau_{x_1} U_1, \ldots, \tau_{x_k} U_k$ are still disjoint and contained in V. It parametrizes the way we can move the open sets without causing overlaps. Let us assume that $\mathrm{Disj}(U_1, \ldots, U_k \mid V)$

has nonempty interior, which happens when the closure of the U_i are disjoint and contained in V.

Let $\mathcal{F}$ be any discretely translation-invariant prefactorization algebra. For each $(x_1, \ldots, x_k) \in \mathrm{Disj}(U_1, \ldots, U_k \mid V)$, we have a multilinear map obtained as a composition

$$m_{x_1,\ldots,x_k} : \mathcal{F}(U_1) \times \cdots \times \mathcal{F}(U_k) \to \mathcal{F}(\tau_{x_1} U_1) \times \cdots \times \mathcal{F}(\tau_{x_k} U_k) \to \mathcal{F}(V),$$

where the second map arises from the inclusion $\tau_{x_1} U_1 \sqcup \cdots \sqcup \tau_{x_k} U_k \hookrightarrow V$.

Definition 4.8.3 A discretely translation-invariant prefactorization algebra $\mathcal{F}$ is *smoothly translation-invariant* if the following conditions hold.

(i) The map $m_{x_1,\ldots,x_k}$ above depends smoothly on $(x_1, \ldots, x_k) \in \mathrm{Disj}$ $(U_1, \ldots, U_k \mid V)$.

(ii) The prefactorization algebra $\mathcal{F}$ is equipped with an action of the Abelian Lie algebra $\mathbb{R}^n$ of translations. If $v \in \mathbb{R}^n$, we will denote the corresponding action maps by

$$\frac{\mathrm{d}}{\mathrm{d}v} : \mathcal{F}(U) \to \mathcal{F}(U).$$

We view this Lie algebra action as an infinitesimal version of the global translation invariance.

(iii) The infinitesimal action is compatible with the global translation invariance in the following sense. For $v \in \mathbb{R}^n$, let $v_i \in (\mathbb{R}^n)^k$ denote the vector $(0, \ldots, v, \ldots, 0)$, with v placed in the i-slot and 0 in the other $k - 1$ slots. If $\alpha_i \in \mathcal{F}(U_i)$, then we require that

$$\frac{\mathrm{d}}{\mathrm{d}v_i} m_{x_1,\ldots,x_k}(\alpha_1, \ldots, \alpha_k) = m_{x_1,\ldots,x_k}\left(\alpha_1, \ldots, \frac{\mathrm{d}}{\mathrm{d}v}\alpha_i, \ldots, \alpha_k\right).$$

When we refer to a translation-invariant prefactorization algebra without further qualification, we will always mean a smoothly translation-invariant prefactorization algebra.

Example: We have already seen that the prefactorization algebra of the free scalar field theory on $\mathbb{R}^n$ is discretely translation invariant and is also equipped with an action of the Abelian Lie algebra $\mathbb{R}^n$ by derivations. It is easy to verify that this prefactorization algebra is smoothly translation-invariant. ◊

Example: This example is a special case of the example in Section 3.7.3.

Let $\mathcal{F}$ be the cohomology of the prefactorization algebra of observables of the free scalar field theory on $\mathbb{R}$ with mass m. We have seen in Section 4.3 that the corresponding associative algebra A is the Weyl algebra, generated by $p, q, \hbar$ with commutation relation $[p, q] = \hbar$.

It is straightforward to verify that $\mathcal{F}$ is a locally constant, smoothly translation invariant prefactorization algebra on $\mathbb{R}$, valued in vector spaces. As the action of $\mathbb{R}$ on A is smooth, it differentiates to an infinitesimal action of the Lie algebra $\mathbb{R}$ on A by derivations. The basis element $\frac{\partial}{\partial x}$ of $\mathbb{R}$ becomes a derivation H of A, called the *Hamiltonian*. The Hamiltonian here is given by $H(a) = \frac{1}{2\hbar}[p^2 - m^2q^2, a]$. $\qquad\qquad\qquad\qquad\qquad\qquad\qquad\qquad\qquad$ $\Diamond$

4.8.2 An Operadic Reformulation

Next, we will explain how to think of the structure of a translation-invariant prefactorization algebra on $\mathbb{R}^n$ in more operadic terms. This description has a lot in common with the E_n algebras familiar from topology.

Let $r_1, \ldots, r_k, s \in \mathbb{R}_{>0}$. Let

$$\mathrm{Discs}_n(r_1, \ldots, r_k \mid s) \subset (\mathbb{R}^n)^k$$

be the (possibly empty) open subset consisting of k-tuples $x_1, \ldots, x_k \in \mathbb{R}^n$ with the property that the closures of the balls $B_{r_i}(x_i)$ are all disjoint and contained in $B_s(0)$. Here, $B_r(x)$ denotes the open ball of radius r around x.

Definition 4.8.4 Let Discs_n be the $\mathbb{R}_{>0}$-colored operad in the category of smooth manifolds whose space of k-ary morphisms $\mathrm{Discs}_n(r_1, \ldots, r_k \mid s)$ between $r_i, s \in \mathbb{R}_{>0}$ is described previously.

Note that a colored operad is the same thing as a multicategory. An $\mathbb{R}_{>0}$-colored operad is thus a multicategory whose set of objects is $\mathbb{R}_{>0}$.

The essential data of the colored operad structure on Discs_n are the following. We have maps

$$\circ_i : \mathrm{Discs}_n(r_1, \ldots, r_k \mid t_i) \times \mathrm{Discs}_n(t_1, \ldots, t_m \mid s)$$
$$\to \mathrm{Discs}_n(t_1, \ldots, t_{i-1}, r_1, \ldots, r_k, t_{i+1}, \ldots, t_m \mid s).$$

This map is defined by inserting the outgoing ball (of radius t_i) of a configuration $x \in \mathrm{Discs}_n(r_1, \ldots, r_k \mid t_i)$ into the ith incoming ball of a point $y \in \mathrm{Discs}_n(t_1, \ldots, t_k \mid s)$. These maps satisfy the natural associativity and commutativity properties of a multicategory.

Next, let $\mathcal{F}$ be a translation-invariant prefactorization algebra on $\mathbb{R}^n$. Let

$$\mathcal{F}_r = \mathcal{F}(B_r(0))$$

denote the cochain complex that $\mathcal{F}$ assigns to a ball of radius r. This notation is reasonable because translation invariance gives us an isomorphism between $\mathcal{F}(B_r(0))$ and $\mathcal{F}(B_r(x))$ for any $x \in \mathbb{R}^n$.

The structure maps for a translation-invariant prefactorization algebra yield, for each configuration $p \in \mathrm{Discs}_n(r_1, \ldots, r_k \mid s)$, a multiplication operation

$$m[p] : \mathcal{F}_{r_1} \times \cdots \times \mathcal{F}_{r_k} \to \mathcal{F}_s.$$

The map $m[p]$ is a smooth multilinear map of differentiable spaces; and furthermore, this map depends smoothly on p.

These operations make the complexes $\mathcal{F}_r$ into an algebra over the $\mathbb{R}_{>0}$-colored operad $\mathrm{Discs}_n(r_1, \ldots, r_k \mid s)$, valued in the multicategory of differentiable cochain complexes. In addition, the complexes $\mathcal{F}_r$ are endowed with an action of the Abelian Lie algebra $\mathbb{R}^n$. This action is by derivations of the Discs_n-algebra $\mathcal{F}$ compatible with the action of translation on Discs_n, as described previously.

Now, let us unravel explicitly what it means to be such a Discs_n algebra.

The first property is that, for each configuration $p \in \mathrm{Discs}_n(r_1, \ldots, r_k \mid s)$, the map $m[p]$ is a multilinear map of cohomological degree 0, compatible with differentials.

Second, let N be a manifold and let $f_i : N \to \mathcal{F}_{r_i}^{d_i}$ be smooth maps into the space $\mathcal{F}_{r_i}^{d_i}$ of elements of degree d_i. The smoothness properties of the map $m[p]$ mean that the map

$$
\begin{array}{ccc}
N \times \mathrm{Discs}_n(r_1, \ldots, r_k \mid s) & \to & \mathcal{F}_s \\
(x, p) & \mapsto & m[p](f_1(x), \ldots, f_k(x))
\end{array}
$$

is smooth. Thus, we can work with smooth families of multiplications.

Next, note that a permutation $\sigma \in S_k$ gives an isomorphism

$$\sigma : \mathrm{Discs}_n(r_1, \ldots, r_k \mid s) \to \mathrm{Discs}_n(r_{\sigma(1)}, \ldots, r_{\sigma(k)} \mid s).$$

We require that for each $p \in \mathrm{Discs}_n(r_1, \ldots, r_k \mid s)$ and collection of elements $\alpha_i \in \mathcal{F}_{r_i}$,

$$m[\sigma(p)](\alpha_{\sigma(1)}, \ldots, \alpha_{\sigma(k)}) = m[p](\alpha_1, \ldots, \alpha_k).$$

Finally, we require that the maps $m[p]$ are compatible with composition, in the following sense. For any configurations $p \in \mathrm{Discs}_n(r_1, \ldots, r_k \mid t_i)$ and $q \in \mathrm{Discs}_n(t_1, \ldots, t_l \mid s)$ and for collections $\alpha_i \in F_{r_i}$ and $\beta_j \in \mathcal{F}_{t_j}$, we require that

$$m[q](\beta_1, \ldots, \beta_{i-1}, m[p](\alpha_1, \ldots, \alpha_k), \beta_{i+1}, \ldots, \beta_l)$$
$$= m[q \circ_i p](\beta_1, \ldots, \beta_{i-1}, \alpha_1, \ldots, \alpha_k, \beta_{i+1}, \ldots, \beta_l).$$

In addition, the action of $\mathbb{R}^n$ on each $\mathcal{F}^r$ is compatible with these multiplication maps, in the manner described in the preceding text.

4.8.3 A Cooperadic Reformulation

Let us give one more equivalent way of rewriting these axioms, which will be useful when we discuss the holomorphic context. These alternative axioms will say that the spaces $C^\infty(\mathrm{Discs}_n(r_1, \ldots, r_k \mid s))$ form an $\mathbb{R}_{>0}$-colored *co*-operad when we use the appropriate completed tensor product. Since we know how to tensor a differentiable vector space with the space of smooth functions on a manifold, it makes sense to talk about an algebra over this colored cooperad in the category of differentiable cochain complexes.

We can rephrase the smoothness axiom for the product map

$$m[p] : \mathcal{F}_{r_1} \otimes \cdots \otimes \mathcal{F}_{r_k} \to \mathcal{F}_s,$$

as p varies in $\mathrm{Discs}_n(r_1, \ldots, r_k \mid s)$, as follows. For any differentiable vector space V and smooth manifold M, we use the notation $V \otimes C^\infty(M)$ interchangeably with the notation $C^\infty(M, V)$; both indicate the differentiable vector space of smooth maps $M \to V$. The smoothness axiom states that the map above extends to a smooth map of differentiable spaces

$$\mu(r_1, \ldots, r_k \mid s) : \mathcal{F}_{r_1} \times \cdots \times \mathcal{F}_{r_k} \to \mathcal{F}_s \otimes C^\infty(\mathrm{Discs}_n(r_1, \ldots, r_k \mid s)).$$

In general, if $V_1, \ldots, V_k, W$ are differentiable vector spaces and if X is a smooth manifold, let

$$C^\infty(X, \mathbb{H}\mathrm{om}_{\mathrm{DVS}}(V_1, \ldots, V_k \mid W))$$

denote the space of smooth multilinear maps $V_1 \times \cdots \times V_k \to C^\infty(X, W)$.

Note that there is a natural gluing map

$$\circ_i : C^\infty(X, \mathbb{H}\mathrm{om}_{\mathrm{DVS}}(V_1, \ldots, V_k \mid W_i)) \times C^\infty(Y, \mathbb{H}\mathrm{om}_{\mathrm{DVS}}(W_1, \ldots, W_l \mid T))$$
$$\to C^\infty(X \times Y, \mathbb{H}\mathrm{om}_{\mathrm{DVS}}(W_1, \ldots, W_{i-1}, V_1, \ldots, V_k, W_{i+1}, \ldots, W_l \mid T)).$$

With this notation in hand, there are elements

$$\mu(r_1, \ldots, r_k \mid s) \in C^\infty\left(\mathrm{Discs}(r_1, \ldots r_k \mid s), \mathbb{H}\mathrm{om}_{\mathrm{DVS}}(\mathcal{F}_{r_1}, \ldots, \mathcal{F}_{r_k} \mid \mathcal{F}_s)\right)$$

with the following properties.

(i) The map $\mu(r_1, \ldots, r_k \mid s)$ is closed under the natural differential, arising from the differentials on the cochain complexes $\mathcal{F}_{r_i}$.

(ii) If $\sigma \in S_k$, then

$$\sigma_* \mu(r_1, \ldots, r_k \mid s) = \mu(r_{\sigma(1)}, \ldots, r_{\sigma(k)} \mid s)$$

where

$$\sigma_* : C^\infty(\text{Discs}(r_1, \ldots r_k \mid s), \text{Hom}_{\text{DVS}}(\mathcal{F}_{r_1}, \ldots, \mathcal{F}_{r_k} \mid \mathcal{F}_s))$$
$$\to C^\infty(\text{Discs}(r_{\sigma(1)}, \ldots r_{\sigma(k)} \mid s), \text{Hom}_{\text{DVS}}$$
$$(\mathcal{F}_{r_{\sigma(1)}}, \ldots, \mathcal{F}_{r_{\sigma(k)}} \mid \mathcal{F}_s))$$

is the natural isomorphism.

(iii) As before, let

$$\circ_i : \text{Discs}_n(r_1, \ldots, r_k \mid t_i) \times \text{Discs}_n(t_1, \ldots, t_m \mid s)$$
$$\to \text{Discs}_n(t_1, \ldots, t_{i-1}, r_1, \ldots, r_k, t_{i+1}, \ldots, t_m \mid s)$$

denote the gluing map. Then we require that

$$\circ_i^* \mu(t_1, \ldots, t_{i-1}, r_1, \ldots, r_k, t_{i+1}, \ldots, t_l) = \mu(r_1, \ldots, r_k \mid t_i) \circ_i \mu$$
$$(t_1, \ldots, t_l \mid s).$$

These elements equip the $\mathcal{F}_r$ with the structure of an algebra over the colored cooperad, as stated earlier.

4.8.4 The Free Scalar Field

Let us write down explicit formulas for these product maps, as a Discs_n algebra, in the case of the free scalar field theory. We have seen in Section 4.6 that the choice of a Green's function G leads to an isomorphism of cochain complexes $W_U : \text{Obs}^{cl}(U)[\hbar] \to \text{Obs}^q(U)[\hbar]$ for every open subset U. This isomorphism allows us to transfer the product (or structure map) in the prefactorization algebra Obs^q to a deformed product $\star_\hbar$ in the prefactorization algebra $\text{Obs}^{cl}[\hbar]$, defined by

$$\alpha \star_\hbar \beta = W_{U \sqcup V}^{-1}(W_U(\alpha) \cdot W_V(\beta)),$$

where U and V are disjoint open sets, $\alpha \in \text{Obs}^{cl}(U)[\hbar]$, $\beta \in \text{Obs}^{cl}(V)[\hbar]$, and the dot $\cdot$ indicates the structure map in the prefactorization algebra Obs^q.

This isomorphism leads to a completely explicit description of the product maps

$$\mu_{r_1, \ldots, r_k}^s : \mathcal{F}_{r_1} \otimes \cdots \otimes \mathcal{F}_{r_k} \to C^\infty(P(r_1, \ldots, r_k \mid s), \mathcal{F}_s)$$

discussed earlier, in the case that $\mathcal{F}$ arises from the prefactorization algebra of quantum observables of a free scalar field theory, or equivalently from the prefactorization algebra $(\text{Obs}^{cl}[\hbar], \star_\hbar)$.

For example, on $\mathbb{R}^2$, let α_1, α_2 be compactly supported smooth functions on discs $B_{r_i}(0)$ of radii r_i around 0. Let us view each α_i as a cohomological degree

0 element of

$$\mathcal{F}_{r_i} = \mathrm{Obs}^{cl}(B_{r_i}(0))[\hbar] = \mathrm{Sym}(C_c^\infty(B_{r_i}(0))[1] \oplus C_c^\infty(B_{r_i}(0)))[\hbar].$$

Let $\tau_x \alpha_i$ denote the translate of α_i to an element of $C_c^\infty(B_{r_i}(x))$.

Then, for $x_1, x_2 \in \mathbb{R}^2$ such that the $B_{r_i}(x_i)$ are disjoint and contained in $B_s(0)$, we have

$$\begin{aligned}
\mu^s_{r_1, r_2}(\alpha_1, \alpha_2) &= \tau_{x_1} \alpha_1 \star_\hbar \tau_{x_2} \alpha_2 \\
&= (\tau_{x_1} \alpha_1) \cdot (\tau_{x_2} \alpha_2) - \hbar \\
&\quad \times \int_{u_1, u_2 \in \mathbb{R}^2} \alpha_1(u_1 + x_1) \alpha_2(u_2 + x_2) \\
&\quad \times \log |u_1 - u_2|.
\end{aligned}$$

Here, we used the Green's function $G(x, y) = -\frac{1}{2\pi} \log |x - y|$.

4.9 States and Vacua for Translation Invariant Theories

We develop notions such as *state* and *vacuum* in this setting of translation-invariant prefactorization algebras. Our running example is the free scalar field.

As should be clear from their construction, the prefactorization algebra of observables only encodes the *local* relationships between the observables. Long-range and global aspects of a physical situation appear in a different way, and this notion of state provides one method by which to introduce them. The examples here exhibit the role of fixing constraints on the asymptotic or boundary behavior of the fields.

Remark: We do not pursue here a comparison with classic axioms for quantum field theory, such as the Wightman or Osterwalder–Schrader axioms, but the familiar reader will recognize the relationship. Both Kazhdan's lectures in Deligne et al. (1999) and Glimm and Jaffe (1987) cover such material and much more. ◇

Definition 4.9.1 Let $\mathcal{F}$ be a smoothly translation-invariant prefactorization algebra on $\mathbb{R}^n$, over a ring R. (In practice, R is $\mathbb{R}$, $\mathbb{C}$ or $\mathbb{R}[[\hbar]]$, $\mathbb{C}[[\hbar]]$).

A *state* for $\mathcal{F}$ is a smooth linear map $\langle - \rangle : H^*(\mathcal{F}(\mathbb{R}^n)) \to R$.

A state $\langle - \rangle$ is *translation invariant* if it commutes with the action of both the group $\mathbb{R}^n$ and of the infinitesimal action of the Lie algebra $\mathbb{R}^n$, where $\mathbb{R}^n$ acts trivially on R.

Recall that "smooth" means the following. The ring R itself is a differentiable vector space: for instance, the real numbers $\mathbb{R}$ as a topological vector space corresponds to the sheaf C^∞ on the site of smooth manifolds. As $\mathcal{F}(\mathbb{R}^n)$ is a differentiable vector space, it is also a sheaf on the site of smooth manifolds, and $H^*(\mathcal{F}(\mathbb{R}^n))$ denotes the cohomology sheaf on the site of smooth manifolds. Unraveling the structures in the preceding definition, we find that for each smooth manifold M and each element $\alpha \in H^*(C^\infty(M, \mathcal{F}(\mathbb{R}^n)))$, the element $\langle \alpha \rangle$ is a smooth R-valued function on M.

A state $\langle - \rangle$ allows us to define correlation functions of observables (even if the complex of fields is not acyclic). If $O_i \in \mathcal{F}(B_{r_i}(0))$ are observables, then we can construct $\tau_{x_i} O_i \in \mathcal{F}(B_{r_i}(x_i))$. If the configuration $(x_1, \ldots, x_k)$ satisfies the condition that the discs $B_{r_i}(x_i)$ are disjoint – that is, $(x_1, \ldots, x_k) \in P(r_1, \ldots, r_k \mid \infty)$ – then we define the *correlation function*

$$\langle O_1(x_1) \cdots O_n(x_n) \rangle$$

by applying the state $\langle - \rangle$ to the product observable $\tau_{x_1} O_1 \cdots \tau_{x_n} O_n \in \mathcal{F}(\mathbb{R}^n)$. Because we have a smoothly translation invariant prefactorization algebra and the state is a map of differentiable cochain complexes, we see that $\langle O_1(x_1) \cdots O_n(x_n) \rangle$ is a smooth function of the points x_i, i.e., an element of $C^\infty(P(r_1, \ldots, r_k \mid \infty), R)$. Further, this function is invariant under simultaneous translation of all the points x_i.

Definition 4.9.2 A translation-invariant state $\langle - \rangle$ is a *vacuum* if it satisfies the *cluster decomposition principle*: in the preceding situation, for any two observables $O_1 \in \mathcal{F}(B_{r_1}(0))$ and $O_2 \in \mathcal{F}(B_{r_2}(0))$, we have

$$\langle O_1(0) \, O_2(x) \rangle - \langle O_1(0) \rangle \langle O_2(0) \rangle \to 0$$

as x goes to infinity. A vacuum is *massive* if $\langle O_1(0) \, O_2(x) \rangle - \langle O_1(0) \rangle \langle O_2(0) \rangle$ tends to zero exponentially fast.

Example: This extended example examines vacua of the free scalar field with mass m.

We have seen that the choice of a Green's function G for the operator $\triangle + m^2$ leads to an isomorphism of cochain complexes $\mathrm{Obs}^{cl}(U)[\hbar] \cong \mathrm{Obs}^q(U)$ for every open subsets $U \subset \mathbb{R}^n$. This produces an isomorphism of prefactorization algebras if we endow $\mathrm{Obs}^{cl}[\hbar]$ with a deformed factorization product $\star_\hbar$, defined using the Green's function.

If $m > 0$, there is a unique Green's function G of the form $G(x, y) = f(x - y)$, where f is a distribution on $\mathbb{R}^n$ that is smooth away from the origin and tends to zero exponentially fast at infinity. For example, if $n = 1$, the function f is $f(x) = \frac{1}{2m} e^{-m|x|}$.

If $m = 0$, we already know

$$G(x, y) = \begin{cases} \frac{1}{4\pi^{d/2}} \Gamma(d/2 - 1) |x - y|^{2-n} & \text{if } n \neq 2 \\ -\frac{1}{2\pi} \log |x - y| & \text{if } n = 2 \end{cases}$$

is the canonically defined Green's function.

It is standard to interpret the Green's function as the two-point correlation function $\langle \delta_x \delta_0 \rangle$. Hence, we see that this two-point function exhibits the necessary behavior: for $m > 0$, it decays exponentially fast, and it decays slowly in the massless case. Strictly speaking, this observable is not in our prefactorization algebra, as it is distributional, but it does exist at the cohomological level. Alternatively, we could work with smeared versions.

Because $\mathrm{Obs}^{cl}(\mathbb{R}^n)$ is, as a cochain complex, the symmetric algebra on the complex $C_c^\infty(\mathbb{R}^n)[1] \xrightarrow{\Delta + m^2} C_c^\infty(\mathbb{R}^n)$, there is a map from $\mathrm{Obs}^{cl}(\mathbb{R}^n)$ to $\mathbb{R}$ that is the identity on Sym^0 and sends $\mathrm{Sym}^{>0}$ to 0. (In other words, we have a natural augmentation map.) This map extends to an $\mathbb{R}[\hbar]$-linear cochain map

$$\langle - \rangle : \mathrm{Obs}^{cl}(\mathbb{R}^n)[\hbar] \to \mathbb{R}[\hbar].$$

Clearly, this map is translation invariant and smooth. Thus, because we have a cochain isomorphism between $\mathrm{Obs}^{cl}(\mathbb{R}^n)[\hbar]$ and $\mathrm{Obs}^q(\mathbb{R}^n)$, we have produced a translation-invariant state.

Lemma 4.9.3 *For $m > 0$, this state is a massive vacuum. For $m = 0$, this state is a vacuum if $n > 2$; for $n \leq 2$, it does not satisfy the cluster decomposition principle.*

Proof Let $F_1 \in C_c^\infty(D(0, r_1))^{\otimes k_1}$ and $F_2 \in C_c^\infty(D(0, r_2))^{\otimes k_2}$. We will view F_1, F_2 as observables in $\mathrm{Obs}^{cl}(B_{r_i}(0))$ by using the natural map from $C_c^\infty(U)^{\otimes k}$ to the coinvariants $\mathrm{Sym}^k C_c^\infty(U)$. Let $\tau_c F_2$ be the translation of F_2 by c, where c is sufficiently large so that the discs $D(0, r_1)$ and $D(c, r_2)$ are disjoint. Explicitly, $\tau_c F_2$ is represented by the function $F_2(y_1 - c, \ldots, y_{k_2} - c)$. We are interested in computing the expected value $\langle F_1 (\tau_c F_2) \rangle$.

In the preceding text we gave an explicit formula for the quantum factorization product $\star_\hbar$ on $\mathrm{Obs}^{cl}[\hbar]$. In this case, it states that $F_1 \star_\hbar \tau_c F_2$ is given by

$$\sum_{r=0}^{\min(k_1, k_2)} \hbar^r \sum_{\substack{1 \leq i_1 < \cdots < i_r \leq k_1 \\ 1 \leq j_1 < \cdots < j_r \leq k_2}} \int_{\mathbf{x} \in \mathbb{R}^{nk_1}} \int_{\mathbf{y} \in \mathbb{R}^{nk_2}} F_1(\mathbf{x}) F_2(\tau_c \mathbf{y}) \prod_{k=1}^{r} G(x_{i_k}, y_{j_k}),$$

where $\mathbf{x} = (x_1, \ldots, x_{k_1})$ and $\tau_c \mathbf{y} = (y_1 - c, \ldots, y_{k_2} - c)$. Note that after performing the integral on the right-hand side, we are left with a function of the

$k_1 + k_2 - 2r$ copies of $\mathbb{R}^n$ that we have not integrated over. This function is then viewed as an observable in $\text{Sym}^{k_1+k_2-2r} C_c^\infty(\mathbb{R}^n)$.

If $k_1, k_2 > 0$, then $\langle F_1 \rangle = 0$ and $\langle F_2 \rangle = 0$. Further, $\langle F_1 (\tau_c F_2) \rangle$ selects the constant term in the expression for $F_1 \star_\hbar \tau_c F_2$. There is only a nonzero constant term if $k_1 = k_2 = k$. In that case, the constant term $\langle F_1, \tau_c F_2 \rangle$ is

$$\hbar^k \int_{x_i, y_i \in \mathbb{R}^n} F_1(x_1, \ldots, x_k) F_2(y_1 - c, \ldots, y_k - c) G(x_1, y_1) \ldots G(x_k, y_k).$$

To check whether the cluster decomposition principle holds, we need to check whether or not

$$\langle F_1 (\tau_c F_2) \rangle = \langle F_1 (\tau_c F_2) \rangle - \langle F_1 \rangle \langle F_2 \rangle$$

tends to zero as $c \to \infty$. If $m > 0$, we know that $G(x, y)$ tends to zero exponentially fast as $x - y \to \infty$. This implies immediately that we have a massive vacuum in this case, because the correlation function involves integrating against powers of G.

If $m = 0$, the Green's function $G(x, y)$ tends to zero like the inverse of a polynomial as long as $n > 2$. In this case, we have a vacuum. If $n = 1$ or $n = 2$, then $G(x, y)$ does not tend to zero, so we don't have a vacuum. $\quad\square$

This lemma simply reduces the full condition on a vacuum down to the Green's function, and hence to the usual perspective. $\qquad\diamond$

Example: We can give a more abstract construction for the vacuum of associated to a massive scalar field theory on $\mathbb{R}^n$. It illuminates how boundary or asymptotic conditions on fields relate to observables.

Let $\mathscr{E} = C^\infty \xrightarrow{\Delta + m^2} C^\infty$ be, as before, the complex of fields. Let $\widehat{\mathscr{E}_c}$ be the Heisenberg central extension $\widehat{\mathscr{E}_c} = \mathscr{E}_c \oplus \mathbb{R} \cdot \hbar$ where $\hbar$ has degree 1. We defined the complex of observables on U as the Chevalley–Eilenberg chain complex of $\widehat{\mathscr{E}_c}(U)$.

Let $\mathscr{E}_S(\mathbb{R}^n)$ be the complex $S(\mathbb{R}^n) \xrightarrow{\Delta + m^2} S(\mathbb{R}^n)$, where $S(\mathbb{R}^n)$ is the space of Schwartz functions on $\mathbb{R}^n$. (Recall that a function is Schwartz if it tends to zero at ∞ faster than the reciprocal of any polynomial.)

A Heisenberg dg Lie algebra can be defined using $\mathscr{E}_S(\mathbb{R}^n)$ instead of $\mathscr{E}_c(\mathbb{R}^n)$. We let $\widehat{\mathscr{E}_S}(\mathbb{R}^n)$ be $\mathscr{E}_S(\mathbb{R}^n) \oplus \mathbb{R} \cdot \hbar$, with bracket defined by $[\phi^0, \phi^1] = \hbar \int \phi^0 \phi^1$. Here ϕ^k denotes a Schwartz function of cohomological degrees $k = 0$ or 1. This bracket makes sense because the product of any two Schwartz function is Schwartz and Schwartz functions are integrable.

Schwartz functions have a natural topology, so we will view them as being a convenient vector space. Since the topology is nuclear Fréchet, a result discussed in Appendix B tells us that the tensor product $\widehat{\otimes}_\beta$ in the category of convenient vector spaces coincides with the completed projective tensor product $\widehat{\otimes}_\pi$ in the category of nuclear Fréchet spaces. A result of Grothendieck (1952) tells us that

$$\mathcal{S}(\mathbb{R}^n) \,\widehat{\otimes}\, \mathcal{S}(\mathbb{R}^m) = \mathcal{S}(\mathbb{R}^{n+m}),$$

and similarly for Schwartz sections of vector bundles.

This allows us to define the Chevalley–Eilenberg chain complex

$$\mathrm{Obs}^q_\mathcal{S}(\mathbb{R}^n) \overset{\mathrm{def}}{=} C_*(\widehat{\mathscr{E}}_\mathcal{S}(\mathbb{R}^n))$$

of the Heisenberg algebra based on Schwartz functions; as usual, we use the tensor product $\widehat{\otimes}_\beta$ when defining the symmetric algebra.

There's a map

$$\mathrm{Obs}^q(\mathbb{R}^n) \to \mathrm{Obs}^q_\mathcal{S}(\mathbb{R}^n)$$

given by viewing a compactly supported function as a Schwartz function. The Schwartz observables are well behaved, as we will show, because zero is the only Schwartz solution to the equations of motion.

Lemma 4.9.4 *The cohomology of* $\mathrm{Obs}^q_\mathcal{S}(\mathbb{R}^n)$ *is* $\mathbb{R}[\hbar]$, *concentrated in degree zero.*

Proof The complex

$$\mathscr{E}_\mathcal{S}(\mathbb{R}^n) = \mathcal{S}(\mathbb{R}^n) \xrightarrow{\triangle+m^2} \mathcal{S}(\mathbb{R}^n)$$

has no cohomology. We can see this using Fourier duality: the Fourier transform is an isomorphism on the space of Schwartz functions, and the Fourier dual of the operator $\triangle+m^2$ is the operator p^2+m^2, where $p^2 = \sum p_i^2$ and p_i are coordinates on the Fourier dual $\mathbb{R}^n$. (Note that our convention is that the Laplacian is $-\sum \frac{\partial}{\partial x_i}^2$.) The operator of multiplication by $p^2 + m^2$ is invertible on the space of Schwartz functions, just because if f is a Schwartz function then so is $(p^2 + m^2)^{-1}f$, as we assumed $m \neq 0$. The inverse is a smooth linear map, so that this complex is smoothly homotopy equivalent to the zero complex.

The fact that the complex $\mathscr{E}_\mathcal{S}(\mathbb{R}^n)$ has no cohomology implies immediately that the Chevalley–Eilenberg chain complex of $\widehat{\mathscr{E}}_\mathcal{S}(\mathbb{R}^n)$ has the same cohomology as the Chevalley–Eilenberg chain complex of the Abelian Lie algebra $\mathbb{R}\hbar[-1]$. But that complex is $\mathbb{R}[\hbar]$, as desired. $\qquad\square$

Thus, we have a translation-invariant state

$$\langle - \rangle : H^*(\mathrm{Obs}^q(\mathbb{R}^n)) \to H^*(\mathrm{Obs}^q_{\mathcal{S}}(\mathbb{R}^n)) = \mathbb{R}[\hbar].$$

This state is characterized uniquely by the fact that it is defined on Schwartz observables. Since the state constructed more explicitly earlier also has this property, we see that these two states coincide, so that this state is a massive vacuum.   ◇

5

Holomorphic Field Theories and
Vertex Algebras

This chapter serves two purposes. On the one hand, we develop several examples that exhibit how to understand the observables of a two-dimensional theory from the point of view of prefactorization algebras and how this approach recovers standard examples of vertex algebras. On the other hand, we provide a precise definition of a prefactorization algebra on $\mathbb{C}^n$ whose structure maps vary holomorphically, much as we defined translation-invariant prefactorization algebras in Section 4.8 in Chapter 4. We then give a proof that when $n = 1$ and the prefactorization algebra possesses a $U(1)$-action, we can extract a vertex algebra. For $n > 1$, the structure we find is a higher-dimensional analog of a vertex algebra. (We leave a detailed analysis of the structure present in the higher-dimensional case for future work). Such higher-dimensional vertex algebras appear, for example, as the prefactorization algebra of observables of partial twists of supersymmetric gauge theories.

5.1 Vertex Algebras and Holomorphic Prefactorization Algebras on $\mathbb{C}$

In mathematics, the notion of a vertex algebra is a standard formalization of the observables of a two-dimensional chiral field theory. In the version of the axioms we consider, we do not ask for the Virasoro algebra to act. Before embarking on our own approach, we recall the definition of a vertex algebra and various properties as given in Frenkel and Ben-Zvi (2004).

Definition 5.1.1 Let V be a vector space. An element $a(z) = \sum_{n \in \mathbb{Z}} a_n z^{-n}$ in End $V[[z, z^{-1}]]$ is a *field* if, for each $v \in V$, there is some N such that $a_j v = 0$ for all $j > N$.

Remark: The usage of the term "field" in the theory of vertex operators often provokes confusion. In this book, the term *field* is used to refer to a configuration in a classical field theory: for example, in a scalar field theory on a manifold M, a field is an element of $C^\infty(M)$. The term "field" as used in the theory of vertex algebras is *not* an example of this usage. Instead, as we will see in the text that follows, a "field" in a vertex algebra corresponds to an observable with support at a point of the surface; it can be understand as a special kind of operator.   ◊

Definition 5.1.2 [Definition 1.3.1 in, Frenkel and Ben-Zvi (2004)] A *vertex algebra* is the following data:

- A vector space V over $\mathbb{C}$ (the *state space*)
- A nonzero vector $|0\rangle \in V$ (the *vacuum vector*)
- A linear map $T : V \to V$ (the *shift operator*)
- A linear map $Y(-, z) : V \to \operatorname{End} V[[z, z^{-1}]]$ sending every vector v to a field (the *vertex operation*)

subject to the following axioms:

- (*Vacuum axiom*) $Y(|0\rangle, z) = \mathrm{id}_V$ and $Y(v, z)|0\rangle \in v + zV[[z]]$ for all $v \in V$
- (*Translation axiom*) $[T, Y(v, z)] = \partial_z Y(v, z)$ for every $v \in V$ and $T|0\rangle = 0$
- (*Locality axiom*) For any pair of vectors $v, v' \in V$, there exists a nonnegative integer N such that $(z - w)^N [Y(v, z), Y(v', w)] = 0$ as an element of $\operatorname{End} V[[z^{\pm 1}, w^{\pm 1}]]$.

The vertex operation is best understood in terms of the following intuition. The vector space V represents the set of pointwise measurements one can make of the fields, and one should imagine labeling each point $z \in \mathbb{C}$ by a copy of V, which we'll denote V_z. Moreover, in a disc D containing the point z, the measurements at z are a dense subspace of the measurements one can make in D (this is a feature of chiral theories). We'll denote the observables on D by V_D. The vertex operation is a way of combining pointwise measurements. Let D be a disc centered on the origin. For $z \neq 0$, we can multiply observables to get a map

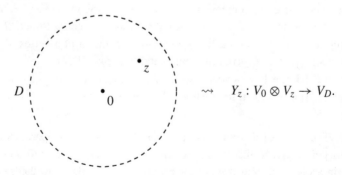

$$\rightsquigarrow \quad Y_z : V_0 \otimes V_z \to V_D.$$

This *vertex operator* should vary holomorphically in $z \in D \setminus \{0\}$. In other words, we should get something with properties like the formal definition $Y(-, z)$ above. (This picture clearly resembles the "pair of pants" product from two-dimensional topological field theories, although we've shrunk the two "incoming" boundary circles to points.)

An appealing aspect of our approach to observables is that this intuition becomes explicit and rigorous. Our procedure provides structure maps from disjoint discs into a larger disc (and also describes how to "multiply" observables supported in more complicated opens). Moreover, it gives the structure maps in a coordinate-free way. By choosing a coordinate z on $\mathbb{C}$, we recover the usual formulas for vertex algebras.

This relationship is seen in the main theorem in this chapter, which we now state. The theorem connects vertex algebras with a certain class of prefactorization algebras on $\mathbb{C}$. The prefactorization algebras of interest are holomorphically translation invariant prefactorization algebra. We will give a precise definition of what this means shortly, but here is a rough definition. Let $\mathcal{F}$ be such a prefactorization algebra. Let $\mathcal{F}_r$ denote the cochain complex $\mathcal{F}(D(0, r))$ that $\mathcal{F}$ assigns to a disc of radius r. Recall, as we explained in Section 4.8 of Chapter 4, if $\mathcal{F}$ is smoothly translation invariant, then we have an operator product map

$$\mu : \mathcal{F}_{r_1} \otimes \cdots \otimes \mathcal{F}_{r_n} \to C^\infty(\mathrm{Discs}(r_1, \ldots, r_k \mid s), \mathcal{F}_s),$$

where $\mathrm{Discs}(r_1, \ldots, r_k \mid s)$ refers to the open subset of $\mathbb{C}^k$ consisting of points $z_1, \ldots, z_k$ such that the discs of radius r_i around z_i are all disjoint and contained in the disc of radius s around the origin. If $\mathcal{F}$ is holomorphically translation invariant, then μ lifts to a cochain map

$$\mu : \mathcal{F}_{r_1} \otimes \cdots \otimes \mathcal{F}_{r_n} \to \Omega^{0,*}(\mathrm{Discs}(r_1, \ldots, r_k \mid s), \mathcal{F}_s),$$

where on the right-hand side we have used the Dolbeault complex of the complex manifold $\mathrm{Discs}(r_1, \ldots, r_k \mid s)$. We also require some compatibility of these lifts with compositions, which we will detail later.

In other words, being holomorphically translation invariant means that the operator product map is holomorphic (up to homotopy) in the location of the discs.

Theorem 5.1.3 *Let $\mathcal{F}$ be a holomorphically translation invariant prefactorization algebra on $\mathbb{C}$. Let $\mathcal{F}$ be equivariant under the action of S^1 on $\mathbb{C}$ by rotation, and let $\mathcal{F}_r^k$ denote the weight k eigenspace of the S^1 action on the complex $\mathcal{F}_r$. Assume that for every $r < s$, the extension map $\mathcal{F}_r^k \to \mathcal{F}_s^k$ associated to the inclusion $D(0, r) \subset D(0, s)$ is a quasi-isomorphism. Finally, we need to* *assume that the S^1 action on each $\mathcal{F}_r$ satisfies a certain technical "tameness" condition.*

Then the vector space

$$V_{\mathcal{F}} = \bigoplus_{k \in \mathbb{Z}} H^*(\mathcal{F}_r^k)$$

has the structure of a vertex algebra. The vertex algebra structure map

$$Y_{\mathcal{F}} : V_{\mathcal{F}} \otimes V_{\mathcal{F}} \to V_{\mathcal{F}}[[z, z^{-1}]]$$

is the Laurent expansion of operator product map

$$H^*\mu : H^*(\mathcal{F}_{r_1}^{k_1}) \otimes H^*(\mathcal{F}_{r_2}^{k_2}) \to \mathrm{Hol}(\mathrm{Discs}(r_1, r_2 \mid s), H^*(\mathcal{F}_s)).$$

On the right-hand side, Hol *denotes the space of holomorphic maps.*

In other words, the intuition that the vertex algebra structure map is the operator product expansion is made precise in our formalism.

This result should be compared with the classic result of Huang (1997), who relates vertex algebras with chiral conformal field theories at genus 0, in the sense used by Segal. As we have seen in Section 1.4 in Chapter 1, the axioms for prefactorization algebras are closely related to Segal's axioms. Our axioms for a holomorphically translation invariant field theory are similarly related to Segal's axioms for a two-dimensional chiral field theory. Although our result is closely related to Huang's, it is a little different because of the technical differences between a prefactorization algebra and a Segal-style chiral conformal field theory.

One nice feature of our definition of holomorphically translation-invariant prefactorization algebra is that it makes sense in any complex dimension. The structure present on the cohomology of a higher-dimensional holomorphically translation-invariant prefactorization algebra is a higher-dimensional version of the axioms of a vertex algebra. This structure is discussed briefly in Costello and Scheimbauer (2015).

Another nice feature of our approach is that our general construction of field theories allows one to construct vertex algebras by perturbation theory, starting with a Lagrangian. This method should lead to the construction of many interesting vertex algebras, and of their higher dimensional analogs.

5.1.1 Organization of this Chapter

We start this chapter by stating and proving our main theorem. The first thing we define, in Section 5.2, is the notion of a holomorphically translation invariant prefactorization algebra on $\mathbb{C}^n$ for every $n \geq 1$. Holomorphic translation invariance guarantees that the operator products are all holomorphic. We will then show to construct a vertex algebra from such an object in dimension 1.

The rest of this chapter is devoted to analyzing examples. In Section 5.4, we discuss the prefactorization algebra associated to a very simple two-dimensional chiral conformal field theory: the free $\beta\gamma$ system. We will show that the vertex algebra associated to the prefactorization algebra of observables of this theory is an object called the $\beta\gamma$ vertex algebra in the literature. Then, in Section 5.5, we will construct a prefactorization algebra encoding the affine Kac–Moody algebra. This prefactorization algebra again encodes a vertex algebra, which is the standard Kac–Moody vertex algebra.

5.2 Holomorphically Translation-Invariant Prefactorization Algebras

In this section we analyze in detail the notion of translation-invariant prefactorization algebras on $\mathbb{C}^n$. On $\mathbb{C}^n$ we can ask for a translation-invariant prefactorization algebra to have a *holomorphic* structure; this implies that all structure maps of the prefactorization algebra are (in a sense we will explain shortly) holomorphic. There are many natural field theories where the corresponding prefactorization algebra is holomorphic: for instance, chiral conformal field theories in complex dimension 1, and minimal twists of supersymmetric field theories in complex dimension 2, as described in Costello (2011b).

5.2.1 The Definition

We now explain what it means for a (smoothly) translation-invariant prefactorization algebra $\mathcal{F}$ on $\mathbb{C}^n$ to be *holomorphically translation-invariant*. For this definition to make sense, we require that $\mathcal{F}$ is defined over $\mathbb{C}$: that is, the vector spaces $\mathcal{F}(U)$ are complex vector spaces and all structure maps are complex-linear.

Recall that such a prefactorization algebra has, as part of its structure, an action of the real Lie algebra $\mathbb{R}^{2n} = \mathbb{C}^n$ by derivations. This action is as a real Lie algebra. Since $\mathcal{F}$ is defined over $\mathbb{C}$, the action extends to an action of the complexified translation Lie algebra $\mathbb{R}^{2n} \otimes_{\mathbb{R}} \mathbb{C}$. We will denote the action maps by

$$\frac{\partial}{\partial z_i}, \frac{\partial}{\partial \bar{z}_j} : \mathcal{F}(U) \to \mathcal{F}(U).$$

Definition 5.2.1 A translation-invariant prefactorization algebra $\mathcal{F}$ on $\mathbb{C}^n$ is *holomorphically translation-invariant* if it is equipped with derivations $\eta_i :$

$\mathcal{F} \to \mathcal{F}$ of cohomological degree -1, for $i = 1 \ldots n$, with the following properties:

$$d\eta_i = \frac{\partial}{\partial \bar{z}_i} \in \mathrm{Der}(\mathcal{F})$$

$$[\eta_i, \eta_j] = 0,$$

$$\left[\eta_i, \frac{\partial}{\partial \bar{z}_i} \right] = 0.$$

Here, d refers to the differential on the dg Lie algebra $\mathrm{Der}(\mathcal{F})$.

We should understand this definition as saying that the vector fields $\frac{\partial}{\partial \bar{z}_i}$ act homotopically trivially on $\mathcal{F}$. In physics terminology, the operators $\frac{\partial}{\partial x_i}$, where x_i are real coordinates on $\mathbb{C}^n = \mathbb{R}^{2n}$, are related to the energy-momentum tensor. We are asking that the components of the energy-momentum tensor in the $\bar{z}_i$ directions are exact for the differential on observables (in physics, this might be called "BRST exact"). This is rather similar to a phenomenon that sometimes appears in the study of topological field theory, in which a topological theory depends on a metric, but the variation of the metric is exact for the BRST differential. See Witten (1988b) for discussion.

5.2.2 An Operadic Reformulation

Now we will interpret holomorphically translation-invariant prefactorization algebras in the language of $\mathbb{R}_{>0}$-colored operads. When we work in complex geometry, it is better to use *polydiscs* instead of balls, as is standard in complex analysis.

Thus, if $z \in \mathbb{C}^n$, let

$$PD_r(z) = \{ w \in \mathbb{C}^n \mid |w_i - z_i| < r \text{ for } 1 \leq i \leq n \}$$

denote the polydisc of radius r around z. Let

$$\mathrm{PDiscs}_n(r_1, \ldots, r_k \mid s) \subset (\mathbb{C}^n)^k$$

denote the set of $z_1, \ldots, z_k \in \mathbb{C}^n$ with the property that the closures of the polydiscs $PD_{r_i}(z_i)$ are disjoint and contained in the polydisc $PD_s(0)$.

The spaces $\mathrm{PDiscs}_n(r_1, \ldots, r_k \mid s)$ form a $\mathbb{R}_{>0}$-colored operad in the category of complex manifolds. We will explain here why a holomorphically translation invariant prefactorization algebra provides an algebra over this colored operad. Proposition 5.2.2 will provide a precise version and proof of this argument.

Now, let $\mathcal{F}$ be a holomorphically translation-invariant prefactorization algebra on $\mathbb{C}^n$. Let $\mathcal{F}_r$ denote the differentiable cochain complex $\mathcal{F}(PD_r(0))$ associated to the polydisc of radius r. For each $p \in \text{PDiscs}_n(r_1, \ldots, r_k \mid s)$, we have a map

$$m[p] : \mathcal{F}_{r_1} \times \cdots \times \mathcal{F}_{r_k} \to \mathcal{F}_s.$$

This map is smooth, multilinear, and compatible with the differential. Further, this map varies smoothly with p.

The fact that $\mathcal{F}$ is a holomorphically translation-invariant prefactorization algebra means that these maps are equipped with extra structure: we have derivations η_j of $\mathcal{F}$ that make the derivations $\frac{\partial}{\partial \bar{z}_j}$ homotopically trivial.

For $i = 1, \ldots, k$ and $j = 1, \ldots, n$, let $z_{ij}, \bar{z}_{ij}$ refer to coordinates on $(\mathbb{C}^n)^k$. We thereby obtain coordinates on the open subset

$$\text{PDiscs}_n(r_1, \ldots, r_k \mid s) \subset (\mathbb{C}^n)^k.$$

Thus, we have operations

$$\frac{\partial}{\partial \bar{z}_{ij}} m[p] : \mathcal{F}_{r_1} \times \cdots \times \mathcal{F}_{r_k} \to \mathcal{F}_s$$

obtained by differentiating the operation $m[p]$, which depends smoothly on p, in the direction $\bar{z}_{ij}$.

Let $m[p] \circ_i \eta_j$ denote the operation

$$m[p] \circ_i \eta_j : \mathcal{F}_{r_1} \times \cdots \times \mathcal{F}_{r_k} \to \mathcal{F}_s$$
$$\alpha_1 \times \cdots \times \alpha_k \mapsto (-1)^{|\alpha_1| + \cdots |\alpha_{i-1}|} m[p](\alpha_1, \ldots, \eta_j \alpha_i, \ldots, \alpha_k),$$

where the sign arises from the usual Koszul rule.

The axioms of a (smoothly) translation invariant prefactorization algebra tell us that

$$\frac{\partial}{\partial \bar{z}_{ij}} m[p] = m[p] \circ_i \frac{\partial}{\partial \bar{z}_j},$$

where $\frac{\partial}{\partial \bar{z}_j}$ is the derivation of the prefactorization algebra $\mathcal{F}$. This equality, together with the fact that $[\mathrm{d}, \eta_i] = \frac{\partial}{\partial \bar{z}_j}$, tells us that

$$\frac{\partial}{\partial \bar{z}_{ij}} m[p] = \big[\mathrm{d}, m[p] \circ_i \eta_j\big]$$

holds. Hence the product map $m[p]$ is holomorphic in p, up to a homotopy given by η_i.

5.2.3 A Cooperadic Reformulation

In the smooth case, we saw that we could also describe the structure as that of an algebra over a $\mathbb{R}_{>0}$-colored cooperad built from smooth functions on the spaces $\mathrm{Discs}_n(r_1, \ldots, r_k \mid s)$. In this section we will see that there is an analogous story in the complex world, where we use the Dolbeault complex of the spaces $\mathrm{PDiscs}_n(r_1, \ldots, r_k \mid s)$.

Because the spaces $\mathrm{PDiscs}_n(r_1, \ldots, r_k \mid s)$ form a colored operad in the category of complex manifolds, their Dolbeault complexes form a colored cooperad in the category of convenient cochain complexes. We use here that the contravariant functor sending a complex manifold to its Dolbeault complex, viewed as a convenient vector space, is symmetric monoidal:

$$\Omega^{0,*}(X \times Y) = \Omega^{0,*}(X) \widehat{\otimes}_\beta \Omega^{0,*}(Y)$$

where $\widehat{\otimes}_\beta$ denotes the symmetric monoidal structure on the category of convenient vector spaces.

Explicitly, the colored cooperad structure is given as follows. The operad structure on the complex manifolds $\mathrm{PDiscs}_n(r_1, \ldots, r_k \mid t_i)$ is given by maps

$$\circ_i : \mathrm{PDiscs}_n(r_1, \ldots, r_k \mid t_i) \times \mathrm{PDiscs}_n(t_1, \ldots, t_m \mid s)$$
$$\to \mathrm{PDiscs}_n(t_1, \ldots, t_{i-1}, r_1, \ldots, r_k, t_{i+1}, \ldots, t_m \mid s).$$

We let

$$\circ_i^* : \Omega^{0,*}(\mathrm{PDiscs}_n(t_1, \ldots, t_{i-1}, r_1, \ldots, r_k, t_{i+1}, \ldots, t_m \mid s))$$
$$\to \Omega^{0,*}(\mathrm{PDiscs}_n(r_1, \ldots, r_k \mid t_i) \times \mathrm{PDiscs}_n(t_1, \ldots, t_m \mid s))$$

be the corresponding pullback map on Dolbeault complexes.

The prefactorization algebras we are interested in take values in the category of differentiable vector spaces. We want to say that if $\mathcal{F}$ is a holomorphically translation invariant prefactorization algebra on $\mathbb{C}^n$, then it defines an algebra over the colored cooperad given by the Dolbeault complex of the spaces PDiscs_n. A priori, this does not make sense, because the category of differentiable vector spaces is not a symmetric monoidal category; only its full subcategory of convenient vector spaces has a tensor structure. However, to make this definition, all we need to be able to do is to tensor differentiable vector spaces with the Dolbeault complexes of complex manifolds, and this we know how to do.

As shown in Appendix B, for any manifold X, $C^\infty(X)$ defines a differentiable vector space (in fact, a commutative algebra in DVS). If V is any differentiable vector space, then there is a differentiable vector space $C^\infty(X, V)$

whose value on a manifold M is $C^\infty(M \times X, V)$. If X is a complex manifold, then $\Omega^{0,*}(X)$ is a differentiable cochain complex, for the same reasons. We define $\Omega^{0,*}(X, V)$ as the tensor product

$$\Omega^{0,*}(X, V) := \Omega^{0,*}(X) \otimes_{C^\infty(X)} C^\infty(X, V)$$

as sheaves of $C^\infty(X)$-modules on the site of smooth manifolds. This is a reasonable thing to do as $\Omega^{0,*}(X)$ is a projective module over $C^\infty(X)$. Note also that the fact that V is a differentiable vector space, and not just a sheaf of C^∞-modules on the site of smooth manifolds, means that differential operators on X act on $C^\infty(X, V)$. This is what allows us to extend the differential $\overline{\partial}$ on the Dolbeault complex $\Omega^{0,*}(X)$ to an operator on $\Omega^{0,*}(X, V)$.

The Dolbeault complex with coefficients in V is functorial both for maps $f^* : \Omega^{0,*}(X) \to \Omega^{0,*}(Y)$ that arise from holomorphic maps $f : Y \to X$ and also for arbitrary smooth maps between differentiable vector spaces. Since the cooperad structure maps in our colored dg cooperad $\Omega^{0,*}(\mathrm{PDiscs}_n)$ arise from maps of complex manifolds, it makes sense to ask for a coalgebra over this cooperad in the category of differentiable vector spaces.

Proposition 5.2.2 *Let $\mathcal{F}$ be a holomorphically translation-invariant prefactorization algebra on $\mathbb{C}^n$. Then $\mathcal{F}$ defines an algebra over the cooperad $\left(\Omega^{0,*}(\mathrm{PDiscs}_n)\right)_{n \in \mathbb{N}}$ in differentiable cochain complexes.*

More precisely, the product maps

$$m[p] : \mathcal{F}_{r_1} \times \cdots \times \mathcal{F}_{r_k} \to \mathcal{F}_s$$

for $p \in \mathrm{PDiscs}_n(r_1, \ldots, r_k \mid s)$ lift to multilinear maps

$$\mu^{\overline{\partial}}(r_1, \ldots, r_k \mid s) : \mathcal{F}_{r_1} \times \cdots \times \mathcal{F}_{r_k} \to \Omega^{0,*}(\mathrm{PDiscs}_n(r_1, \ldots, r_k \mid s), \mathcal{F}_s))$$

that are compatible with differentials and that satisfy the properties needed to define a coalgebra over a cooperad.

In other words, we have closed elements

$$\mu^{\overline{\partial}}(r_1, \ldots, r_k \mid s) \in \Omega^{0,*}(\mathrm{PDiscs}_n(r_1, \ldots, r_k \mid s), \mathbb{H}\mathrm{om}_{\mathrm{DVS}}(\mathcal{F}_{r_1}, \ldots, \mathcal{F}_{r_k} \mid \mathcal{F}_s))$$

satisfying the following properties.

(i) The element $\mu^{\overline{\partial}}(r_1, \ldots, r_k \mid s)$ is closed under the natural differential on

$$\Omega^{0,*}(\mathrm{PDiscs}_n(r_1, \ldots, r_k \mid s), \mathbb{H}\mathrm{om}_{\mathrm{DVS}}(\mathcal{F}_{r_1}, \ldots, \mathcal{F}_{r_k} \mid \mathcal{F}_s)),$$

which incorporates the Dolbeault differential as well as the internal differentials on the complexes $\mathcal{F}_{r_i}, \mathcal{F}_s$. Explicitly, the differential

$$\left((d_{\mathcal{F}} + \overline{\partial})\mu^{\overline{\partial}}(r_1, \ldots, r_k \mid s)\right)(f_1, \ldots, f_k)$$

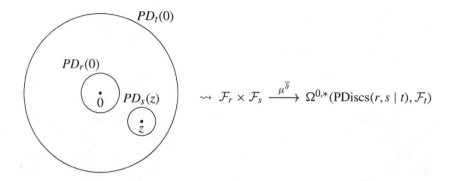

Figure 5.1. Configurations of discs and the operator product map.

is given by the sum

$$\sum \pm \mu^{\overline{\partial}}(r_1, \ldots, r_k \mid s)(f_1, \ldots, \mathrm{d}_{\mathcal{F}} f_i, \ldots, f_k).$$

(ii) Let $\sigma \in S_k$. As in the smooth case,

$$\sigma_* \mu^{\overline{\partial}}(r_1, \ldots, r_k \mid s) = \mu^{\overline{\partial}}(r_{\sigma(1)}, \ldots, r_{\sigma(k)} \mid s)$$

where σ_* is induced by the isomorphism of complex manifolds

$$\sigma : \mathrm{PDiscs}_n(r_1, \ldots, r_k \mid s) \to \mathrm{PDiscs}_n(r_{\sigma(1)}, \ldots, r_{\sigma(k)} \mid s).$$

(iii) For all $1 \leq i \leq m$,

$$o_i^* \mu^{\overline{\partial}}(t_1, \ldots, t_{i-1}, r_1, \ldots, r_k, t_{i+1}, \ldots, t_m \mid s)$$
$$= \mu^{\overline{\partial}}(t_1, \ldots, t_m \mid s) \circ_i \mu^{\overline{\partial}}(r_1, \ldots, r_k \mid t_i).$$

That is, we have associativity of composition.

Remark: For similar axiom systems in the context of topological field theory, the interested reader should consult, for example, Getzler (1994) and Costello (2007b). In Segal (2004) there is a related system of axioms for chiral conformal field theories. This construction is also closely related to the construction of "descendants" in the physics literature on topological field theory (see, e.g., Witten (1988b, 1992)).

Here is a brief explanation of the relationship. Suppose one has a field theory that depends on a Riemannian metric, but where the variation of the original action functional with respect to the metric is exact for the BRST operator (which corresponds to the differential on observables in our language). A metric-dependent functional making the variation BRST exact can be viewed as a 1-form on the moduli space of metrics. Higher homotopies yield forms on the moduli space of metrics, or, in two dimensions, on the moduli of conformal classes of metrics (i.e., on the moduli of Riemann surfaces).

In our approach, because we are working with holomorphic instead of topological theories, we find elements of the Dolbeault complex of the appropriate moduli spaces of complex manifolds. In the simple situation considered here, these moduli spaces are the spaces PDiscs_n.

For the factorization algebras associated to holomorphic twists of supersymmetric field theories, these elements of the Dolbeault complex of PDiscs_n come from the operator product of supersymmetric descendents of BPS operators. $\Diamond$

Proof of the proposition Giving a smooth multilinear map

$$\phi : \mathcal{F}_{r_1} \times \cdots \times \mathcal{F}_{r_k} \to \Omega^{0,*}(\mathrm{PDiscs}_n(r_1,\ldots,r_k \mid s), \mathcal{F}_s)$$

compatible with the differentials is equivalent to giving an element of

$$\phi \in \Omega^{0,*}\left(\mathrm{PDiscs}_n(r_1,\ldots,r_k \mid s), \mathbb{H}\mathrm{om}_{\mathrm{DVS}}(\mathcal{F}_{r_1},\ldots,\mathcal{F}_{r_k} \mid \mathcal{F}_s)\right)$$

that is closed with respect to the differential $\overline{\partial} + \mathrm{d}_{\mathcal{F}}$, where $\mathrm{d}_{\mathcal{F}}$ refers to the natural differential on the differentiable cochain complex of smooth multilinear maps $\mathbb{H}\mathrm{om}_{\mathrm{DVS}}(\mathcal{F}_{r_1},\ldots,\mathcal{F}_{r_k} \mid \mathcal{F}_s)$.

We will produce the desired element

$$\mu^{\overline{\partial}}(r_1,\ldots,r_k \mid s) \in \Omega^{0,*}\left(\mathrm{PDiscs}_n(r_1,\ldots,r_k \mid s), \mathbb{H}\mathrm{om}_{\mathrm{DVS}}(\mathcal{F}_{r_1},\ldots,\mathcal{F}_{r_k} \mid \mathcal{F}_s)\right)$$

starting from the operations

$$\mu^0(r_1,\ldots,r_k \mid s) \in C^\infty(\mathrm{PDiscs}(r_1,\ldots,r_k \mid s), \mathbb{H}\mathrm{om}_{\mathrm{DVS}}(\mathcal{F}_{r_1},\ldots,\mathcal{F}_{r_k} \mid s))$$

provided by the hypothesis that $\mathcal{F}$ is a smoothly translation-invariant prefactorization algebra.

First, we need to introduce some notation. Recall that

$$\mathrm{PDiscs}_n(r_1,\ldots,r_k \mid s) \subset \left(\mathbb{C}^n\right)^k$$

is an open (possibly empty) subset. Thus,

$$\Omega^{0,*}\left(\mathrm{PDiscs}_n(r_1,\ldots,r_k \mid s)\right) = \Omega^{0,0}\left(\mathrm{PDiscs}_n(r_1,\ldots,r_k \mid s)\right) \otimes \mathbb{C}[\mathrm{d}\overline{z}_{ij}]$$

where the $\mathrm{d}\overline{z}_{ij}$ are commuting variables of cohomological degree 1, with $i = 1,\ldots,k$ and $j = 1,\ldots,n$. We let $\frac{\partial}{\partial(\mathrm{d}\overline{z}_{ij})}$ denote the graded derivation that removes $\mathrm{d}\overline{z}_{ij}$.

As before, let $\eta_j : \mathcal{F}_r \to \mathcal{F}_r$ denote the derivation that cobounds the derivation $\frac{\partial}{\partial \overline{z}_j}$. We can compose any element

$$\alpha \in \Omega^{0,*}(\mathrm{PDisc}_n(r_1,\ldots,r_k \mid s), \mathrm{Hom}(\mathcal{F}_{r_1},\ldots,\mathcal{F}_{r_k} \mid s)) \tag{$\dagger$}$$

with η_j acting on $\mathcal{F}_{r_i}$, to get

$$\alpha \circ_i \eta_j \in \Omega^{0,*}(\mathrm{PDisc}_n(r_1, \ldots, r_k \mid s), \mathrm{Hom}(\mathcal{F}_{r_1}, \ldots, \mathcal{F}_{r_k} \mid s)).$$

We use the shorthand notations:

$$l_{ij}(\alpha) = \alpha \circ_i \eta_j,$$
$$r_{ij}(\alpha) = \alpha \circ_i \frac{\partial}{\partial \bar{z}_j},$$

where $\frac{\partial}{\partial \bar{z}_j}$ refers to the derivation acting on $\mathcal{F}_{r_j}$.
    Note that

$$[\mathrm{d}_{\mathcal{F}}, l_{ij}] = r_{ij},$$

where $\mathrm{d}_{\mathcal{F}}$ is the differential on the graded vector space (†) above arising from the differentials on the spaces $\mathcal{F}_{r_i}, \mathcal{F}_s$. Further, the operators l_{ij}, r_{ij} all commute with each other in the graded sense.
    In what follows, for concision, we will write $\mu^{\bar{\partial}}$ instead of $\mu^{\bar{\partial}}(r_1, \ldots, r_k \mid s)$, and similarly for μ^0.
    The cochains $\mu^{\bar{\partial}}$ are defined by

$$\mu^{\bar{\partial}} = \exp(-\sum_{i,j} \mathrm{d}\bar{z}_{ij} l_{ij}) \mu^0.$$

Here $\mathrm{d}\bar{z}_{ij} l_{ij}$ denotes the operation of wedging with $\mathrm{d}\bar{z}_{ij}$ after applying the operator l_{ij}. Note that these two operators graded-commute.
    Next, we need to verify that $(\bar{\partial} + \mathrm{d}_{\mathcal{F}}) \mu^{\bar{\partial}} = 0$, where $\bar{\partial} + \mathrm{d}_{\mathcal{F}}$ is the differential on graded vector space (†) above that arises by combining the $\bar{\partial}$ operator with the differential arising from the complexes $\mathcal{F}_{r_i}, \mathcal{F}_s$.
    We use the following identities:

$$[\mathrm{d}_{\mathcal{F}}, \sum_{i,j} \mathrm{d}\bar{z}_{ij} l_{ij}] = \sum_{i,j} \mathrm{d}\bar{z}_{ij} r_{ij},$$
$$r_{ij} \mu^0 = \frac{\partial}{\partial \bar{z}_{ij}} \mu^0,$$
$$\mathrm{d}_{\mathcal{F}} \mu^0 = 0.$$

The second and third identities are part of the axioms for a smoothly translation-invariant prefactorization algebra.

These identities allows us to calculate that

$$
\begin{aligned}
(\bar{\partial} + d_{\mathcal{F}})\mu^{\bar{\partial}} &= (\bar{\partial} + d_{\mathcal{F}}) \exp(-\sum d\bar{z}_{ij} l_{ij})\mu^0 \\
&= -\exp\left(-\sum d\bar{z}_{ij} l_{ij}\right)\left(\sum d\bar{z}_{ij} r_{ij}\right)\mu^0 \\
&\quad + \exp\left(-\sum d\bar{z}_{ij} l_{ij}\right)(\bar{\partial} + d_{\mathcal{F}})\mu^0 \\
&= \exp\left(-\sum d\bar{z}_{ij} l_{ij}\right)\left(-\sum d\bar{z}_{ij} \frac{\partial}{\partial \bar{z}_{ij}} + \bar{\partial}\right)\mu^0 \\
&= 0.
\end{aligned}
$$

Thus, $\mu^{\bar{\partial}}$ is closed.

It is straightforward to verify that the elements $\mu^{\bar{\partial}}$ are compatible with composition and with the symmetric group actions. $\qquad\square$

5.3 A General Method for Constructing Vertex Algebras

In this section we prove that the cohomology of a holomorphically translation invariant prefactorization algebra on $\mathbb{C}$ with a compatible circle action gives rise to a vertex algebra. Together with the central theorem of the second volume, which allows one to construct prefactorization algebras by obstruction theory starting from the Lagrangian of a classical field theory, this gives a general method to construct vertex algebras.

5.3.1 The Circle Action

Recall from Section 3.7 in Chaper 3 the definition of a smoothly G-equivariant prefactorization algebra on a manifold M with smooth action of a Lie group G. The case of interest here is the action of the isometry group $S^1 \ltimes \mathbb{C}$ of $\mathbb{C}$ acting on $\mathbb{C}$ itself.

Definition 5.3.1 A *holomorphically translation-invariant prefactorization algebra $\mathcal{F}$ on $\mathbb{C}$ with a compatible S^1 action* is a smoothly $S^1 \ltimes \mathbb{R}^2$-invariant prefactorization algebra $\mathcal{F}$, defined over the base field of complex numbers, together with an extension of the action of the complex Lie algebra

$$
\mathrm{Lie}_{\mathbb{C}}(S^1 \ltimes \mathbb{R}^2) = \mathbb{C}\{\partial_\theta, \partial_z, \partial_{\bar{z}}\},
$$

where ∂_θ is a basis of $\mathrm{Lie}_{\mathbb{C}}(S^1)$, to an action of the dg Lie algebra

$$
\mathbb{C}\{\partial_\theta, \partial_z, \partial_{\bar{z}}\} \oplus \mathbb{C}\{\eta\},
$$

where η is of cohomological degree -1 and the differential is

$$d\eta = \partial_{\bar{z}}.$$

In this dg Lie algebra, all commutators involving η vanish except for

$$[\partial_\theta, \eta] = -\eta.$$

Note that, in particular, $\mathcal{F}$ is a holomorphically translation invariant prefactorization algebra on $\mathbb{C}$.

Our prefactorization algebras take values in differentiable cochain complexes. Here we will work with such vector spaces over the complex numbers; in other words, every differentiable vector space E discussed here will assign a complex vector space $E(M)$ to a manifold M, the restriction maps are complex-linear, and so on.

By their very definition, we know how to differentiate "smooth maps to a differentiable vector space." Thus, if M is a complex manifold and E is a differentiable cochain complex, we have, for every open $U \subset M$, a cochain map

$$\bar{\partial} : C^\infty(U, E) \to \Omega^{0,1}(U, E).$$

These maps form a map of sheaves of cochain complexes on M. Taking the cohomology sheaves of E, we obtain a graded differentiable vector space $H^*(E)$ and we have a map of sheaves of graded vector spaces

$$\bar{\partial} : C^\infty(U, H^*(E)) \to \Omega^{0,1}(U, H^*(E)).$$

The kernel of this map defines a sheaf on M, whose sections we denote by $\mathrm{Hol}(U, H^*(E))$. We call these the *holomorphic sections* of $H^*(E)$ on M.

The theorem on vertex algebras will require an extra hypothesis regarding the S^1-action on the prefactorization algebra. Given any compact Lie group G, we will formulate a concept of *tameness* for a G-action on a differentiable vector space. We will require that the S^1-action on the spaces $\mathcal{F}_r$ in our factorization algebra is tame.

Note that for any compact Lie group G, the space $\mathcal{D}(G)$ of distributions on G is an algebra under convolution. Since spaces of distributions are naturally differentiable vector spaces, and the convolution product

$$* : \mathcal{D}(G) \times \mathcal{D}(G) \to \mathcal{D}(G)$$

is smooth, $\mathcal{D}(G)$ forms an algebra in the category of differentiable vector spaces. There is a map

$$\delta : G \to \mathcal{D}(G)$$

sending an element g to the δ-distribution at g. It is a smooth map and a homomorphism of monoids.

Definition 5.3.2 Let E be a differentiable vector space and G be a compact Lie group. A *tame* action of G on E is a smooth action of the algebra $\mathcal{D}(G)$ on E. Note that this is, in particular, an action of G on E via the smooth map of groups $G \to \mathcal{D}(G)^{\times}$ sending g to δ_g.

If E is a differentiable cochain complex, a tame action is an action of $\mathcal{D}(G)$ on E that commutes with the differential on E.

Let us now specialize to the case when $G = S^1$, which is the case relevant for the theorem on vertex algebras. For each integer k, there is an irreducible representation ρ_k of S^1 given by the function $\rho_k : \lambda \mapsto \lambda^k$. (Here we use λ to denote a complex number by viewing S^1 as a subset of $\mathbb{C}$.) We can view ρ_k as a map of sheaves on the site of smooth manifolds from the sheaf represented by the manifold S^1 to the differentiable vector space C^∞, which assigns complex-valued smooth functions to each manifold. It lifts naturally to a representation $\widetilde{\rho}_k$ of the algebra $\mathcal{D}(S^1)$, defined by

$$\widetilde{\rho}_k(\phi) = \langle \phi, \rho_k \rangle,$$

where $\langle -, - \rangle$ indicates the pairing between distributions and functions. We will abusively denote this representation by ρ_k, for simplicity.

If E is a differentiable vector space equipped with such a smooth action of $\mathcal{D}(S^1)$, we use $E_k \subset E$ to denote the subspace on which $\mathcal{D}(S^1)$ acts by ρ_k. We call this the *weight k eigenspace* for the S^1-action on E.

In the algebra $\mathcal{D}(S^1)$, the element ρ_k, viewed as a distribution on S^1, is an idempotent. If we denote the action of $\mathcal{D}(S^1)$ on E by $*$, then the map $\rho_k * - : E \to E$ defines a projection from E onto E_k. We will denote this projection by π_k.

Remark: Of course, all this holds for a general compact Lie group where instead of these ρ_k, we use the characters of irreducible representations. $\Diamond$

5.3.2 The Statement of the Main Theorem

Now we can state the main theorem of this section.

Theorem 5.3.3 *Let $\mathcal{F}$ be a unital S^1-equivariant holomorphically translation invariant prefactorization algebra on $\mathbb{C}$ valued in differentiable vector spaces. Assume that, for each disc $D(0, r)$ around the origin, the action of S^1 on $\mathcal{F}(D(0, r))$ is tame. Let $\mathcal{F}_k(D(0, r))$ be the weight k eigenspace of the S^1 action on $\mathcal{F}(D(0, r))$.*

We make the following additional assumptions.

(i) *Assume that, for $r < r'$, the structure map*

$$\mathcal{F}_k(D(0,r)) \to \mathcal{F}_k(D(0,r'))$$

is a quasi-isomorphism.

(ii) *For $k \gg 0$, the vector space $H^*(\mathcal{F}_k(D(0,r)))$ is zero.*

(iii) *For each k and r, we require that $H^*(\mathcal{F}_k(D(0,r)))$ is isomorphic, as a sheaf on the site of smooth manifolds, to a countable sequential colimit of finite-dimensional graded vector spaces.*

Let $V_k = H^*(\mathcal{F}_k(D(0,r)))$, and let $V = \bigoplus_{k\in\mathbb{Z}} V_k$. (This space is independent of r by assumption.) Then V has the structure of a vertex algebra, determined by the structure maps of $\mathcal{F}$.

The construction is functorial: it will be manifest that a map of factorization algebras respecting all the equivariance conditions produces a map between the associated vertex algebras.

We give the proof in Section 5.3.4. We spell out the data of the vertex algebra – the vacuum vector, translation operator, vertex operation, and so on – as we construct it. First, though, we discuss issues around applying the theorem.

Remark: If V is not concentrated in cohomological degree 0, then it will have the structure of a vertex algebra valued in the symmetric monoidal category of graded vector spaces. That is, the Koszul rule of signs will apear in the axioms.   ◊

Remark: We will often deal with prefactorization algebras $\mathcal{F}$ equipped with a complete decreasing filtration $F^i\mathcal{F}$, so that $\mathcal{F} = \lim \mathcal{F}/F^i\mathcal{F}$. In this situation, to construct the vertex algebra we need that the properties listed in the theorem hold on each graded piece $\mathrm{Gr}^i \mathcal{F}$. This condition implies, by a spectral sequence, that they hold on each $\mathcal{F}/F^i\mathcal{F}$, allowing us to construct an inverse system of vertex algebras associated to the prefactorization algebra $\mathcal{F}/F^i\mathcal{F}$. The inverse limit of this system of vertex algebras is the vertex algebra associated to $\mathcal{F}$.   ◊

5.3.3 Using the Theorem

The conditions of the theorem are always satisfied in practice by prefactorization algebras arising from quantizing a holomorphically translation invariant classical field theory. We describe in great detail the example of the free $\beta\gamma$ system in Section 5.4.

There is also the problem of recognizing the vertex algebra produced by such a prefactorization algebra. Thankfully, there is a useful "reconstruction" theorem that provides simple criteria to uniquely construct a vertex algebra given "generators and relations." We will exploit this theorem to verify that we have recovered standard vertex algebras in our examples later in this chapter.

Theorem 5.3.4 (Reconstruction Theorem 4.4.1 in Frenkel and Ben-Zvi (2004)) *Let V be a complex vector space equipped with a nonzero vector $|0\rangle$, an endomorphism T, a countable ordered set $\{a^\alpha\}_{\alpha \in S}$ of vectors, and fields*

$$a^\alpha(z) = \sum_{n \in \mathbb{Z}} a^\alpha_{(n)} z^{-n-1} \in \operatorname{End}(V)[[z, z^{-1}]]$$

such that

(i) *For all α, $a^\alpha(z)|0\rangle = a^\alpha + O(z)$.*
(ii) *$T|0\rangle = 0$ and $[T, a^\alpha(z)] = \partial_z a^\alpha(z)$ for all α.*
(iii) *All fields $a^\alpha(z)$ are mutually local.*
(iv) *V is spanned by the vectors*

$$a^{\alpha_1}_{(j_1)} \cdots a^{\alpha_m}_{(j_m)} |0\rangle$$

with the $j_i < 0$.

Then, using the formula

$$Y(a^{\alpha_1}_{(j_1)} \cdots a^{\alpha_m}_{(j_m)} |0\rangle, z) = \frac{1}{(-j_1 - 1)! \cdots (-j_m - 1)!} : \partial_z^{-j_1 - 1}$$
$$a^{\alpha_1}(z) \cdots \partial_z^{-j_m - 1} a^{\alpha_m}(z) :$$

to define a vertex operation, we obtain a well-defined and unique vertex algebra $(V, |0\rangle, T, Y)$ such that $Y(a^\alpha, z) = a^\alpha(z)$.

Here $: a(z)b(w) :$ denotes the *normally ordered product* of fields, defined as

$$: a(z)b(w) := a(z)_+ b(w) + b(w)a(z)_-$$

where

$$a(z)_+ = \sum_{n \geq 0} a_n z^n \text{ and } a(z)_- = \sum_{n < 0} a_n z^n.$$

Normal ordering eliminates various "divergences" that appear in naively taking products of fields.

5.3.4 The Proof of the Main Theorem

The strategy of the proof is as follows. We will analyze the structure on V given to us by the axioms of a translation-invariant prefactorization algebra.

The prefactorization product will become the operator product expansion or state-field map. The locality axiom of a vertex algebra will follow from the associativity axioms of the prefactorization algebra.

The Vector Space Structure

Let us begin by describing important features of V and its natural completion $\overline{V}$.

Because our prefactorization algebra $\mathcal{F}$ is translation invariant, the cochain complex $\mathcal{F}(D(z, r))$ associated to a disc of radius r is independent of z. We therefore use the notation $\mathcal{F}(r)$ for $\mathcal{F}(D(z, r))$ and $\mathcal{F}(\infty)$ to denote $\mathcal{F}(\mathbb{C})$.

Let $\mathcal{F}_k(r)$ denote the weight k space of the S^1 action on $\mathcal{F}$. The S^1 action on each $\mathcal{F}(r)$ extends to an action of $\mathcal{D}(S^1)$. The projection map $\mathcal{F}(r) \to \mathcal{F}_k(r)$ onto the weight k space induces a map at the level of cohomology that we also denote

$$\pi_k : H^*(\mathcal{F}(r)) \to H^*(\mathcal{F}_k(r)) = V_k.$$

It is a map of differentiable vector spaces, splitting the natural inclusion. Furthermore, the extension map $H^*(\mathcal{F}_k(r)) \to H^*(\mathcal{F}_k(r'))$ associated to the inclusion $D(0, r) \hookrightarrow D(0, r')$ is the identity on V.

By assumption, the differentiable vector space $V_k = H^*(\mathcal{F}_k(r))$ is the colimit of its finite-dimensional subspaces. This means that a section of V_k on the manifold M is given, locally on M, by a smooth map (in the ordinary sense) to a finite dimensional subspace of V_k.

By construction there is a canonical map

$$V = \bigoplus_k V_k \to H^*(\mathcal{F}(r))$$

for every $r > 0$. We emphasize that V need not be the limit $\lim_{r \to 0} H^*(\mathcal{F}(r))$. Instead, it is an algebraically tractable vector space mapping into that limit (so to speak, the finite sums of modes).

We introduce now a natural partner to V into which all the $H^*(\mathcal{F}(r))$ map. Let

$$\overline{V} = \prod_k V_k,$$

where the product is taken in the category of differentiable vector spaces. There is a canonical map

$$\prod_k \pi_k : H^*(\mathcal{F}(r)) \to \overline{V}$$

given by the product of all the projection maps. We emphasize that $\overline{V}$ need not be the colimit $\mathrm{colim}_{r \to \infty} H^*(\mathcal{F}(r))$. Instead, it is an algebraically tractable vector space receiving a map from that colimit. Note that $\overline{V}$ is a completion of V in the sense that we now allow arbitrary, infinite sums of the modes.

The Operator Product

Let us start to analyze the structure on V given by the prefactorization algebra structure on $\mathcal{F}$. Our approach will be to use instead the structure as an algebra over the cooperad $\Omega^{0,*}(\mathrm{PDiscs}_1)$. (Since a one-dimensional polydisc is simply a disc, we will simply write disc from hereon.) We will call the structure maps for this algebra the *operator product*. Recall that we use

$$\mu^{\overline{\partial}}(r_1,\ldots,r_k \mid \infty) : \mathcal{F}(r_1) \times \cdots \times \mathcal{F}(r_k) \to \Omega^{0,*}(\mathrm{Discs}(r_1,\ldots,r_k \mid \infty), \mathcal{F}(\infty))$$

to denote such a multilinear cochain map.

We want to use this operator product to produce a map

$$m_{z_1,\ldots,z_k} : V^{\otimes k} \to \overline{V}$$

for every configuration $(z_1,\ldots,z_k)$ of k ordered distinct points in $\mathbb{C}$. This map will vary holomorphically over that configuration space.

Note that for any complex manifold X there is a truncation map

$$(-)^0 : H^* \left(\Omega^{0,*}(X, \mathcal{F}(\infty)) \right) \to \mathrm{Hol}(X, H^*(\mathcal{F}(\infty))).$$

If $\alpha \in \Omega^{0,*}(X, \mathcal{F}(\infty))$ is a cocycle, this map is defined by first extracting the component α^0 of α in $\Omega^{0,0}(X, \mathcal{F}(\infty))$. The fact that α is closed means that α^0 is closed for the differential $D_{\mathcal{F}}$ on $\mathcal{F}(\infty)$ and that $\overline{\partial}\alpha^0$ is exact. Thus, the cohomology class $[\alpha^0]$ is a section of $\mathcal{F}(\infty)$ on X, and $\overline{\partial}[\alpha^0] = 0$ so that $[\alpha^0]$ is holomorphic.

Hence, at the level of cohomology, the operator product produces a smooth multilinear map

$$m_{z_1,\ldots,z_k} : H^*(\mathcal{F}(r_1)) \times \cdots \times H^*(\mathcal{F}(r_k))$$
$$\to \mathrm{Hol}(\mathrm{Discs}(r_1,\ldots,r_k \mid \infty), H^*(\mathcal{F}(\infty))).$$

Here z_i indicate the positions of the centers of the discs in $\mathrm{Discs}(r_1,\ldots,r_k \mid \infty)$.

Consider what happens if we shrink the discs. If $r_i' < r_i$, then

$$\mathrm{Discs}(r_1,\ldots,r_k \mid \infty) \subset \mathrm{Discs}(r_1',\ldots,r_k' \mid \infty).$$

We also have the extension map $\mathcal{F}(r') \to \mathcal{F}(r)$, from the inclusion $D(0, r') \hookrightarrow D(0, r)$. Note that the following diagram commutes:

$$
\begin{array}{ccc}
H^*(\mathcal{F}(r'_1)) \otimes \cdots \otimes H^*(\mathcal{F}(r'_k)) & \longrightarrow & \mathrm{Hol}(\mathrm{Discs}(r'_1, \ldots, r'_k \mid \infty), H^*(\mathcal{F}(\infty))) \\
\downarrow & & \downarrow \\
H^*(\mathcal{F}(r_1)) \otimes \cdots \otimes H^*(\mathcal{F}(r_k)) & \longrightarrow & \mathrm{Hol}(\mathrm{Discs}(r_1, \ldots, r_k \mid \infty), H^*(\mathcal{F}(\infty))).
\end{array}
$$

Because V maps to $\lim_{r \to 0} H^*(\mathcal{F}(r))$, the operator product, when restricted to V, gives a map

$$
m^{H^*(\mathcal{F}(\infty))}_{z_1, \ldots, z_k} : V^{\otimes k} \to \lim_{r \to 0} \mathrm{Hol}(\mathrm{Discs}(r, \ldots, r \mid \infty), H^*(\mathcal{F}(\infty))),
$$

but

$$
\lim_{r \to 0} \mathrm{Hol}(\mathrm{Discs}(r, \ldots, r \mid \infty), H^*(\mathcal{F}(\infty))) = \mathrm{Hol}(\mathrm{Conf}_k(\mathbb{C}), H^*(\mathcal{F}(\infty))),
$$

where $\mathrm{Conf}_k(\mathbb{C})$ is the configuration space of k ordered distinct points in $\mathbb{C}$. Note that if the z_i lie in a disc $D(0, r)$, this map provides a map

$$
m^{H^*(\mathcal{F}(r))}_{z_1, \ldots, z_k} : V^{\otimes k} \to \mathrm{Hol}(\mathrm{Conf}_k(D(0, r)), H^*(\mathcal{F}(r))).
$$

Composing the prefactorization product map $m^{H^*(\mathcal{F}(\infty))}_{z_1, \ldots, z_k}$ with the map

$$
\prod_n \pi_n : H^*(\mathcal{F}(\infty)) \to \overline{V} = \prod_n V_n
$$

gives a map

$$
m_{z_1, \ldots, z_k} : V^{\otimes k} \to \mathrm{Hol}(\mathrm{Conf}_k(\mathbb{C}), \overline{V}) = \prod_n \mathrm{Hol}(\mathrm{Conf}_k(\mathbb{C}), V_n).
$$

This map does not involve the spaces $\mathcal{F}(r)$ anymore, only the space V and its natural completion $\overline{V}$. If there is potential confusion, we will refer to this version of the operator product map by $m^V_{z_1, \ldots, z_k}$ instead of just $m_{z_1, \ldots, z_k}$.

The vertex operator will, of course, be constructed from the maps $m_{z_1, \ldots, z_k}$. Consider the map

$$
m_{z,0} : V \otimes V \to \prod_n \mathrm{Hol}(\mathbb{C}^\times, V_n),
$$

where we have restricted the map $m^V_{z,w}$ to the locus where $w = 0$. Since each space V_n is a discrete vector space (i.e., a colimit of finite dimensional vector spaces), we can form the ordinary Laurent expansion of an element in $\mathrm{Hol}(\mathbb{C}^\times, V_n)$ to get a map

$$
\mathcal{L}_z m^V_{z,0} : V \otimes V \to \overline{V}[[z, z^{-1}]].
$$

It has the following important property.

Lemma 5.3.5 *The image of* $\mathcal{L}_z m^V_{z,0}$ *is in the subspace* $V((z))$.

Proof The map $m^V_{z,0}$ is S^1-equivariant, where S^1 acts on V and $\mathbb{C}^\times$ in the evident way. Therefore, so is $\mathcal{L}_z m^V_{z,0}$. Since every element in $V \otimes V$ is in a finite sum of the S^1-eigenspaces, the image of $\mathcal{L}_z m^V_{z,0}$ is in the subspace of $\overline{V}[[z, z^{-1}]]$ spanned by finite sums of eigenvectors. An element of $\overline{V}[[z, z^{-1}]]$ is in the weight k eigenspace of the S^1 action if it is of the form

$$\sum_n z^{k-n} v_n,$$

where $v_n \in V_n$. Since $V_n = 0$ for $n \gg 0$, every such element is in $V((z))$. □

The Vertex Algebra Structure

Let us now define the structures on V that will correspond to the vertex algebra structure.

(i) **The vacuum element** $|0\rangle \in V$: By assumption, $\mathcal{F}$ is a unital prefactorization algebra. Therefore, the commutative algebra $\mathcal{F}(\emptyset)$ has a unit element $|0\rangle$. The prefactorization structure map $\mathcal{F}(\emptyset) \to \mathcal{F}(D(0, r))$ for any r gives an element $|0\rangle \in \mathcal{F}(r)$. This element is automatically S^1-invariant, and therefore in V_0.

(ii) **The translation map** $T : V \to V$: The structure of a holomorphically translation prefactorization algebra on $\mathcal{F}$ includes a derivation $\frac{\partial}{\partial z}$ corresponding to infinitesimal translation in the (complex) direction z. The fact that $\mathcal{F}$ has a compatible S^1 action means that, for all r, the map $\frac{\partial}{\partial z}$ maps $\mathcal{F}_k(r)$ to $\mathcal{F}_{k-1}(r)$. Therefore, passing to cohomology, it becomes a map $\frac{\partial}{\partial z} : V_k \to V_{k-1}$. Let $T : V \to V$ be the map that is $\frac{\partial}{\partial z}$ on V_k.

(iii) **The state-field map** $Y : V \to \text{End}(V)[[z, z^{-1}]]$: We let

$$Y(v, z)(v') = \mathcal{L}_z m_{z,0}(v, v') \in V((z)).$$

Note that $Y(v, z)$ is a field in the sense used in the axioms of a vertex algebra, because $Y(v, z)(v')$ has only finitely many negative powers of z.

It remains to check the axioms of a vertex algebra. We need to verify:

(i) **Vacuum axiom**: $Y(|0\rangle, z)(v) = v$.

(ii) **Translation axiom**:

$$Y(Tv_1, z)(v_2) = \frac{\partial}{\partial z} Y(v_1, z)(v_2)$$

and $T |0\rangle = 0$.

(iii) **Locality axiom**:

$$(z_1 - z_2)^N [Y(v_1, z_1), Y(v_2, z_2)] = 0$$

for N sufficiently large.

The vacuum axiom follows from the fact that the unit $|0\rangle \in \mathcal{F}(\emptyset)$, viewed as an element of $\mathcal{F}(D(0, r))$, is a unit for the prefactorization product.

The translation axiom follows immediately from the corresponding axiom of prefactorization algebras, which is built into our definition of holomorphic translation invariance: namely,

$$\frac{\partial}{\partial z} m_{z,0}(v_1, v_2) = m_{z,0}(\partial_z v_1, v_2).$$

The fact that $T|0\rangle = 0$ follows from the fact that the derivation ∂_z of $\mathcal{F}$ gives a derivation of the commutative algebra $\mathcal{F}(\emptyset)$, which must therefore send the unit to zero.

It remains to prove the locality axiom. Essentially, it follows from the associativity property of prefactorization algebras.

5.3.5 Proof of the Locality Axiom

Let us observe some useful properties of the operator product maps

$$m_{z_1, \ldots, z_k} : V^{\otimes k} \to \mathrm{Hol}(\mathrm{Conf}_k(\mathbb{C}), \overline{V}).$$

First, the map $m_{z_1, \ldots, z_k}$ is S^1-equivariant, where we use the diagonal S^1 action on $V^{\otimes k}$ on the left-hand side and on the right-hand side we use the rotation action of S^1 on $\mathrm{Conf}_k(\mathbb{C})$ coupled to the S^1 action on $\overline{V}$ coming from $\mathcal{F}$'s S^1-equivariance. The operator product is also S_k-equivariant, where S_k acts on $V^{\otimes k}$ and on $\mathrm{Conf}_k(\mathbb{C})$ in the evident way. It is also invariant under translation, in the sense that for arbitrary $v_i \in V$,

$$m_{z_1 + \lambda, \ldots, z_k + \lambda}(v_1, \ldots, v_k) = \tau_\lambda \left(m_{z_1, \ldots, z_k}(v_1, \ldots, v_k) \right) \in \overline{V}$$

where τ_λ denotes the action of $\mathbb{C}$ on $\overline{V}$ that integrates the infinitesimal translation action of the Lie algebra $\mathbb{C}$.

We write

$$\mathcal{L}_z m_{z,0}(v_1, v_2) = \sum_k z^k m^k(v_1, v_2)$$

where $m^k(v_1, v_2) \in V$. As we have seen, $m^k(v_1, v_2)$ is zero for $k \ll 0$.

The key proposition in proving the locality axiom is the following bit of complex analysis.

Proposition 5.3.6 *Let $U_{ij} \subset \mathrm{Conf}_k(\mathbb{C})$ be the open subset of configurations for which $|z_j - z_i| < |z_j - z_l|$ for $l \neq i, j$.*
Then, for $(z_1, \ldots, z_k) \in U_{ij}$, we have the following identity in $\mathrm{Hol}(U_{ij}, \overline{V})$:

$$m_{z_1 \cdots z_k}(v_1, \ldots, v_k) = \sum (z_i - z_j)^n m_{z_1 \cdots \widehat{z_j} \cdots z_k}$$
$$\times (v_1, \ldots, v_{i-1}, m^n(v_i, v_j), \ldots, \widehat{v_j}, \ldots, v_k).$$

In particular, the sum on the right-hand side converges. (In this expression, the hats $\widehat{z_j}$ and $\widehat{v_j}$ indicate that we skip these entries.)

When $k = 3$, this identity allows us to expand m_{z_1, z_2, z_3} in two different ways. We find that

$$m_{z_1, z_2, z_3}(v_1, v_2, v_3) = \begin{cases} \sum (z_2 - z_3)^k m_{z_1, z_3}(m^k(v_1, v_2), v_3) & \text{if } |z_2 - z_3| < |z_1 - z_3| \\ \sum (z_1 - z_3)^k m_{z_2, z_3}(v_2, m^k(v_1, v_3)) & \text{if } |z_2 - z_3| > |z_1 - z_3|. \end{cases}$$

This formula should be compared to the "associativity" property in the theory of vertex algebras (see, e.g., Theorem 3.2.1 of Frenkel and Ben-Zvi (2004)).

Proof By symmetry, we can reduce to the case when $i = 1$ and $j = 2$.

Since the operator product is induced from the structure maps of $H^*\mathcal{F}$, the argument will proceed by analyzing the operator product for discs of small radius and then showing that the relevant property extends to the case where the radii go to zero.

Fix a configuration $(z_1, \ldots, z_k) \in \mathrm{Conf}_k(\mathbb{C})$. There is an $\varepsilon > 0$ such that the closures of every disc $D(z_i, \varepsilon)$ are disjoint. As we are interested in the region U_{12}, we can find $\delta > \varepsilon$ (possibly after shrinking ε) such that $D(z_1, \varepsilon) \subset D(z_2, \delta)$ and the closure of $D(z_2, \delta)$ is disjoint from the closure of $D(z_m, \varepsilon)$ for $m > 2$. There is an open neighborhood U' of the configuration $(z_1, \ldots, z_k)$ in U_{12} where these conditions hold.

For configurations in U', the axioms of a prefactorization algebra tell us that the following associativity condition holds:

$$m_{z_1, z_2, \ldots, z_k}(v_1, \ldots, v_k) = m_{z_2, z_3, \ldots, z_k}(m_{z_1, z_2}(v_1, v_2), v_3, \ldots, v_k). \qquad (\dagger)$$

Here, we view $m_{z_1, z_2}(v_1, v_2)$ as an element of $H^*(\mathcal{F}(D(z_2, \delta)))$ that depends holomorphically on points (z_1, z_2) that lie in the open set where $D(z_1, \varepsilon)$ is disjoint from $D(z_2, \varepsilon)$ and contained in $D(z_2, \delta)$. By translating z_2 to 0, we identify $H^*(\mathcal{F}(D(z_2, \delta)))$ with $H^*(\mathcal{F}(\delta))$. This associativity property is an immediate consequence of the axioms of a prefactorization algebra.

By taking ε and δ sufficiently small, we can cover U_{12} by sets of the form U'. Hence, locally on U_{12}, we have the associativity we need. We will continue to work with our open U' described earlier.

We now assume, without loss of generality, that each v_i is homogeneous of weight $|v_i|$ under the S^1 action on V.

Let $\pi_k : H^*(\mathcal{F}(\delta)) \to V_k$ be the projection onto the eigenspace V_k. Recall that the algebra $\mathcal{D}(S^1)$ of distributions on S^1 acts on $\mathcal{F}(\delta)$ and so on $H^*(\mathcal{F}(\delta))$. If this action is denoted by $*$, and if $\lambda \in S^1 \subset \mathbb{C}^\times$ denotes an element of the circle viewed as a complex number, then

$$\pi_k(f) = \lambda^{-k} * f.$$

Note that S^1-equivariance of the operator product map means that we can write the Laurent expansion of $m_{z,0}(v_1, v_2)$ as

$$m_{z,0}(v_1, v_2) = \sum_k z^k m^k(v_1, v_2) = \sum_k \pi_{|v_1|+|v_2|-k} m_{z,0}(v_1, v_2).$$

That is, $z^k m^k(v_1, v_2)$ is the projection of $m_{z,0}(v_1, v_2)$ onto the weight $|v_1 + v_2| - k$ eigenspace of $\overline{V}$.

We want to show that

$$\sum_n (z_1 - z_2)^n m_{z_2, z_3, \dots, z_k}(m^n(v_1, v_2), \dots, v_k) = m_{z_1, \dots, z_k}(v_1, \dots, v_k) \quad (\ddagger)$$

where the z_i lie in our open U'. This identity is equivalent to showing that

$$\sum_n m_{z_2, z_3, \dots, z_k}(\pi_n m_{z_1, z_2}(v_1, v_2), \dots, v_k) = m_{z_1, \dots, z_k}(v_1, \dots, v_k).$$

Indeed, the Laurent expansion of $m_{z_1, z_2}(v_1, v_2)$ and the expansion in terms of eigenspaces of the S^1 action on $\overline{V}$ differ only by a reordering of the sum.

Fix $v_1, \dots, v_k \in U'$. Define a map

$$\Phi : \quad \mathcal{D}(S^1) \quad \to \quad \mathrm{Hol}(U', \overline{V})$$
$$\alpha \quad \mapsto \quad m_{z_2, z_3, \dots, z_k}(\alpha * m_{z_1, z_2}(v_1, v_2), v_3, \dots, v_k).$$

Here $\alpha * -$ refers to the action of $\mathcal{D}(S^1)$ on $H^*(\mathcal{F}(\delta))$. Note that Φ is a smooth map.

Let δ_1 denote the delta-function on S^1 with support at the identity 1. The associativity identity $(\dagger)$ of prefactorization algebras implies that

$$\Phi(\delta_1) = m_{z_1, \dots, z_k}(v_1, \dots, v_k).$$

The point here is that $\delta_1 * $ is the identity on $H^*(\mathcal{F}(\delta))$.

To prove the identity (‡), it now suffices to prove that

$$\Phi\left(\sum_{n\in\mathbb{Z}}\lambda^n\right) = \Phi(\delta_1).$$

In the space $\mathcal{D}(S^1)$ with its natural topology the sum $\sum_{n\in\mathbb{Z}}\lambda^n$ converges to δ_1, as it is simply the Fourier expansion of the delta function. So, to prove the proposition, it suffices to prove that Φ is continuous, and not just smooth, where the spaces $\mathcal{D}(S^1)$ and $\mathrm{Hol}(U^\delta_{12}, \overline{V})$ are endowed with their natural topologies. (The topology on $\mathrm{Hol}(U^\delta_{ij}, \overline{V})$ is given by saying that a sequence converges if its projection to each V_k converges uniformly, with all derivatives, on compact sets of U^δ_{ij}.)

The spaces $\mathcal{D}(S^1)$ and $\mathrm{Hol}(U^\delta_{ij}, \overline{V})$ both lie in the essential image of the functor from locally convex topological vector spaces to differentiable vector spaces. A result of Kriegl and Michor (1997) tells us that smooth linear maps between topological vector spaces are bounded. Therefore, the map Φ is a bounded linear map of topological vector spaces.

In Lemma B.5.5 in Appendix B we show, using results of Kriegl and Michor (1997), that the space of compactly supported distributions on any manifold has the *bornological* property, meaning that a bounded linear map from it to any topological vector space is the same as a continuous linear map. It follows that Φ is continuous, thus completing the proof. □

As a corollary, we find the following.

Corollary 5.3.7 *For $v_1, \ldots, v_k \in V$, $m_{z_1,\ldots,z_k}(v_1, \ldots, v_k)$ has finite order poles on every diagonal in $\mathrm{Conf}_k(\mathbb{C})$. That is, for some N sufficiently large, the function*

$$(\prod_{i<j}(z_i - z_j)^N)m_{z_1,\ldots,z_k}(v_1, \ldots, v_k)$$

extends to an element of $\mathrm{Hol}(\mathbb{C}^k, \overline{F})$.

Proof This corollary follows immediately from the previous proposition. □

We are finally ready to prove the locality axiom.

Proposition 5.3.8 *The locality axiom holds: for any $v_1, v_2 \in V$,*

$$(z_1 - z_2)^N[Y(v_1, z_1), Y(v_2, z_2)] = 0$$

for $N \gg 0$.

Proof For any holomorphic function $F(z_1, \ldots, z_k)$ of variables $z_1, \ldots, z_k \in \mathbb{C}^\times$, we let $\mathcal{L}_{z_i} F$ denote the Laurent expansion of F in the variable z_i. This expansion converges when $|z_i| < |z_j|$ for all j. It can be defined by fixing the values of z_j with $j \neq i$, then viewing F as a function of z_i on the punctured disc where $0 < |z_i| < \min_{j \neq i} |z_j|$, and taking the usual Laurent expansion. We can also define iterated Laurent expansions. For example, if F is a function of $z_1, z_2 \in \mathbb{C}^\times$, we define

$$\mathcal{L}_{z_2} \mathcal{L}_{z_1} F \in \mathbb{C}[[z_1^{\pm 1}, z_2^{\pm 1}]]$$

by first taking the Laurent expansion with respect to z_1, yielding a series in z_1 whose coefficients are holomorphic functions of $z_2 \in \mathbb{C}^\times$, and then applying the Laurent expansion with respect to z_2 to each of the coefficient functions of the expansion with respect to z_1.

Recall that we define

$$Y(v_1, z)(v_2) = \mathcal{L}_z m_{z,0}(v_1, v_2) \in V((z)).$$

We define $m_k(v_1, v_2)$ so that

$$\mathcal{L}_z m_{z,0}(v_1, v_2) = \sum_k z^k m_k(v_1, v_2).$$

Note that, by definition,

$$Y(v_1, z_1) Y(v_2, z_2)(v_3) = \mathcal{L}_{z_1} m_{z_1,0}\left(v_1, \mathcal{L}_{z_2} m_{z_2,0}(v_2, v_3)\right)$$
$$= \mathcal{L}_{z_1} \sum_k z_2^k m_{z_1,0}\left(v_1, m_n(v_2, v_3)\right) \in V[[z_1^{\pm 1}, v_2^{\pm 1}]].$$

Proposition 5.3.6 tells us that

$$m_{z_1,z_2,0}(v_1, v_2, v_3) = \sum_n z_2^n m_{z_1,0}\left(v_1, m^n(v_2, v_3)\right)$$

as long as $|z_2| < |z_1|$. Thus,

$$Y(v_1, z_1) Y(v_2, z_2)(v_3) = \mathcal{L}_{z_1} \mathcal{L}_{z_2} m_{z_1,z_2,0}(v_1, v_2, v_3).$$

Similarly,

$$Y(v_2, z_2) Y(v_1, z_1)(v_3) = \mathcal{L}_{z_2} \mathcal{L}_{z_1} m_{z_2,z_2,0}(v_2, v_1, v_3)$$
$$= \mathcal{L}_{z_2} \mathcal{L}_{z_1} m_{z_1,z_2,0}(v_1, v_2, v_3).$$

Therefore,

$$[Y(v_1, z_1), Y(v_2, z_2)](v_3) = \left(\mathcal{L}_{z_1} \mathcal{L}_{z_2} - \mathcal{L}_{z_2} \mathcal{L}_{z_1}\right) m_{z_1,z_2,0}(v_1, v_2, v_3).$$

Since $\mathcal{L}_{z_1}$ and $\mathcal{L}_{z_2}$ are maps of $\mathbb{C}[z_1, z_2]$ modules, we have

$$(z_1 - z_2)^N \left(\mathcal{L}_{z_1} \mathcal{L}_{z_2} - \mathcal{L}_{z_2} \mathcal{L}_{z_1} \right) m_{z_1, z_2, 0}(v_1, v_2, v_3)$$
$$= \left(\mathcal{L}_{z_1} \mathcal{L}_{z_2} - \mathcal{L}_{z_2} \mathcal{L}_{z_1} \right) (z_1 - z_2)^N m_{z_1, z_2, 0}(v_1, v_2, v_3).$$

Finally, we know that for N sufficiently large, $(z_1 - z_2)^N z_1^N z_2^N m_{z_1, z_2, 0}$ has no poles, and so extends to a function on $\mathbb{C}^2$. It follows from the fact that partial derivatives commute that

$$\left(\mathcal{L}_{z_1} \mathcal{L}_{z_2} - \mathcal{L}_{z_2} \mathcal{L}_{z_1} \right) (z_1 - z_2)^N z_1^N z_2^N m_{z_1, z_2, 0}(v_1, v_2, v_3) = 0.$$

Since Laurent expansion is a map of $\mathbb{C}[z_1, z_2]$-modules, and since z_1, z_2 act invertibly on $\mathbb{C}[[z_1^{\pm 1}, z_2^{\pm 1}]]$, the result follows. $\qquad\square$

Remark: In the vertex algebra literature, a heuristic justification of the locality axiom is frequently given by unpacking the consequences of pretending that $Y(v_1, z_1) Y(v_2, z_2)(v_3)$ and $Y(v_2, z_2) Y(v_1, z_1)(v_3)$ arise as expansions of a holomorphic function of $z_1, z_2 \in \mathbb{C}^\times$ in the regions when $|z_1| < |z_2|$ and $|z_2| < |z_1|$. Our approach makes this idea rigorous. $\qquad\diamondsuit$

5.4 The $\beta\gamma$ System and Vertex Algebras

This section focuses on one of the simplest holomorphic field theories, the free $\beta\gamma$ system. Our goal is to study it just as we studied the free particle in Section 4.2 in Chapter 4. Following the methods developed there, we will construct the prefactorization algebra for this theory, show that it is holomorphically translation-invariant, and finally show that the associated vertex algebra is what is known in the vertex algebra literature as the $\beta\gamma$ system. Along the way, we will compute the simplest operator product expansions for the theory using purely homological methods.

5.4.1 The $\beta\gamma$ System

Let $M = \mathbb{C}$ and let $\mathscr{E} = \left(\Omega_M^{0,*} \oplus \Omega_M^{1,*}, \overline{\partial} \right)$ be the Dolbeault complex resolving holomorphic functions and holomorphic 1-forms as a sheaf on M. Following the convention of physicists, we denote by γ an element of $\Omega^{0,*}$ and by β an element of $\Omega^{1,*}$. The pairing $\langle -, - \rangle$ is

$$\langle -, - \rangle : \qquad \mathscr{E}_c \otimes \mathscr{E}_c \qquad \rightarrow \qquad \mathbb{C},$$
$$(\gamma_0 + \beta_0) \otimes (\gamma_1 + \beta_1) \quad \mapsto \quad \int_{\mathbb{C}} \gamma_0 \wedge \beta_1 + \beta_0 \wedge \gamma_1.$$

Thus we have the data of a free BV theory. The action functional for the theory is

$$S(\gamma, \beta) = \langle \gamma + \beta, \overline{\partial}(\gamma + \beta) \rangle = 2 \int_M \beta \wedge \overline{\partial}\gamma.$$

The Euler–Lagrange equation is simply $\overline{\partial}\gamma = 0 = \overline{\partial}\beta$. One should think of $\mathscr{E}$ as the "derived space of holomorphic functions and 1-forms on M."

Remark: Note that this theory is well defined on any Riemann surface, and one can study how it varies over the moduli space of curves. In fact, there are many variants of this theory. Let Σ be a Riemann surface and $\mathcal{V}$ a holomorphic vector bundle on Σ. Define a free BV theory on Σ with fields $\mathscr{E} = \Omega^{0,*}(\mathcal{V}) \oplus \Omega^{1,*}(\mathcal{V}^\vee)$ and with pairing given by "fiberwise evaluate duals and then integrate."

For instance, if one adds d copies of $\mathscr{E}$ from above (equivalently, tensor $\mathscr{E}$ with $\mathbb{C}^d$) and lets S_d be the d-fold sum of the action S on each copy, then the Euler–Lagrange equations for S_d pick out holomorphic maps γ from M to $\mathbb{C}^d$ and holomorphic sections β of $\Omega^1_M(\gamma^* T_{\mathbb{C}^d})$. $\Diamond$

5.4.2 The Quantum Observables of the $\beta\gamma$ System

To construct the quantum observables, following Section 4.2 in Chapter 4, we start by defining a certain graded Heisenberg Lie algebra and then take its Chevalley–Eilenberg complex for Lie algebra homology.

For each open $U \subset \mathbb{C}$, we set

$$\mathcal{H}(U) = \Omega_c^{0,*}(U) \oplus \Omega_c^{1,*}(U) \oplus (\mathbb{C}\hbar)[-1],$$

where $\mathbb{C}\hbar$ is situated in cohomological degree 1. The Lie bracket is simply

$$[\mu, \nu] = \hbar \int_U \mu \wedge \nu,$$

so $\mathcal{H}$ is a central extension of the abelian dg Lie algebra given by all the Dolbeault forms (with $\overline{\partial}$ as differential).

The prefactorization algebra Obs^q of quantum observables assigns to each open $U \subset \mathbb{C}$, the cochain complex $C_*(\mathcal{H}(U))$, which we will write as

$$\mathrm{Obs}^q(U) = \left(\mathrm{Sym}\left(\Omega_c^{1,*}(U)[1] \oplus \Omega_c^{0,*}(U)[1] \right)[\hbar], \overline{\partial} + \hbar\Delta \right).$$

The differential has a component $\overline{\partial}$ arising from the underlying cochain complex of $\mathcal{H}$ and a component arising from the Lie bracket, which we'll denote Δ. It is the *BV Laplacian* for this theory.

In the text that follows we will unpack the information Obs^q that encodes by examining some simple open sets and the cohomology $H^* \mathrm{Obs}^q$ on those open sets. As usual, the meaning of a complex is easiest to garner through its cohomology. The main theorem of this section is that the vertex algebra extracted from Obs^q is the well-known $\beta\gamma$ vertex algebra.

5.4.3 An Isomorphism of Vertex Algebras

Our goal is to demonstrate that the vertex algebra constructed by Theorem 5.3.3 from the quantum observables of the $\beta\gamma$-system is isomorphic to a vertex algebra considered in the physics literature called the $\beta\gamma$ vertex algebra.

The $\beta\gamma$ Vertex Algebra

We follow Frenkel and Ben-Zvi (2004), notably chapters 11 and 12, to make the relationship clear. Let W denote the space of polynomials $\mathbb{C}[a_n, a_m^*]$ generated by variables a_n, a_m^* $n < 0$ and $m \leq 0$.

Definition 5.4.1 The $\beta\gamma$ *vertex algebra* has state space W, vacuum vector 1, translation operator T the map

$$a_i \mapsto -ia_{i-1},$$
$$a_i^* \mapsto -(i-1)a_{i-1}^*,$$

and the vertex operator

$$Y(a_{-1}, z) = \sum_{n<0} a_n z^{-1-n} + \sum_{n\geq 0} \frac{\partial}{\partial a_{-n}^*} z^{-1-n}$$

and

$$Y(a_0^*, z) = \sum_{n\leq 0} a_n^* z^{-n} - \sum_{n>0} \frac{\partial}{\partial a_{-n}} z^{-n}.$$

By Theorem 5.3.4, these determine a vertex algebra.

The main theorem of this section is the following.

Theorem 5.4.2 *Let* $V_{\hbar=2\pi i}$ *denote the vertex algebra constructed from quantum observables of the* $\beta\gamma$ *system, specialized to* $\hbar = 2\pi i$. *There is an* S^1-*equivariant isomorphism of vertex algebras*

$$V_{\hbar=2\pi i} \cong W.$$

The circle S^1 *acts on* W *by giving* a_i, a_j^* *weights* i, j *respectively.*

The proof of this theorem breaks up into a few stages. First, we need to verify that we can apply Theorem 5.3.3. In particular, we need to verify that Obs^q is holomorphically translation invariant and S^1-equivariant. We then need to compute the weight spaces for the S^1 action. Finally, we need to demonstrate that the vertex operator arising from Obs^q agrees with that of the $\beta\gamma$ vertex algebra.

These arguments are divided over several subsections. The techniques we develop here make it possible to prove theorems of this flavor for many other field theories on $\mathbb{C}$ whose action functional is holomorphic in flavor.

5.4.4 The Criteria to Obtain a Vertex Algebra

The $\beta\gamma$ field theory is manifestly translation-invariant, so it remains to verify that the action of $\frac{\partial}{\partial \bar{z}}$ is homotopically trivial. Consider the operator

$$\eta = \frac{\mathrm{d}}{\mathrm{d}(\mathrm{d}\bar{z})},$$

which acts on the space of fields $\mathscr{E}$. The operator η maps $\Omega^{1,1}$ to $\Omega^{1,0}$ and $\Omega^{0,1}$ to $\Omega^{0,0}$. Then we see that

$$[\bar{\partial} + \hbar\Delta, \eta] = \frac{\mathrm{d}}{\mathrm{d}\bar{z}},$$

so that the action of $\mathrm{d}/\mathrm{d}\bar{z}$ is homotopically trivial, as desired.

We would like to apply the result of Theorem 5.3.3, which shows that a holomorphically translation invariant prefactorization algebra with certain additional conditions gives rise to a vertex algebra. We need to check the following conditions.

(i) The prefactorization algebra must have an action of S^1 covering the action on $\mathbb{C}$ by rotation.

(ii) For every disc $D(0, r) \subset \mathbb{C}$, including $r = \infty$, the S^1 action on Obs^q $(D(0, r))$ must extend to an action of the algebra $\mathcal{D}(S^1)$ of distributions on S^1.

(iii) If $\mathrm{Obs}^q_k(D(0, r))$ denotes the weight k eigenspace of the S^1 action, then we require that the extension map

$$H^*(\mathrm{Obs}^q_k(D(0, r))) \to H^*(\mathrm{Obs}^q_k(D(0, s)))$$

is an isomorphism for $r < s$.

(iv) Finally, we require that the space $H^*(\mathrm{Obs}^q_k(D(0, r)))$ is a discrete vector space, that is, a colimit of finite dimensional vector spaces.

The first condition is obvious in our example: the S^1 action arises from the natural action of S^1 on $\Omega^{0,*}_c(\mathbb{C})$ and $\Omega^{1,*}_c(\mathbb{C})$. The second condition is also easy to check: if $f \in C_c^\infty(D(0, r))$ then the expression

$$z \mapsto \int_{\lambda \in S^1} \phi(\lambda) f(\lambda z)$$

makes sense for any distribution $\phi \in \mathcal{D}(S^1)$, and it defines a continuous and hence smooth map

$$\rho : \mathcal{D}(S^1) \times C_c^\infty(D(0, r)) \to C_c^\infty(D(0, r)).$$

To check the remaining two conditions, we need to analyze $H^*(\mathrm{Obs}^q(D(0, r)))$ more explicitly.

5.4.5 Analytic Preliminaries

In Section 4.6 in Chapter 4 we showed that if a free field theory possesses a Green's function for the differential defining the elliptic complex of fields, then there is an isomorphism of differentiable cochain complexes

$$\mathrm{Obs}^{cl}(U)[\hbar] \cong \mathrm{Obs}^q(U)$$

for any open set U. In our example, we want to understand $H^*(\mathrm{Obs}^q(D(0, r)))$ as a differentiable cochain complex and also its decomposition into eigenspaces for the action of S^1. This result about the Green's function shows that it suffices to understand the cohomology of the corresponding complex of classical observables (see Eq. (5.4.6.1) for the Green's function). We describe the classical observables in this subsection.

Note that the classical observables are particularly simple to analyze because computing their cohomology breaks up into a collection of easier problems. To be explicit, we recall that

$$\mathrm{Obs}^{cl}(U) = \left(\mathrm{Sym}\left(\Omega_c^{1,*}(U)[1] \oplus \Omega_c^{0,*}(U)[1]\right), \bar{\partial}\right),$$

and $\bar{\partial}$ preserves the symmetric powers. Hence, we can analyze each symmetric power

$$\left(\mathrm{Sym}^n\left(\Omega_c^{1,*}(U)[1] \oplus \Omega_c^{0,*}(U)[1]\right), \bar{\partial}\right)$$

separately. The nth symmetric power can then be studied as an elliptic complex on the complex manifold U^n.

Recollections

We remind readers of some facts from the theory of several complex variables (references for this material are Gunning and Rossi (1965), Forster (1991), and Serre (1953)). We then use these facts to describe the cohomology of the observables.

Proposition 5.4.3 *Every open set $U \subset \mathbb{C}$ is Stein. As the product of Stein manifolds is Stein, every product $U^n \subset \mathbb{C}^n$ is Stein.*

Remark: Behnke and Stein (1949) proved that every noncompact Riemann surface is Stein, so the arguments we develop here extend farther than we exploit them.   ◊

We need a particular instance of Cartan's theorem B about coherent analytic sheaves. See Gunning and Rossi (1965).

Theorem 5.4.4 (Cartan's Theorem B) *Let X be a Stein manifold and let E be a holomorphic vector bundle on X. Then,*

$$H^k(\Omega^{0,*}(X,E), \bar{\partial}) = \begin{cases} 0, & k \neq 0 \\ \mathrm{Hol}(X,E), & k = 0, \end{cases}$$

where $\mathrm{Hol}(X,E)$ denotes the holomorphic sections of E on X.

We now use a corollary noted by Serre (1953); it is a special case of the Serre duality theorem. (Nowadays, people normally talk about the Serre duality theorem for compact complex manifolds, but in Serre's original paper he proved it for noncompact manifolds too, under some additional hypothesis that will be satisfied on Stein manifolds).

Note that we use the Fréchet topology on $\mathrm{Hol}(X,E)$, obtained as a closed subspace of $C^\infty(X,E)$. We let $E^!$ be the holomorphic vector bundle $E^\vee \otimes K_X$ where K_X is the canonical bundle of X.

Corollary 5.4.5 *For X a Stein manifold of complex dimension n, the* compactly supported *Dolbeault cohomology is*

$$H^k(\Omega^{0,*}_c(X,E), \bar{\partial}) = \begin{cases} 0, & k \neq n \\ (\mathrm{Hol}(X,E^!))^\vee, & k = n, \end{cases}$$

where $(\mathrm{Hol}(X,E^!))^\vee$ denotes the continuous linear dual to $\mathrm{Hol}(X,E^!)$.

Proof The Atiyah–Bott lemma (see Lemma D) shows that the inclusion

$$(\Omega^{0,*}_c(X,E), \bar{\partial}) \hookrightarrow (\overline{\Omega}^{0,*}_c(X,E), \bar{\partial})$$

is a chain homotopy equivalence. (Recall that the bar denotes "distributional sections.") As $\overline{\Omega}^{0,k}_c(X,E)$ is the continuous linear dual of $\Omega^{0,n-k}(X,E^!)$, it suffices to prove the desired result for the continuous linear dual complex.

Consider the acyclic complex

$$0 \to \mathrm{Hol}(X, E) \overset{i}{\hookrightarrow} C^\infty(X, E) \overset{\bar\partial}{\to} \Omega^{0,1}(X, E) \to \cdots \to \Omega^{0,n}(X) \to 0.$$

Our aim is to show that the linear dual of this complex is also acyclic. Note that this is a complex of Fréchet spaces. The result is then a consequence of the following lemma.   □

Lemma 5.4.6 *If V^* is an acyclic cochain complex of Fréchet spaces, then the dual complex $(V^*)^\vee$ is also acyclic.*

Proof Let $d_i : V^i \to V^{i+1}$ denote the differential. We need to show that the sequence

$$(V^{i+1})^\vee \to (V^i)^\vee \to (V^{i-1})^\vee$$

is exact in the middle. That is, we need to show that if $\alpha : V^i \to \mathbb{C}$ is a continuous linear map, and if $\alpha \circ d_{i-1} = 0$, then there exists some $\beta : V^{i+1} \to \mathbb{C}$ such that $\alpha = \beta \circ d_i$.

Note that α is zero on $\mathrm{Im}\, d_{i-1} = \mathrm{Ker}\, d_i$, so that α descends to a linear map

$$\bar\alpha : V^i / \mathrm{Im}\, d_{i-1} \to \mathbb{C}.$$

Since the complex is acyclic, $\mathrm{Im}\, d_{i-1} = \mathrm{Ker}\, d_i$ as vector spaces. However, it is *not* automatically true that they are the same as *topological* vector spaces, where we view $\mathrm{Im}\, d_{i-1}$ as a quotient of V^i and $\mathrm{Ker}\, d_i$ as a subspace of V^i. Here is where we use the Fréchet hypothesis: the open mapping theorem holds for Fréchet spaces, and it tells us that any surjective map between Fréchet spaces is open. Since $\mathrm{Ker}\, d_i$ is a closed subspace of V^i, it is a Fréchet space. The map $d_{i-1} : V^{i-1} \to \mathrm{Ker}\, d_i$ is surjective and therefore open. It follows that $\mathrm{Im}\, d_{i-1} = \mathrm{Ker}\, d_i$ as topological vector spaces.

From this, we see that our $\alpha : V^i \to \mathbb{C}$ descends to a continuous linear functional on $\mathrm{Ker}\, d_{i+1}$. As it is a closed subspace of V^{i+1}, the Hahn–Banach theorem tells us that it extends to a continuous linear functional on V^{i+1}.   □

The Setting of Differentiable Vector Spaces

These lemmas allow us to understand the cohomology of classical observables just as a vector space. Since we treat classical observables as a *differentiable* vector space, however, we are really interested in its cohomology as a sheaf on the site of smooth manifolds. It turns out (perhaps surprisingly) that for X a Stein manifold of dimension n, the isomorphism

$$H^n(\Omega_c^{0,*}(X, E)) = \mathrm{Hol}(X, E^!)^\vee$$

from above is *not* an isomorphism of sheaves on the site of smooth manifolds. (Here we define a smooth map from a manifold M to $\mathrm{Hol}(X, E^!)^\vee$ to be a continuous linear map $\mathrm{Hol}(X, E) \to C^\infty(M)$, as discussed in Appendix B.) Thankfully, there is a different and very useful description of compactly supported Dolbeault cohomology as a differentiable vector space.

We will present this description for polydiscs, although it works more generally. If $0 < r_1, \ldots, r_n \leq \infty$, let $D_r \subset \mathbb{C}$ be the disc of radius r, and let

$$D_{r_1, \ldots, r_n} = D_{r_1} \times \cdots \times D_{r_k} \subset \mathbb{C}^n$$

be the corresponding polydisc. We can view D_r as an open subset in $\mathbb{P}^1$.

In general, if X is a complex manifold and $C \subset X$ is a closed subset, we define

$$\mathrm{Hol}(C) = \operatorname*{colim}_{C \subset U} \mathrm{Hol}(U)$$

to be the germs of holomorphic functions on C. (The colimit is over opens U containing C.) The space $\mathrm{Hol}(C)$ has a natural structure of differentiable vector space, where we view $\mathrm{Hol}(U)$ as a differentiable vector space and take the colimit in the category of differentiable vector spaces. The same definition holds for the space $\mathrm{Hol}(C, E)$ of germs on C of holomorphic sections of a holomorphic vector bundle on E.

We have the following theorem, describing compactly supported Dolbeault cohomology as a differentiable vector space. Let $\mathscr{O}(-1)$ denote the holomorphic line bundle on $\mathbb{P}^1$ consisting of functions vanishing at $\infty \in \mathbb{P}^1$. Let $\mathscr{O}(-1)^{\boxtimes n}$ denote the line bundle on $(\mathbb{P}^1)^n$ consisting of functions that vanish at infinity in each variable.

Theorem 5.4.7 *For a polydisc $D_{r_1, \ldots, r_n} \subset \mathbb{C}^n$, we have a natural isomorphism of differentiable vector spaces*

$$H^n(\Omega_c^{0,*}(D_{r_1, \ldots, r_n})) \cong \mathrm{Hol}((\mathbb{P}^1 \setminus D_{r_1}) \times \cdots \times (\mathbb{P}^1 \times D_{r_n}), \mathscr{O}(-1)^{\boxtimes n}).$$

Further, all other cohomology groups of $\Omega_c^{0,*}(D_{r_1, \ldots, r_n})$ are zero as differentiable vector spaces.

This isomorphism is invariant under holomorphic symmetries of the polydisc $D_{r_1, \ldots, r_k}$, under the actions of S^1 by rotation in each coordinate, and under the action of the symmetric group (when the r_i are all the same).

Before we prove this theorem, we need a technical result.

Proposition 5.4.8 *For any complex manifold X and holomorphic vector bundle E on X and for any manifold M, the cochain complex $C^\infty(M, \Omega^{0,*}(X, E))$*

is a fine resolution of the sheaf on $M \times X$ consisting of smooth sections of the bundle $\pi_X^* E$ that are holomorphic along X.

Further, if we assume that $H^i(\Omega^{0,*}(X,E)) = 0$ for $i > 0$, then

$$H^i(C^\infty(M, \Omega^{0,*}(X,E))) = \begin{cases} C^\infty(M, \mathrm{Hol}(X,E)) & \text{if } i = 0, \\ 0 & \text{if } i > 0. \end{cases}$$

Proof The first statement follows from the second statement. Locally on X, the Dolbeault–Grothendieck lemma tells us that the sheaf $\Omega^{0,*}(X,E)$ has no higher cohomology, and the sheaves $C^\infty(M, \Omega^{0,i}(X,E))$ are certainly fine.

To prove the second statement, consider the exact sequence

$$0 \to \mathrm{Hol}(X,E) \to \Omega^{0,0}(X,E) \to \Omega^{0,1}(X,E) \cdots \to \Omega^{0,n}(X,E) \to 0.$$

It is an exact sequence of Fréchet spaces. Proposition 3 of section I.1.2 in Grothendieck (1955) tells us that the completed projective tensor product of nuclear Fréchet spaces is an exact functor, that is, takes exact sequences to exact sequences. Another result of Grothendieck (1952) tells us that for any complete locally convex topological vector space F, $C^\infty(M,F)$ is naturally isomorphic to $C^\infty(M) \widehat{\otimes}_\pi F$, where $\widehat{\otimes}_\pi$ denotes the completed projective tensor product. The result then follows. $\square$

Proof of the theorem We need to produce an isomorphism of differentiable vector spaces, i.e., an isomorphism of sheaves. In other words, we need to find an isomorphism between their sections on each manifold M,

$$C^\infty(M, H^n(\Omega_c^{0,*}(D_{r_1,\dots,r_n}))) \cong C^\infty(M, \mathrm{Hol}((\mathbb{P}^1 \setminus D_{r_1})$$
$$\times \dots (\mathbb{P}^1 \times D_{r_n}), \mathcal{O}(-1)^{\boxtimes n})),$$

and this isomorphism must be natural in M. We also need to show that the sections $C^\infty(M, H^i(\Omega_c^{0,*}(D_{r_1,\dots,r_n})))$ of the other cohomology sheaves, with $i < n$, are zero.

The first thing we prove is that, for any complex manifold X and holomorphic vector bundle E on X, and any open subset $U \subset X$, there is an exact sequence

$$0 \to C^\infty(M, \Omega_c^{0,*}(U,E)) \to C^\infty(M, \Omega^{0,*}(X,E))$$
$$\to C^\infty(M, \Omega^{0,*}(X \setminus U, E)) \to 0 \tag{$\ddagger$}$$

of the sections on any smooth manifold M.

We will do this by working with these differentiable cochain complexes $\Omega^{0,*}(X,E)$ simply as sheaves on M, not on the whole site of smooth manifolds. To start, we will prove that we have an exact sequence of sheaves on M. As these complexes are built out of topological vector spaces, they are fine sheaves on M and hence their global sections will provide the exact sequence (‡).

Pick an exhausting family $\{K_i\}$ of compact subsets of U. Define

$$\Omega^{0,*}_{K_i}(U, E) = \mathrm{Ker}\left(\Omega^{0,*}(U, E) \xrightarrow{\mathrm{res}} \Omega^{0,*}(U \setminus K_i, E)\right).$$

This vector space coincides with $\Omega^{0,*}_{K_i}(X, E)$, since a section that vanishes outside of K_i in X must vanish outside of U. The sequence

$$0 \to \Omega^{0,*}_{K_i}(X, E) \to \Omega^{0,*}(X, E) \to \Omega^{0,*}(X \setminus K_i, E)$$

is therefore exact, but it is not necessarily exact on the right.

Consider the colimit of this sequence as i goes to ∞ in the category of sheaves on M. We then obtain an exact sequence

$$0 \to \Omega^{0,*}_c(U, E) \to \Omega^{0,*}(X, E) \to \Omega^{0,*}(X \setminus U, E)$$

as sheaves on M. Now, we claim that this sequence is exact on the right, as sheaves. The point is that, locally on M, every smooth function on $M \times (X \setminus U)$ extends to a smooth function on some $M \times (X \setminus K_i)$ and by applying a bump function that is 1 on a neighborhood of $M \times (X \setminus U)$ and zero on the interior of K_i, we can extend it to a smooth function on $M \times X$.

Let's apply this exact sequence to the case when $X = (\mathbb{P}^1)^n$, $U = D_{r_1,\dots,r_n}$ and $E = \mathcal{O}(-1)^{\boxtimes n}$. Note that in this case E is trivialized on U, so that we find, on taking cohomology of the Dolbeault complexes, a long exact sequence of sheaves on M:

$$\cdots \to H^i(\Omega^{0,*}_c(D_{r_1,\dots,r_n})) \to H^i(\Omega^{0,*}((\mathbb{P}^1)^n, (\mathcal{O}(-1))^{\boxtimes n}))$$
$$\to H^i(\Omega^{0,*}((\mathbb{P}^1)^n \setminus D_{r_1,\dots,r_n}, \mathcal{O}(-1)^{\boxtimes n})) \to \cdots.$$

Note that the Dolbeault cohomology of $(\mathbb{P}^1)^n$ with coefficients in $\mathcal{O}(-1)^{\boxtimes n}$ vanishes. Hence, the middle term in this exact sequence vanishes by the preceding proposition, and so we get an isomorphism

$$C^\infty(M, H^i(\Omega^{0,*}_c(D_{r_1,\dots,r_n}))) \cong C^\infty(M, H^{i-1}(\Omega^{0,*}((\mathbb{P}^1)^n \setminus D_{r_1,\dots,r_n}, \mathcal{O}(-1)^{\boxtimes n}))).$$

For the purposes of the theorem, we now need to compute the right-hand side. The complex appearing on the right-hand side is the colimit, as $\varepsilon \to 0$, of

$$C^\infty(M, \Omega^{0,*}((\mathbb{P}^1)^n \setminus \overline{D}_{r_1-\varepsilon,\dots,r_n-\varepsilon}, \mathcal{O}(-1)^{\boxtimes n})).$$

Here $\overline{D}$ indicates the closed disc.

We now have a sheaf on $M \times \left((\mathbb{P}^1)^n \setminus \overline{D}_{r_1-\varepsilon,\dots,r_n-\varepsilon}\right)$ that sends an open set $U \times V$ to $C^\infty(U, \Omega^{0,*}(V, \mathcal{O}(-1)^{\boxtimes n}))$. It is a cochain complex of fine sheaves.

We are interested in the cohomology of global sections. We can compute this sheaf cohomology using the local-to-global spectral sequence associated to a Čech cover of $(\mathbb{P}^1)^n \setminus \overline{D}_{r_1-\varepsilon,\dots,r_n-\varepsilon}$. Consider the cover $\{U_i\}_{1 \le i \le n}$ given by

$$U_i = \mathbb{P}^1 \times \cdots \times (\mathbb{P}^1 \setminus \overline{D}_{r_i-\varepsilon}) \times \cdots \times \mathbb{P}^1.$$

We take the corresponding cover of $M \times \left((\mathbb{P}^1)^n \setminus \overline{D}_{r_1-\varepsilon,\dots,r_n-\varepsilon}\right)$ given by the opens $\{M \times U_i\}_{1 \le i \le n}$.

Note that, for $k < n$, we have

$$H^*(\Omega^{0,*}(U_{i_1,\dots,i_k}, \mathcal{O}(-1)^{\boxtimes n})) = 0,$$

where $U_{i_1,\dots,i_k}$ denotes the intersection of all the U_{i_j}. The previous proposition implies that we also have

$$H^*(C^\infty(M, \Omega^{0,*}(U_{i_1,\dots,i_k}, \mathcal{O}(-1)^{\boxtimes n}))) = 0.$$

The local-to-global spectral sequence then tells us that we have a natural isomorphism identifying

$$H^*(C^\infty(M, \Omega^{0,*}((\mathbb{P}^1)^n \setminus \overline{D}_{r_1-\varepsilon,\dots,r_n-\varepsilon}, \mathcal{O}(-1)^{\boxtimes n})))$$

with

$$C^\infty(M, \mathrm{Hol}(U_{1,\dots,n}, \mathcal{O}(-1)^{\boxtimes n}))[-n].$$

That is, all cohomology groups on the left-hand side of this equation are zero, except for the top cohomology, which is equal to the vector space on the right.

Note that

$$U_{1,\dots,n} = (\mathbb{P}^1 \setminus \overline{D}_{r_1-\varepsilon}) \times \cdots \times (\mathbb{P}^1 \setminus \overline{D}_{r_n-\varepsilon}).$$

Taking the colimit as $\varepsilon \to 0$, combined with our previous calculations, gives the desired result. Note that this colimit must be taken in the category of sheaves on M. We are also using the fact that sequential colimits commute with formation of cohomology. $\square$

5.4.6 A Description of Observables on a Disc

This analytic discussion allows us to understand both classical and quantum observables of the $\beta\gamma$ system. Let us first define some basic classical observables.

Definition 5.4.9 On any disc $D(x, r)$ centered at the point x, let $c_n(x)$ denote the linear classical observable

$$c_n(x) : \gamma \in \Omega^{0,0}(D(x,r)) \mapsto \frac{1}{n!}(\partial_z^n \gamma)(x).$$

Likewise, for $n > 0$, let $b_n(x)$ denote the linear functional

$$b_n(x) : \beta \, dz \in \Omega^{1,0}(D(x,r)) \mapsto \frac{1}{(n-1)!} (\partial_z^{n-1} \beta)(x).$$

These observables descend to elements of $H^0(\mathrm{Obs}^{cl}(D(x,r)))$.

Let us introduce some notation to deal with products of these. For a multi-index $K = (k_1, \ldots, k_n)$ or $L = (l_1, \ldots, l_m)$, we let

$$b_K(x) = b_{k_1}(x) \cdots b_{k_n}(x),$$
$$c_L(x) = c_{l_1}(x) \cdots c_{l_m}(x).$$

Of course these products make sense only if $k_i > 0$ for all i. Note that under the natural S^1 action on $H^0(\mathrm{Obs}^{cl}(D(x,r)))$, the elements $b_K(x)$ and $c_L(x)$ are of weight $-|K|$ and $-|L|$, where $|K| = \sum_i k_i$ and similarly for L.

Lemma 5.4.10 *These observables generate the cohomology of the classical observables in the following sense.*

(i) *For $i \neq 0$ the cohomology groups $H^i(\mathrm{Obs}^{cl}(D(x,r)))$ vanish as differentiable vector spaces.*

(ii) *The monomials $\{b_K(x)c_L(x)\}_{|K|+|L|=n}$ form a basis for the weight n space $H^0(\mathrm{Obs}^{cl}(D(x,r)))_n$. Further, they form a basis in the sense of differentiable vector spaces, meaning that any smooth section*

$$f : M \to H_n^0(\mathrm{Obs}^{cl}(D(x,r)))$$

can be expressed uniquely as a sum

$$f = \sum_{|K|+|L|=n} f_{KL} b_K(x) c_L(x),$$

where $f_{KL} \in C^\infty(M)$ and locally on M all but finitely many of the f_{KL} are zero.

(iii) *More generally, any smooth section*

$$f : M \to H^0(\mathrm{Obs}^{cl}(D(x,r)))$$

can be expressed uniquely as a sum

$$f = \sum f_{KL} b_k(x) c_L(x)$$

that is convergent in a natural topology on $C^\infty(M, H^0(\mathrm{Obs}^{cl}(D(x,r))))$, with the coefficient functions f_{KL} satisfying the following properties:

(a) *Locally on M, $f_{k_1,\ldots,k_n,l_1,\ldots,l_m} = 0$ for $n + m \gg 0$.*

(b) *For fixed n and m, the sum*

$$\sum_{K,L} f_{k_1,\ldots,k_n,l_1,\ldots,l_m} z_1^{-k_1} \cdots z_n^{-k_n} w_1^{-l_1-1} \cdots w_1^{-l_m-1}$$

is absolutely convergent when $|z_i| \geq r$ and $|w_j| \geq r$, in the natural topology on $C^\infty(M)$.

This lemma gives an explicit description of $C^\infty(M, H^0(\mathrm{Obs}^{cl}(D(x,r))))$.

Proof We will set $x = 0$. By definition, $\mathrm{Obs}^{cl}(D(0,r))$ is a direct sum, as differentiable cochain complexes,

$$\mathrm{Obs}^{cl}(U) = \bigoplus_n \left(\Omega_c^{0,*}(D(0,r)^n, E^{\boxtimes n})[n] \right)_{S_n},$$

where E is the holomorphic vector bundle $\mathcal{O} \oplus K$ on $\mathbb{C}$.

It follows from Theorem 5.4.7 that, as differentiable vector spaces,

$$H^i(\Omega^{0,*}(D(0,r)^n, E^{\boxtimes n}) = 0$$

unless $i = n$, and that

$$H^n(\Omega^{0,*}(D(0,r)^n, E^{\boxtimes n}) = \mathrm{Hol}((\mathbb{P}^1 \setminus D(0,r)), (\mathcal{O}(-1) \oplus \mathcal{O}(-1)\mathrm{d}z)^{\boxtimes n}).$$

Theorem 5.4.7 was stated only for the trivial bundle, but the bundle E is trivial on $D(x,r)$. It is not, however, S^1-equivariantly trivial. The notation $\mathcal{O}(-1)\mathrm{d}z$ indicates how we have changed the S^1-action on $\mathcal{O}(-1)$.

It follows that we have an isomorphism, of S^1-equivariant differentiable vector spaces

$$H^0(\mathrm{Obs}^{cl}(D(0,r))) \cong \bigoplus_n H^0((\mathbb{P}^1 \setminus D(0,r))^n, (\mathcal{O}(-1) \oplus \mathcal{O}(-1)\mathrm{d}z)^{\boxtimes n}))_{S_n}.$$

Under this isomorphism, the observable $b_K(0)c_L(0)$ goes to the function

$$\frac{1}{(2\pi i)^{n+m}} z_1^{-k_1}\mathrm{d}z_1 \cdots z_n^{-k_n}\mathrm{d}z_n w_1^{-l_1-1} \cdots w_m^{-l_m-1}.$$

Everything in the statement is now immediate. On $H^0(\mathrm{Obs}^{cl}(D(0,r)))$ we use the topology that is the colimit of the topologies on holomorphic functions on $(\mathbb{P}^1 \setminus \overline{D}(0, r - \varepsilon))^n$ as $\varepsilon \to 0$. $\qquad\square$

Lemma 5.4.11 *We have*

$$H^*(\mathrm{Obs}^q(U)) = H^*(\mathrm{Obs}^{cl}(U))[\hbar]$$

as S^1-equivariant differentiable vector spaces. This isomorphism is compatible with maps induced from inclusions $U \hookrightarrow V$ of open subsets of $\mathbb{C}$. It is not compatible with the prefactorization product map.

Proof This lemma follows, as explained in Section 4.6 in Chapter 4, from the existence of a Green's function for the $\bar{\partial}$ operator, namely

$$G(z_1, z_2) = \frac{dz_1 - dz_2}{z_1 - z_2} \in \mathscr{E}(\mathbb{C}) \otimes \mathscr{E}(\mathbb{C}) \qquad (5.4.6.1)$$

where $\mathscr{E}$ denotes the complex of fields of the $\beta\gamma$ system.   □

We will use the notation $b_K(x)c_L(x)$ for the quantum observables on $D(x, r)$ that arise from the classical observables discussed above, using the isomorphism given by this lemma.

Corollary 5.4.12 *The properties listed in Lemma 5.4.10 also hold for quantum observables. As a result, all the conditions of Theorem 5.3.3 are satisfied, so that the structure of prefactorization algebra leads to a $\mathbb{C}[\hbar]$-linear vertex algebra structure on the space*

$$V = \bigoplus_n H^*(\mathrm{Obs}^q(D(0, r)))_n$$

where $H^*(\mathrm{Obs}^q(D(0, r)))_n$ indicates the weight n eigenspace of the S^1 action.

Proof The only conditions we have not yet checked are

(i) The inclusion maps

$$H^*(\mathrm{Obs}^q(D(0, r)))_n \to H^*(\mathrm{Obs}^q(D(0, s)))_n$$

for $r < s$ are quasi-isomorphisms, and

(ii) the differentiable vector spaces $V_n = H^*(\mathrm{Obs}^q(D(0, r)))_n$ are countable colimits of finite-dimensional vector spaces in the category of differentiable vector spaces.

Both of these conditions follow immediately from the analog of Lemma 5.4.10 that applies to quantum observables.   □

5.4.7 The Proof of Theorem 5.4.2

We now finish the proof of the main theorem. We have shown that we obtain some vertex algebra, because we have verified the criteria to apply Theorem 5.3.3. What remains is to exhibit an explicit isomorphism with the $\beta\gamma$ vertex algebra.

Proof Note that $V_{\hbar=2\pi i}$ is the polynomial algebra on the generators b_n, c_m where $n \geq 1$ and $m \geq 0$. Also b_n, c_m have weights $-n, -m$ respectively, under the S^1-action. We define an isomorphism $V_{\hbar=2\pi i}$ to W by sending b_n to a_{-n} and

c_m to a^*_{-m}, and extending it to be an isomorphism of commutative algebras. By the reconstruction theorem, it suffices to calculate $Y(b_1, z)$ and $Y(c_0, z)$.

By the way we defined the vertex algebra associated to the prefactorization algebra of quantum observables in Theorem 5.3.3, we have

$$Y(b_1, z)(\alpha) = \mathcal{L}_z m_{z,0}(b_1, \alpha) \in V_{\hbar = 2\pi i}((z)),$$

where $\mathcal{L}_z$ denotes Laurent expansion, and

$$m_{z,0} : V_{\hbar = 2\pi i} \otimes V_{\hbar = 2\pi i} \to \overline{V}_{\hbar = 2\pi i}$$

is the map associated to the prefactorization product coming from the inclusion of the disjoint discs $D(z, r)$ and $D(0, r)$ into $D(0, \infty)$ (where r can be taken to be arbitrarily small).

We are using the Green's function for the $\overline{\partial}$ operator to identify classical and quantum observables. Let us recall how the Green's function leads to an explicit formula for the prefactorization product.

Let $\mathcal{E}$ denote the sheaf on $\mathbb{C}$ of fields of our theory, so that

$$\mathcal{E}(U) = \Omega^{0,*}(U, \mathcal{O} \oplus K).$$

It is the sheaf of smooth sections of a graded bundle E on $\mathbb{C}$. Let $\mathcal{E}^! = \mathcal{E}[1]$ denote the sheaf of smooth sections of the vector bundle $E^\vee \otimes \wedge^2 T^*$, so that a section is a fiberwise linear functional on $\mathcal{E}$ valued in 2-forms. Let $\overline{\mathcal{E}}$ denote the sheaf of distributional sections and $\mathcal{E}_c$ denote compactly supported sections.

The propagator, or Green's function, is

$$P = \frac{dz_1 \otimes 1 - 1 \otimes dz_2}{2\pi i(z_1 - z_2)}.$$

It is an element of

$$\overline{\mathcal{E}}(\mathbb{C}) \widehat{\otimes}_\pi \overline{\mathcal{E}}(\mathbb{C}) = \mathcal{D}(\mathbb{C}^2, E \boxtimes E),$$

where $\mathcal{D}$ denotes the space of distributional sections.

We can also view at as a symmetric and smooth linear map

$$P : \mathcal{E}^!_c(\mathbb{C}) \widehat{\otimes}_\beta \mathcal{E}^!_c(\mathbb{C}) = C^\infty_c(\mathbb{C}^2, (E^!)^{\boxtimes 2}) \to \mathbb{C}. \tag{†}$$

Here $\widehat{\otimes}_\beta$ denotes the completed bornological tensor product on the category of convenient vector spaces.

Recall that we identify

$$\mathrm{Obs}^{cl}(U) = \mathrm{Sym}(\mathcal{E}^!_c(U)) = \mathrm{Sym}(\mathcal{E}_c(U)[1]),$$

where the symmetric algebra is defined using the completed tensor product on the category of convenient vector spaces.

From P we construct an order two differential operator on this symmetric algebra:

$$\partial_P : \text{Obs}^{cl}(U) \to \text{Obs}^{cl}(U).$$

This operator is characterized by the fact that it is a smooth (or, equivalently, continuous) order two differential operator, vanishes on $\text{Sym}^{\leq 1}$, and on Sym^2 is determined by the map in (†).

For example, if $x, y \in U$, then

$$\partial_P(b_i(x)c_j(y)) = \frac{1}{(i-1)!j!} \frac{\partial^{i-1}}{\partial^{i-1}x} \frac{\partial^j}{\partial^j y} \frac{1}{2\pi i(x-y)}.$$

To compute this, note that $b_i(x)c_j(y)$ is quadratic and we can pick a representative for this observable in $\mathscr{E}_c^!(\mathbb{C}) \widehat{\otimes}_\beta \mathscr{E}_c^!(\mathbb{C})$. We then apply P to obtain the number on the right-hand side. Explicitly, $b_i(x)$ and $c_j(y)$ are derivatives of delta functions and we apply them to the Green's function.

We can identify $\text{Obs}^q(U)$ as a graded vector space with $\text{Obs}^{cl}(U)[\hbar]$. The quantum differential is $d = d_1 + \hbar d_2$, a sum of two terms, where d_1 is the differential on $\text{Obs}^{cl}(U)$ and d_2 is the differential arising from the Lie bracket in the shifted Heisenberg Lie algebra whose Chevalley–Eilenberg chain complex defines $\text{Obs}^q(U)$.

It is easy to check that

$$[\hbar \partial_P, d_1] = d_2.$$

This identity follows immediately from the fact that $(1,1)$-current $\overline{\partial}P$ on $\mathbb{C}^2$ is the delta current on the diagonal.

As we explained in Section 4.6 in Chapter 4, we get an isomorphism of cochain complexes

$$\begin{aligned} W: \quad \text{Obs}^{cl}(U)[\hbar] &\mapsto \text{Obs}^q(U) \\ \alpha &\mapsto e^{\hbar \partial_P} \alpha \end{aligned}.$$

Further, for U_1, U_2 disjoint opens in V, the prefactorization product map

$$\star_\hbar : \text{Obs}^{cl}(U_1)[\hbar] \times \text{Obs}^{cl}(U_2)[\hbar] \to \text{Obs}^{cl}(V)[\hbar],$$

arising from that on Obs^q under the identification W, is given by the formula

$$\alpha \star_\hbar \beta = e^{-\hbar \partial_P}\left(\left(e^{\hbar \partial_P}\alpha\right) \cdot \left(e^{\hbar \partial_P}\beta\right)\right).$$

Here $\cdot$ refers to the commutative product on classical observables.

Let us apply this formula to $\alpha = b_1(z)$ and β in the algebra generated by $b_i(0)$ and $c_j(0)$. First, note that since $b_1(z)$ is linear, $\hbar \partial_P b_1(z) = 0$. Note also that $[\partial_P, b_1(z)]$ commutes with ∂_P. Thus, we find that

$$b_1(z) \star_\hbar \beta = e^{-\hbar\partial_P} \left(b_1(z) e^{\hbar\partial_P} \beta \right)$$
$$= b_1(z)\beta - [\hbar\partial_P, b_1(z)]\beta.$$

Note that $[\partial_P, b_1(z)]$ is an order one operator, and so a derivation. So it suffices to calculate what it does on generators. We find that

$$[\partial_P, b_1(z)]c_j(0) = \frac{1}{2\pi i}\frac{1}{j!}\left(\frac{\partial^j}{\partial^j w}\frac{1}{z-w}\right)\Bigg|_{w=0}$$
$$= \frac{1}{2\pi i}z^{-j-1},$$

and

$$[\partial_P, b_1(z)]b_j(0) = 0.$$

In other words,

$$[\partial_P, b_1(z)] = \frac{1}{2\pi i}\sum_{j=0}^{\infty} z^{-j-1}\frac{\partial}{\partial c_j(0)}.$$

Note as well that for $|z| < r$, we can expand the cohomology class $b_1(z)$ in $H^0(\mathrm{Obs}^{cl}(D(0,r)))$ as a sum

$$b_1(z) = \sum_{n=0}^{\infty} b_{n+1}(0)z^n.$$

Indeed, for a classical field $\gamma \in \Omega^1_{hol}(D(0,r))$ satisfying the equations of motion, we have

$$b_1(z)(\gamma) = \gamma(z)$$
$$= \sum_{n=0}^{\infty} z^n \frac{1}{n!}\gamma^{(n)}(0)$$
$$= \sum_{n=0}^{\infty} z^n b_{n_1}(0)(\gamma).$$

Putting all these computations together, we find that, for β in the algebra generated by $c_j(0), b_i(0)$, we have

$$b_1(z) \star_\hbar \beta = \left(\sum_{n=0}^{\infty} b_{n+1}(0)z^n + \frac{\hbar}{2\pi i}\sum_{m=0}^{\infty}\frac{\partial}{\partial c_m(0)}z^{-m-1}\right)$$
$$\times \beta \in H^0(\mathrm{Obs}^{cl}(D(0,r)))[\hbar].$$

Thus, if we set $\hbar = 2\pi i$, we see that the operator product on the space $V_{\hbar=2\pi i}$ matches the one on W if we sent $b_n(0)$ to a_{-n} and $c_n(0)$ to a^*_{-n}. A similar calculation of the operator product with $c_0(z)$ completes the proof.   □

5.5 Kac–Moody Algebras and Factorization Envelopes

In this section, we will construct a holomorphically translation invariant prefactorization algebra whose associated vertex algebra is the affine Kac–Moody vertex algebra. This construction is an example of the twisted prefactorization envelope construction, which also produces the prefactorization algebras for free field theories (see Section 3.6 in Chapter 3). As we explained there, it is our version of the chiral envelope construction from Beilinson and Drinfeld (2004).

Our work here recovers the Heisenberg vertex algebra, the free fermion vertex algebra, and the affine Kac–Moody vertex algebras. These methods can be applied, however, to any dg Lie algebra with an invariant pairing, so there is a plethora of unexplored prefactorization algebras provided by this construction.

5.5.1 The Context

The input data are the following:

- A Riemann surface Σ.
- A Lie algebra $\mathfrak{g}$ (for simplicity, we stick to ordinary Lie algebras such as $\mathfrak{sl}_2$).
- A $\mathfrak{g}$-invariant symmetric pairing $\kappa : \mathfrak{g}^{\otimes 2} \to \mathbb{C}$.

From these data, we obtain a cosheaf on Σ,

$$\mathfrak{g}^\Sigma : U \mapsto (\Omega_c^{0,*}(U) \otimes \mathfrak{g}, \overline{\partial}),$$

where U denotes an open in Σ. Note that $\mathfrak{g}^\Sigma$ is a cosheaf of dg vector spaces and merely a precosheaf of dg Lie algebras. When κ is nontrivial (though not necessarily nondegenerate), we obtain a -1-shifted central extension on each open:

$$\mathfrak{g}_\kappa^\Sigma : U \mapsto (\Omega_c^{0,*}(U) \otimes \mathfrak{g}, \overline{\partial}) \oplus \underline{\mathbb{C}} c,$$

where $\underline{\mathbb{C}}$ denotes the locally constant cosheaf on Σ and c is a central element of cohomological degree 1. The bracket is defined by

$$[\alpha \otimes X, \beta \otimes Y]_\kappa := \alpha \wedge \beta \otimes [X, Y] - \frac{1}{2\pi i} \left(\int_U \partial\alpha \wedge \beta \right) \kappa(X, Y) c,$$

with $\alpha, \beta \in \Omega^{0,*}(U)$ and $X, Y \in \mathfrak{g}$. (These constants are chosen to match with the use of κ for the affine Kac–Moody algebra in the text that follows.)

Remark: Every dg Lie algebra $\mathfrak{g}$ has a geometric interpretation as a *formal moduli space* $B\mathfrak{g}$, a theme we develop further in Volume 2. This dg Lie algebra $\mathfrak{g}^{\Sigma}(U)$ in fact possesses a natural geometric interpretation: it describes "deformations *with compact support in U* of the trivial G-bundle on Σ." Equivalently, it describes the moduli space of holomorphic G-bundles on U which are trivialized outside of a compact set. For U a disc, it is closely related to the affine Grassmannian of G. The affine Grassmannian is defined to be the space of algebraic bundles on a formal disc trivialized away from a point, whereas our formal moduli space describes G-bundles on an actual disc trivialized outside a compact set.

The choice of κ has the interpretation of a line bundle on the formal moduli problem $B\mathfrak{g}^{\Sigma}(U)$ for each U. In general, -1-shifted central extensions of a dg Lie algebra $\mathfrak{g}$ are the same as L_{∞}-maps $\mathfrak{g} \to \mathbb{C}$, that is, as rank-one representations. Rank-one representations of a group are line bundles on the classifying space of the group. In the same way, rank-one representations of a Lie algebra are line bundles on the formal moduli problem $B\mathfrak{g}$. ◇

As explained in Section 3.6 in Chapte 3, we can form the twisted prefactorization envelope of $\mathfrak{g}^{\Sigma}$. Concretely, this prefactorization algebra assigns to an open subset $U \subset \Sigma$, the complex

$$\mathcal{F}^{\kappa}(U) = C_*(\mathfrak{g}_{\kappa}^{\Sigma}(U))$$
$$= (\mathrm{Sym}(\Omega_c^{0,*}(U) \otimes \mathfrak{g}[1])[c], d_{CE}),$$

where c now has cohomological degree 0 in the Lie algebra homology complex. It is a prefactorization algebra in modules for the algebra $\mathbb{C}[c]$, generated by the central parameter. We should therefore think of it as a family of prefactorization algebras depending on the central parameter c.

Remark: Given a dg Lie algebra $(\mathfrak{g}, d)$, we interpret $C_*\mathfrak{g}$ as the "distributions with support on the closed point of the formal space $B\mathfrak{g}$." Hence, our prefactorization algebras $\mathcal{F}^{\kappa}(\Sigma)$ describes the κ-twisted distributions supported at the point in $Bun_G(\Sigma)$ given by the trivial bundle on Σ.

This description is easier to understand in its global form, particularly when Σ is a closed Riemann surface. Each point of $P \in Bun_G(\Sigma)$ has an associated dg Lie algebra $\mathfrak{g}_P$ describing the formal neighborhood of P. This dg Lie algebra, in the case of the trivial bundle, is precisely the global sections over Σ of $\mathfrak{g}^{\Sigma}$. For a nontrivial bundle P, the Lie algebra $\mathfrak{g}_P$ is also global sections of a natural cosheaf, and we can apply the enveloping construction to this cosheaf to obtain a prefactorization algebras. By studying families of such bundles, we recognize that our construction $C_*\mathfrak{g}_P$ should recover differential operators on $Bun_G(\Sigma)$. When we include a twist κ, we should recover κ-twisted differential operators. When the twist is integral, the twist corresponds to a line bundle on $Bun_G(\Sigma)$ and the twisted differential operators are precisely differential operators for that line bundle.

It is nontrivial to properly define differential operators on the stack $Bun_G(\Sigma)$, and we will not attempt it here. At the formal neighbourhood of a point, however, there are no difficulties and our statements are rigorous. $\diamond$

5.5.2 The Main Result

Note that if we take our Riemann surface to be $\mathbb{C}$, the prefactorization algebra $\mathcal{F}^\kappa$ is holomorphically translation invariant because the derivation $\frac{\partial}{\partial \bar{z}}$ is homotopically trivial, via the homotopy given by $\frac{\partial}{\partial d\bar{z}}$. It follows that we are in a situation where we might be able to apply Theorem 5.3.3. The main result of this section is the following.

Theorem 5.5.1 *The holomorphically translation invariant prefactorization algebra $\mathcal{F}^\kappa$ on $\mathbb{C}$ satisfies the conditions of Theorem 5.3.3, and so defines a vertex algebra. This vertex algebra is isomorphic to the affine Kac–Moody vertex algebra.*

Before we can prove this statement, we of course need to describe the affine Kac–Moody vertex aigebra.

Recall that the Kac–Moody Lie algebra is the central extension of the loop algebra $L\mathfrak{g} = \mathfrak{g}[t, t^{-1}]$,

$$0 \to \mathbb{C} \cdot c \to \widehat{\mathfrak{g}}_\kappa \to L\mathfrak{g} \to 0.$$

As vector spaces, we have $\widehat{\mathfrak{g}}_\kappa = \mathfrak{g}[t, t^{-1}] \oplus \mathbb{C} \cdot c$, and the Lie bracket is given by the formula

$$[f(t) \otimes X, g(t) \otimes Y]_\kappa := f(t)g(t) \otimes [X, Y] + \left(\oint f \partial g \right) \kappa(X, Y)$$

for $X, Y \in \mathfrak{g}$ and $f, g \in \mathbb{C}[t, t^{-1}]$. Here, c has cohomological degree 0 and is central. The notation $\oint$ denotes an algebraic version of integration around the unit circle (*aka* the residue pairing), and so $\oint t^n dt = 2\pi i \delta_{n,-1}$. In particular, we have $\oint t^n \partial t^m = 2\pi i m \delta_{m+n,0}$.

Observe that $\mathfrak{g}[t]$ is a sub-Lie algebra of the Kac–Moody algebra. The *vacuum module W* for the Kac–Moody algebra is the induced representation from the trivial rank one representation of $\mathfrak{g}[t]$. The induction-restriction adjunction provides a natural map $\mathbb{C} \to W$ of $\mathfrak{g}[t]$-modules, where $\mathbb{C}$ is the trivial representation of $\mathfrak{g}[t]$. Let $|0\rangle \in W$ denote the image of $1 \in \mathbb{C}$.

There is a useful, explicit description of W as a vector space. Consider the sub Lie algebra $\mathbb{C} \cdot c \oplus t^{-1}\mathfrak{g}[t^{-1}]$ of $\widehat{\mathfrak{g}}_\kappa$, which is complementary to $\mathfrak{g}[t]$. The action of this Lie algebra on the vacuum element $|0\rangle \in W$ gives a canonical isomorphism

$$U(\mathbb{C} \cdot c \oplus t^{-1}\mathfrak{g}[t^{-1}]) \cong W$$

as vector spaces.

The vacuum module W is also a $\mathbb{C}[c]$ module in a natural way, because $\mathbb{C}[c]$ is inside the universal enveloping algebra of $\widehat{\mathfrak{g}}_\kappa$.

Definition 5.5.2 The Kac–Moody vertex algebra is defined as follows. It is a vertex algebra structure over the base ring $\mathbb{C}[c]$ on the vector space W. (Working over the base ring $\mathbb{C}[c]$ simply means all maps are $\mathbb{C}[c]$-linear.) By the reconstruction theorem 5.3.4, to specify the vertex algebra structure it suffices to specify the state-field map on a subset of elements of W that generate all of W (in the sense of the reconstruction theorem). The following state-field operations define the vertex algebra structure on W.

(i) The vacuum element $|0\rangle \in W$ is the unit for the vertex algebra, that is, $Y(|0\rangle, z)$ is the identity.

(ii) If $X \in t^{-1}\mathfrak{g} \subset \widehat{\mathfrak{g}}_\kappa$, we have an element $X|0\rangle \in W$. We declare that

$$Y(X|0\rangle, z) = \sum_{n \in \mathbb{Z}} X_n z^{-1-n}$$

where $X_n = t^n X \in \widehat{\mathfrak{g}}_\kappa$, and we are viewing elements of $\widehat{\mathfrak{g}}_\kappa$ as endomorphisms of W.

5.5.3 Verification of the Conditions to Define a Vertex Algebra

We need to verify that $\mathcal{F}^\kappa$ satisfies the conditions listed in Theorem 5.3.3 guaranteeing that we can construct a vertex algebra. Our situation here is entirely parallel to that of the $\beta\gamma$ system, so we will be brief. The first thing to check is that the natural S^1 action on $\mathcal{F}^\kappa(D(0,r))$ extends to an action of the algebra $\mathcal{D}(S^1)$ of distributions on the circle, which is easy to see by the same methods as with $\beta\gamma$.

Next, we need to check that, if $\mathcal{F}^\kappa_n(D(0,r))$ denotes the weight n eigenspace for the S^1-action, then the following properties hold.

(i) The inclusion $\mathcal{F}^\kappa_n(D(0,r)) \to \mathcal{F}^\kappa_n(D(0,s))$ for $r < s$ is a quasi-isomorphism.

(ii) For $n \gg 0$, the cohomology $H^*(\mathcal{F}^\kappa_l(D(0,r))$ vanishes as a sheaf on the site of smooth manifolds.

(iii) The differentiable vector spaces $H^*(\mathcal{F}_n^\kappa(D(0,r))$ are countable sequential colimits of finite-dimensional vector spaces in DVS.

Note that

$$\mathcal{F}^\kappa(D(0,r)) = \mathrm{Sym}\left(\Omega_c^{0,*}(D(0,r),\mathfrak{g})[1] \oplus \mathbb{C}\cdot c\right)$$

with differential d_{CE} the Chevalley–Eilenberg differential for $\mathfrak{g}_\kappa^{\mathbb{C}}$. Give $\mathcal{F}^\kappa(D(0,r)$ an increasing filtration, by degree of the symmetric power. This filtration is compatible with the action of S^1 and of $\mathcal{D}(S^1)$. In the associated graded, the differential is just that from the differential $\overline{\partial}$ on $\Omega_c^{0,*}(D(0,r))$.

It follows that there is a spectral sequence of differentiable cochain complexes

$$H^*\left(\mathrm{Gr}^* \mathcal{F}_l^\kappa(D(0,r))\right) \Rightarrow H^*\left(\mathcal{F}_l^\kappa(D(0,r))\right).$$

The analytic results we proved in Section 5.4.5 concerning compactly supported Dolbeault cohomology immediately imply that $H^*\left(\mathrm{Gr}^* \mathcal{F}_l^\kappa(D(0,r))\right)$ satisfy properties $(1)-(3)$ in the preceding. It follows that these properties also hold for $H^*(\mathcal{F}_l^\kappa(D(0,r))$.

5.5.4 Proof of the Theorem

Let us now prove that the vertex algebra associated to $\mathcal{F}_\kappa$ is isomorphic to the Kac–Moody vertex algebra. The proof will be a little different from the proof of the corresponding result for the $\beta\gamma$ system.

We first prove a statement concerning the behavior of the prefactorization algebra $\mathcal{F}_\kappa$ on annuli. Consider the radial projection map

$$\rho: \quad \mathbb{C}^\times \quad \rightarrow \quad \mathbb{R}_{>0},$$
$$z \quad \mapsto \quad |z|.$$

We define the pushforward prefactorization algebra $\rho_* \mathcal{F}_\kappa$ on $\mathbb{R}_{>0}$ that assigns to any open subset $U \subset \mathbb{R}_{>0}$, the cochain complex $\mathcal{F}_\kappa(\rho^{-1}(U))$. In particular, this prefactorization algebra assigns to an interval (a,b), the space $\mathcal{F}_\kappa(A(a,b))$, where $A(a,b)$ indicates the annulus of those z with $a < |z| < b$. The product map for $\rho_* \mathcal{F}$ associated to the inclusion of two disjoint intervals in a larger one arises from the product map for $\mathcal{F}$ from the inclusion of two disjoint annuli in a larger one.

Recall from the example in Section 3.1.1 in Chapter 3 that any associative algebra A gives rise to a prefactorization algebra A^{fact} on $\mathbb{R}$ which assigns to the interval (a,b) the vector space A and whose product map is the multiplication in A. The first result we will show is the following.

Proposition 5.5.3 *There is an injective map of* $\mathbb{C}[c]$-*linear prefactorization algebras on* $\mathbb{R}_{>0}$

$$i : U(\widehat{\mathfrak{g}}_\kappa)^{fact} \to H^*(\rho_* \mathcal{F}^\kappa)$$

whose image is a dense subspace.

This map i is controlled by the following property. Observe that, for every open subset $U \subset \mathbb{C}$, the subspace of linear elements

$$\Omega_c^{0,*}(U, \mathfrak{g})[1] \oplus \mathbb{C} \cdot c \subset C_*(\Omega_c^{0,*}(U, \mathfrak{g}) \oplus \mathbb{C} \cdot c[1]) = \mathcal{F}^\kappa(U)$$

is, in fact, a subcomplex. Applying this fact to $U = \rho^{-1}(I)$ for an interval $I \subset \mathbb{R}_{>0}$ and taking cohomology, we obtain a linear map

$$H^1(\Omega_c^{0,*}(\rho^{-1}(I))) \otimes \mathfrak{g}) \oplus \mathbb{C} \cdot c \to H^0(\mathcal{F}^\kappa(I)).$$

Note that we have a natural identification

$$H^1(\Omega_c^{0,*}(\rho^{-1}(I))) = \Omega_{hol}^1(\rho^{-1}(I))^\vee,$$

where $\vee$ indicates continuous linear dual. Thus, the linear elements of $H^* \mathcal{F}^\kappa$, which generate the factorization algebra in a certain sense that we'll identify below, are $\mathfrak{g}$-valued linear functionals on holomorphic 1-forms.

The following notation will be quite useful. Fix a circle $\{|z| = r\}$ where $r \in I$. Performing a contour integral around this circle defines a linear function on $\Omega_{hol}^1(\rho^{-1}(I))$, and so an element of $H^1(\Omega_c^{0,*}(\rho^{-1}(I)))$. Denote it by $\phi(1)$. Likewise, for each integer n, performing a contour integral against z^n around this circle defines an element of $H^1(\Omega_c^{0,*}(\rho^{-1}(I)))$ that we call $\phi(z^n)$.

The map

$$i : \widehat{\mathfrak{g}}_\kappa \to H^*(\mathcal{F}^\kappa(\rho^{-1}(I)))$$

constructed by the theorem factors through the map

$$
\begin{array}{rcl}
i : & \widehat{\mathfrak{g}}_\kappa & \to & H^1(\Omega_c^{0,*}(\rho^{-1}(I))) \otimes \mathfrak{g} \oplus \mathbb{C} \cdot c \\
& X z^n & \mapsto & \phi(z^n) \otimes X \\
& c & \mapsto & c
\end{array}.
$$

Note that Cauchy's integral theorem ensures that, in a sense, it does not matter what circle we use for the contour integrals that define ϕ. Hence, this map should produce a map from the universal enveloping algebra of $\widehat{\mathfrak{g}}_\kappa$, viewed as a locally constant prefactorization algebra, to $H^* \mathcal{F}^\kappa$.

Proof In Chapter 3, Section 3.4, we showed how the universal enveloping algebra of any Lie algebra $\mathfrak{a}$ arises as a prefactorization envelope. Let $U^{fact}(\mathfrak{a})$

denote the prefactorization algebra on $\mathbb{R}$ that assigns to an interval I, the cochain complex

$$U(\mathfrak{a})^{fact}(I) = C_*(\Omega_c^*(I, a)).$$

We showed that the cohomology of $U(\mathfrak{a})^{fact}$ is locally constant and corresponds to the ordinary universal enveloping algebra $U\mathfrak{a}$.

Let us apply this construction to $\mathfrak{a} = \widehat{\mathfrak{g}}_\kappa$. To prove the theorem, we will produce a map of prefactorization algebras on $\mathbb{R}_{>0}$

$$i : U(\widehat{\mathfrak{g}}_\kappa)^{fact} \to \rho_* \mathcal{F}^\kappa.$$

Since both sides are defined as the Chevalley–Eilenberg chains of local Lie algebras, it suffices to produce such a map at the level of dg Lie algebras.

We will produce such a map in a homotopical sense. To be explicit, we will introduce several precosheaves of Lie algebras and quasi-isomorphisms

$$\Omega_c^* \otimes \widehat{\mathfrak{g}}_\kappa \xrightarrow{\simeq} \mathcal{L}_1 \leftrightarrows \mathcal{L}_1' \xrightarrow{\simeq} \mathcal{L}_2. \tag{†}$$

In the middle, between $\mathcal{L}_1$ and $\mathcal{L}_1'$, there will be a cochain homotopy equivalence. At the level of cohomology, we then obtain an isomorphism

$$\widehat{\mathfrak{g}}_\kappa = H^*(\Omega_c^* \otimes \widehat{\mathfrak{g}}_\kappa) \xrightarrow{\cong} H^* \mathcal{L}_2$$

that proves the theorem.

Let us remark on a small but important point. These precosheaves will satisfy a stronger condition that ensures the functor C_* of Chevalley–Eilenberg chains produces a prefactorization algebra. Consider the symmetric monoidal category whose underlying category is dg Lie algebras but whose symmetric monoidal structure is given by direct sum as cochain complexes. (The coproduct in dg Lie algebras is *not* this direct sum.) All four precosheaves from the sequence (†) are prefactorization algebras valued in this symmetric monoidal category. In general, a precosheaf $\mathcal{L}$ of dg Lie algebras is such a prefactorization algebra if it has the property that for any two disjoint opens U, V in W, the elements in $\mathcal{L}(W)$ coming from $\mathcal{L}(U)$ commute with those coming from $\mathcal{L}(V)$.

We now describe these prefactorization dg Lie algebras.

Let $\mathcal{L}_1$ be the prefactorization dg Lie algebras on $\mathbb{R}$ that assigns to an interval I the dg Lie algebra

$$\mathcal{L}_1(I) = \left(\Omega_c^*(I) \otimes \mathfrak{g}[z, z^{-1}] \right) \oplus \mathbb{C} \cdot c[-1].$$

Thus, $\mathcal{L}_1$ is a central extension of $\Omega_c^* \otimes \mathfrak{g}[z, z^{-1}]$. The cocycle defining the central extension on the interval I is

$$\alpha X z^n \otimes \beta Y z^m \mapsto \left(\int_I \alpha \wedge \beta \right) \kappa(X, Y) \left(\oint z^n \partial_z z^m \right),$$

where $\alpha, \beta \in \Omega_c^*(U)$.

Note that there is a natural map of such prefactorization dg Lie algebras

$$j : \Omega_c^* \otimes \widehat{\mathfrak{g}_\kappa} \rightarrow \mathcal{L}_1$$

that is the identity on $\Omega_c^* \otimes \mathfrak{g}[z, z^{-1}]$ but on the central element is given by

$$j(\alpha c) = \left(\int_I \alpha \right) c,$$

where $\alpha \in \Omega_c^*(I)$. This map is clearly a quasi-isomorphism when I is an interval. Hence, we have produced the first map in the sequence (†).

It follows immediately that the map

$$C_* j : U(\widehat{\mathfrak{g}_\kappa})^{fact} = C_*(\Omega_c^* \otimes \widehat{\mathfrak{g}_\kappa}) \rightarrow C_* \mathcal{L}_1$$

is a map of prefactorization algebras that is a quasi-isomorphism on intervals. Therefore, the cohomology prefactorization algebra of $C_* \mathcal{L}_1$ assigns to an interval $U(\widehat{\mathfrak{g}_\kappa})$, and the prefactorization product is just the associative product on this algebra.

Let $\mathcal{L}_2$ be the prefactorization dg Lie algebra $\rho_* \mathfrak{g}_\kappa^{\mathbb{C}}$. In other words, it assigns to an interval I the dg Lie algebra

$$\mathcal{L}_2(I) = \left(\Omega_c^{0,*}(\rho^{-1}(I)) \otimes \mathfrak{g} \right) \oplus \mathbb{C} \cdot c[-1],$$

with the central extension inherited from $\mathfrak{g}_\kappa^{\mathbb{C}}$. Hence,

$$C_* \mathcal{L}_2 = \rho_* \mathcal{F}^\kappa,$$

so that the last element in the sequence (†) is what we need for the theorem.

Finally, we define $\mathcal{L}_1'$ to be a central extension of $\Omega_c^* \otimes \mathfrak{g}[z, z^{-1}]$, but where the cocycle defining the central extension is

$$\alpha X z^n \otimes \beta Y z^m \mapsto \left(\int_I \alpha \wedge \beta \right) \kappa(X, Y) \left(\oint z^n \partial_z z^m \right) + \pi \kappa(X, Y) \delta_{n+m,0} \left(\int_I \alpha r \frac{\partial}{\partial r} \beta \right),$$

where the vector field $r \frac{\partial}{\partial r}$ acts by the Lie derivative on the form $\beta \in \Omega_c^*(I)$. It is easy to verify that this cochain is closed and so defines a central extension. Note that $\mathcal{L}_1'$ looks like $\mathcal{L}_1$ except with a small addition to the cocycle giving the central extension.

Remark: This central extension, as well as the ones defining $\mathcal{L}_1$ and $\mathcal{L}_2$, are local central extensions of local dg Lie algebras in the sense of Definition 3.6.3 of Chapter 3. This concept is studied in more detail in Volume 2. ◇

To prove the proposition, we will do the following.

(i) Prove that $\mathcal{L}_1$ and $\mathcal{L}'_1$ are homotopy equivalent prefactorization dg Lie algebras.
(ii) Construct a map of precosheaves of dg Lie algebras $\mathcal{L}'_1 \to \mathcal{L}_2$.

For the first point, note that the extra term

$$\pi\kappa(X,Y)\delta_{n+m,0}\int_U \alpha r\frac{\partial}{\partial r}\beta$$

in the coycle for $\mathcal{L}'_1$ is an exact cocycle. It is cobounded by the cochain

$$\pi\kappa(X,Y)\delta_{n+m,0}\int_U \alpha\iota_{\frac{\partial}{\partial r}}\beta,$$

where $\iota_{\frac{\partial}{\partial r}}$ indicates contraction. The fact that this expression cobounds follows from the Cartan homotopy formula for the Lie derivative of vector fields on differential forms. Since the cobounding cochain is also local, $\mathcal{L}_1$ and $\mathcal{L}'_1$ are homotopy equivalent as prefactorization dg Lie algebras.

Now we will produce the desired map $\Phi : \mathcal{L}'_1 \to \mathcal{L}_2$. We use the following notation:

- r denotes both the coordinate on $\mathbb{R}_{>0}$ and also the radial coordinate in $\mathbb{C}^\times$,
- θ is the angular coordinate on $\mathbb{C}^\times$, and
- X will denote an element of $\mathfrak{g}$.

We will view $\widehat{\mathfrak{g}}_\kappa$ as $\mathfrak{g}[z,z^{-1}] \oplus \mathbb{C}\cdot c$.

On an open $I \subset \mathbb{R}_{>0}$, the map Φ is

$$\begin{aligned}
f(r)Xz^n &\mapsto & f(r)z^nX, \\
f(r)\mathrm{d}r\, Xz^n &\mapsto & \tfrac{1}{2}e^{i\theta}f(r)z^n\mathrm{d}\bar{z}\,X, \\
c &\mapsto & c,
\end{aligned}$$

with $f \in \Omega_c^0(U)$. It is a map of precosheaves but we need to verify it is a map of dg Lie algebras. Compatibility with the differential follows from the formula for $\frac{\partial}{\partial\bar{z}}$ in polar coordinates:

$$\frac{\partial}{\partial\bar{z}} = \frac{1}{2}e^{i\theta}\left(\frac{\partial}{\partial r} - \frac{1}{ir}\frac{\partial}{\partial\theta}\right).$$

Only the central extension might cause incompatibility with the Lie bracket.
Consider the following identity:

$$\int_{\rho^{-1}(I)} f(r)z^n\partial\left(z^mg(r)\frac{1}{2}e^{i\theta}\right)\mathrm{d}\bar{z} = 2\pi\, im\delta_{n+m,0}\int_I f(r)g(r)\,\mathrm{d}r$$
$$+ \pi\delta_{n+m,0}\int_I f(r)r\frac{\partial}{\partial r}g(r)\,\mathrm{d}r,$$

where $I \subset \mathbb{R}_{>0}$ is an interval. The expression on the right-hand side gives the central extension term in the Lie bracket on $\mathcal{L}'_1$, whereas that on the left is the central extension term for $\mathcal{L}_2$ applied to the elements $\Phi(f(r)Xz^n)$ and $\Phi(g(r)\mathrm{d}r\,Yz^m)$.

Applying Chevalley chains, we get a map of prefactorization algebras

$$C_* \Phi : C_* \mathcal{L}'_1 \to C_* \mathcal{L}_2 = \rho_* \mathcal{F}_\kappa.$$

Since $C_* \mathcal{L}'_1$ is homotopy equivalent to $C_* \mathcal{L}_1$ (although we have not given an equivalence explicitly), we get the desired map of prefactorization algebras on $\mathbb{R}_{>0}$:

$$U(\widehat{\mathfrak{g}}_\kappa)^{fact} \to H^*(\rho_* \mathcal{F}_\kappa).$$

The analytical results about compactly supported Dolbeault cohomology in Section 5.4.5 imply immediately this map has dense image.

Finally, we check that this map has the properties discussed just after the statement of the theorem. Let $I \subset \mathbb{R}_{>0}$ be an interval. Then, under the isomorphism

$$i : U(\widehat{\mathfrak{g}}_\kappa) \xrightarrow{\cong} H^*(C_*(\mathcal{L}'_1(I))),$$

the element Xz^n is represented by the element

$$f(r)z^n X\mathrm{d}r \in \mathcal{L}'_1(U)[1],$$

where f has compact support and is chosen so that $\int f(r)\,\mathrm{d}r = 1$.

To check the desired properties, we need to verify that if α is a holomorphic 1-form on $\rho^{-1}(I)$, then

$$\int_{\rho^{-1}(I)} \alpha f(r) z^n \tfrac{1}{2} e^{i\theta} \mathrm{d}\bar{z} = \oint_{|z|=1} \alpha z^n.$$

But this identity follows immediately from Stokes' theorem and the equation

$$f(r)\tfrac{1}{2}e^{i\theta}\mathrm{d}\bar{z} = \bar{\partial}(h(r)),$$

where $h(r) = \int_\infty^r f(r)\,\mathrm{d}r$. □

This theorem shows how to relate observables on an annulus to the universal enveloping algebra of the affine Kac–Moody Lie algebra. Recall that the Kac–Moody vertex algebra is the vacuum representation equipped with a vertex algebra structure. The next result will show that the vertex algebra associated to the prefactorization algebra $\mathcal{F}^\kappa$ is also a vertex algebra structure on the vacuum representation.

More precisely, we will show the following.

Proposition 5.5.4 *Let*

$$V = \bigoplus_n H^*(\mathcal{F}_n^\kappa(D(0,\varepsilon))$$

be the cohomology of the direct sum of the weight n eigenspaces of the S^1 action on $\mathcal{F}_\kappa(D(0,\varepsilon))$. Let $A(r,r')$ be the annulus $\{r < |z| < r'\}$.
　　The map

$$i : U(\widehat{\mathfrak{g}}_\kappa) \to H^*(\mathcal{F}^\kappa(A(r,r')))$$

constructed in the previous proposition induces an action of $U(\widehat{\mathfrak{g}}_\kappa)$ on V. Moreover, there is a unique isomorphism of $U(\widehat{\mathfrak{g}}_\kappa)$-modules from V to the vacuum module W, which sends the unit observable $1 \in V$ to the vacuum element $|0\rangle \in W$.

Proof If $\varepsilon < r < r'$, then the prefactorization product gives a map

$$\mathcal{F}^\kappa(D(0,\varepsilon)) \times \mathcal{F}^\kappa(A(r,r')) \to \mathcal{F}^\kappa(D(0,r'))$$

of cochain complexes. Passing to cohomology, and using the relationship between $U(\widehat{\mathfrak{g}}_\kappa)$ and $\mathcal{F}^\kappa(A(r,r'))$, we get a map

$$V \otimes U(\widehat{\mathfrak{g}}_\kappa) \to H^*(\mathcal{F}^\kappa(D(0,r'))).$$

Note that on the left-hand side, every element is a finite sum of elements in S^1-eigenspaces. Since the map is S^1-equivariant, its image is the subspace in $H^*(\mathcal{F}(D(0,r')))$ that consists of finite sums of S^1-eigenvectors. This subspace is V. We therefore find a map

$$V \otimes U(\widehat{\mathfrak{g}}_\kappa) \to V. \qquad\qquad (\dagger)$$

To see that it makes V a module over $U(\widehat{\mathfrak{g}}_\kappa)$, combine the following observations. First, the map $i : U(\widehat{\mathfrak{g}}_\kappa) \to H^*(\mathcal{F}^\kappa(A(r,r')))$ takes the asociative product on $U(\widehat{\mathfrak{g}}_\kappa)$ to the prefactorization product map

$$H^*(\mathcal{F}^\kappa(A(r,r'))) \otimes H^*(\mathcal{F}^\kappa(A(s,s'))) \to H^*(\mathcal{F}^\kappa(A(r,s')))$$

for $r < r' < s < s'$. Second, the diagram of prefactorization product maps

$$\mathcal{F}^\kappa(D(0,\varepsilon)) \otimes \mathcal{F}^\kappa(A(r,r')) \otimes \mathcal{F}^\kappa(A(s,s')) \longrightarrow \mathcal{F}^\kappa(D(0,\varepsilon)) \otimes \mathcal{F}^\kappa(A(r,s'))$$

$$\downarrow \qquad\qquad\qquad\qquad\qquad\qquad\qquad \downarrow$$

$$\mathcal{F}^\kappa(D(0,r')) \otimes \mathcal{F}^\kappa(A(s,s')) \longrightarrow \mathcal{F}^\kappa(D(0,s'))$$

commutes.
　　We need to show that this action identifies V with the vacuum representation W. The putative map from W to V sends the vacuum element $|0\rangle$ to the unit

observable $1 \in V$. To show that this map is well defined, we need to show that $1 \in V$ is annihilated by the elements $z^n X \in \widehat{\mathfrak{g}}_\kappa$ for $n \geq 0$.

The unit axiom for prefactorization algebras implies that the following diagram commutes:

$$
\begin{array}{ccc}
U(\widehat{\mathfrak{g}}_\kappa) & \longrightarrow & H^*(\mathcal{F}^\kappa(A(s,s'))) \\
\downarrow & & \downarrow \\
V & \longrightarrow & H^*(\mathcal{F}^\kappa(D(0,s'))).
\end{array}
$$

The left vertical arrow is given by the action of $U(\widehat{\mathfrak{g}}_\kappa)$ on the unit element $1 \in V$, and the right vertical arrow is the map arising from the inclusion $A(r, r') \subset D(0, r')$. The bottom right arrow is the inclusion into the direct sum of S^1-eigenspaces, which is injective.

The proof of Theorem 5.5.3 gives an explicit representative for the element $X z^n \in \widehat{\mathfrak{g}}_\kappa$ in $F^\kappa(A(s, s'))$. Namely, let $f(r)$ be a function supported in the interval (s, s') and with $\int f(r)\,dr = 1$. Then $X z^n$ is represented by

$$
\tfrac{1}{2} e^{i\theta} f(r) d\bar{z} z^n X \in \Omega_c^{0,1}(A(s, s')) \otimes \mathfrak{g}.
$$

We will show that this element is exact when it is viewed as an element of $\Omega_c^{0,1}(D(0, s')) \otimes \mathfrak{g}$, so that $X z^n$ vanishes when going right and then down in the square above.

Set

$$
h(r) = \int_\infty^r f(t)\,dt,
$$

so $h(r) = 0$ for $r \gg s'$ and $h(r) = 1$ for $r < s$. Also, $\frac{\partial}{\partial r} h = f$. The polar-coordinate representation of $\frac{\partial}{\partial \bar{z}}$ tells us that

$$
\bar{\partial} h(r) z^n X = \tfrac{1}{2} e^{i\theta} f(r)\, d\bar{z} z^n X.
$$

Thus, we have shown that the elements $X z^n \in \widehat{\mathfrak{g}}_\kappa$, where $n \geq 0$, act by zero on the element $1 \in V$. We therefore have a unique map of $U(\widehat{\mathfrak{g}}_\kappa)$-modules from W to V sending $|0\rangle$ to 1.

It remains to show that this map is an isomorphism. Note that every object we are discussing is filtered. The universal enveloping algebra $U(\widehat{\mathfrak{g}}_\kappa)$ is filtered, as usual, by setting $F^i U(\widehat{\mathfrak{g}}_\kappa)$ to be the subspace spanned by products of $\leq i$ elements of $\widehat{\mathfrak{g}}_\kappa$. (This filtration is inherited from the tensor algebra.) Similarly, the space

$$
\mathcal{F}^\kappa(U) = C_*(\Omega_c^{0,*}(U, \mathfrak{g}) \oplus \mathbb{C} \cdot c[-1])
$$

is filtered by saying that $F^i \mathcal{F}^\kappa(U)$ is the subcomplex $C_{\leq i}$. All the maps we have been discussing are compatible with these increasing filtrations.

The associated graded of $U(\widehat{\mathfrak{g}}_\kappa)$ is the symmetric algebra of $\widehat{\mathfrak{g}}_\kappa$, and the associated graded of $\mathcal{F}^\kappa(U)$ is the appropriate symmetric algebra on $\Omega_c^{0,*}(U,\mathfrak{g})[1]\oplus \mathbb{C}\cdot c$ in differentiable cochain complexes. On taking associated graded, the maps

$$\operatorname{Gr} U(\widehat{\mathfrak{g}}_\kappa) \to \operatorname{Gr} H^*\mathcal{F}^\kappa(A(s,s')),$$
$$\operatorname{Gr} H^*(\mathcal{F}^\kappa(A(s,s'))) \to \operatorname{Gr} H^*\mathcal{F}^\kappa(D(0,s'))$$

are maps of commutative algebras. It follows that the map

$$\operatorname{Gr} U(\widehat{\mathfrak{g}}_\kappa) \to \operatorname{Gr} V \subset H^*\operatorname{Sym}\left(\Omega_c^{0,*}(D(0,s),\mathfrak{g})[1]\oplus\mathbb{C}\cdot c\right)$$

is a map of commutative algebras. Moreover, $\operatorname{Gr} V$ is the direct sum of the S^1-eigenspaces in the space on the right side of this map. The direct sum of the S^1-eigenspaces in $H^1(\Omega^{0,1}(D(0,s))$ is naturally identified with $z^{-1}\mathbb{C}[z^{-1}]$. Thus, we find the associated graded of the map $U(\widehat{\mathfrak{g}}_\kappa)$ to V is a map of commutative algebras

$$\left(\operatorname{Sym}\mathfrak{g}[z,z^{-1}]\right)[c] \to \left(\operatorname{Sym} z^{-1}\mathfrak{g}[z^{-1}]\right)[c].$$

We have already calculated that on the generators of the commutative algebra, it arises from the natural projection map

$$\mathfrak{g}[z,z^{-1}]\oplus\mathbb{C}\cdot c \to z^{-1}\mathfrak{g}[z^{-1}]\oplus\mathbb{C}\cdot c.$$

It follows immediately that the map $\operatorname{Gr} W \to \operatorname{Gr} V$ is an isomorphism, as desired. $\qquad\square$

To complete the proof that the vertex algebra associated to the prefactorization algebra $\mathcal{F}^\kappa$ is isomorphic to the Kac–Moody vertex algebra, we need to identify the operator product expansion map

$$Y : V \otimes V \to V((z)).$$

Recall that this map is defined, in terms of the prefactorization algebra $\mathcal{F}^\kappa$, as follows. Consider the map

$$m_{z,0} : V \otimes V \to H^*(\mathcal{F}^\kappa(D(0,\infty))$$

defined by restricting the prefactorization product map

$$H^*(\mathcal{F}^\kappa(D(z,\varepsilon)) \times H^*(\mathcal{F}^\kappa(D(0,\varepsilon)) \to H^*(\mathcal{F}^\kappa(D(0,\infty))$$

to the subspace

$$V \subset H^*(\mathcal{F}^\kappa(D(z,\varepsilon)) = H^*(\mathcal{F}^\kappa(D(0,\varepsilon)).$$

Composing the map $m_{z,0}$ with the map

$$H^*(\mathcal{F}^\kappa(D(0, \infty))) \to \overline{V} = \prod_k V_k$$

given by the product of the projection maps to the S^1-eigenspaces, we get a map

$$m_{z,0} : V \otimes V \to \overline{V}.$$

This map depends holomorphically on z. The operator product map is obtained as the Laurent expansion of $m_{z,0}$.

Our aim is to calculate the operator product map and identify it with the vertex operator map in the Kac–Moody vertex algebra. We use the following notation. If $X \in \mathfrak{g}$, let $X_i = z^i X \in \widehat{\mathfrak{g}}_\kappa$. We denote the action of $\widehat{\mathfrak{g}}_\kappa$ on V by the symbol $\cdot$. Then we have the following.

Proposition 5.5.5 *For all $v \in V$,*

$$m_{z,0}(X_{-1} \cdot 1, v) = \sum_{i \in \mathbb{Z}} z^{-i-1}(X_i \cdot v) \in \overline{V}$$

where the sum on the right-hand side converges.

Before we prove this proposition, let us observe that it proves our main result.

Corollary 5.5.6 *This isomorphism of $U(\widehat{\mathfrak{g}}_\kappa)$-modules from the vacuum representation W to V is an isomorphism of vertex algebras, where V is given the vertex algebra structure arising from the prefactorization algebra $\mathcal{F}^\kappa$, and W is given the Kac–Moody vertex algebra structure defined in 5.5.2.*

Proof This corollary follows immediately from the reconstruction theorem 5.3.4. □

Proof of the proposition The inclusion of a disc into an annulus induces a structure map of the prefactorization algebra. Each element v of V thus determines an element of the prefactorization algebra on that annulus, and hence a map from V to $\overline{V}$. The formula will follow by applying our earlier analysis of the structure maps.

As before, we let $X_i \in \widehat{\mathfrak{g}}_\kappa$ denote Xz^i for $X \in \mathfrak{g}$. As we explained in the discussion following Theorem 5.5.3, we can view the element

$$X_{-1} \cdot 1 \in V$$

as being represented at the cochain level by X times the linear functional

$$
\begin{aligned}
\Omega^1_{\mathrm{hol}}(D(0,s)) &\to & \mathbb{C} \\
\alpha &\mapsto & \int_{|z|=r} z^{-1}\alpha.
\end{aligned}
$$

Cauchy's theorem tells us that this linear functional sends $h(z)\,dz$, when h is holomorphic, to $2\pi i h(0)$.

Now fix $z_0 \in A(s,s')$, and let

$$
\iota_{z_0} : V \to H^*(\mathcal{F}^\kappa(A(s,s')))
$$

denote the map arising from the restriction to V of the structure map

$$
H^*(\mathcal{F}^\kappa(D(0,\varepsilon))) = H^*(\mathcal{F}^\kappa(D(z_0,\varepsilon))) \to H^*(\mathcal{F}^\kappa(A(s,s')))
$$

arising from the inclusion of the disc $D(z_0,\varepsilon)$ into the annulus $A(s,s')$. It is clear from the definition of $m_{z_0,0}$ and the axioms of a prefactorization algebra that the following diagram commutes:

$$
\begin{array}{ccc}
V \otimes V \otimes \mathrm{Id} & \xrightarrow{\ m_{z_0,0}\ } & \overline{V} \\
{\scriptstyle \iota_{z_0}}\big\downarrow & & \big\uparrow \\
H^*(\mathcal{F}^\kappa(A(s,s'))) \otimes V & \longrightarrow & H^*(\mathcal{F}^\kappa(D(0,\infty))).
\end{array}
$$

Here the bottom right arrow is the restriction to V of the prefactorization structure map

$$
H^*(\mathcal{F}^\kappa(A(s,s'))) \otimes H^*(\mathcal{F}^\kappa(D(0,\varepsilon))) \to H^*(\mathcal{F}^\kappa(D(0,\infty))).
$$

It therefore suffices to show that

$$
\iota_{z_0}(X_{-1} \cdot 1) = \sum_{i\in\mathbb{Z}} X_i z_0^{-i-i} \in H^*(\mathcal{F}^\kappa(A(s,s')))
$$

where we view $X_i \in \widehat{\mathfrak{g}}_\kappa$ as elements of $H^0(\mathcal{F}^\kappa(A(s,s')))$ via the map

$$
U(\widehat{\mathfrak{g}}^\kappa) \to H^0(\mathcal{F}^\kappa(A(s,s')))
$$

constructed in Theorem 5.5.3.

It is clear from the construction of $\mathcal{F}^\kappa$ that $\iota_{z_0}(X_{-1} \cdot 1)$ is in the image of the natural map

$$
\Omega^{0,*}_c(A(s,s')) \otimes \mathfrak{g} \to \mathcal{F}^\kappa(A(s,s')).
$$

Recall that the cohomology of $\Omega^{0,*}_c(A(s,s'))$ is the linear dual of the space of holomorphic 1-forms on the annulus $A(s,s')$. The element $\iota_{z_0}(X_{-1} \cdot 1)$ can thus be represented by the continuous linear map

$$\Omega^1_{\mathrm{hol}}(D(0,s)) \quad \to \qquad \mathfrak{g}$$
$$\alpha \qquad \mapsto \quad \left(\textstyle\int_{|z|=r} z^{-1}\alpha\right) X.$$

Similarly, the elements $X_i \in H^*(\mathcal{F}^\kappa(A(s,s')))$ are represented by the linear maps

$$h(z)dz \mapsto \left(\int_{|z|=r} z^i h(z)dz\right) X.$$

It remains to show that, for all holomorphic functions $h(z)$ on the annulus $A(s,s')$, we have

$$2\pi i h(z_0) = \sum_{i \in \mathbb{Z}} z_0^i \left(\oint_{|z|=s+\varepsilon} z^{-i-1} h(z)\, dz\right).$$

This equality is proved using Cauchy's theorem. We know

$$2\pi i h(z_0) = \oint_{|z|=s+\varepsilon} \frac{h(z)}{z-z_0}\, dz - \oint_{|z|=s'-\varepsilon} \frac{h(z)}{z-z_0}\, dz.$$

Expanding $(z-z_0)^{-1}$ in the regions when $|z| < |z_0|$ (relevant for the first integral) and when $|z| > |z_0|$ (relevant for the second integral) gives the desired expression. $\qquad\square$

PART III

Factorization algebras

6

Factorization Algebras: Definitions
and Constructions

Our definition of a prefactorization algebra is closely related to that of a pre-cosheaf or of a presheaf. Mathematicians have found it useful to refine the axioms of a presheaf to those of a sheaf: a sheaf is a presheaf whose value on a large open set is determined, in a precise way, by values on arbitrarily small subsets. In this chapter we describe a similar "descent" axiom for prefactorization algebras. We call a prefactorization algebra satisfying this axiom a *factorization algebra*.

After defining this axiom, our next task is to verify that the examples we have constructed so far, such as the observables of a free field theory, satisfy it. This we do in Sections 6.5 and 6.6.

Philosophically, our descent axiom for factorization algebras is important: a prefactorization algebra satisfying descent (i.e., a factorization algebra) is built from local data, in a way that a general prefactorization algebra need not be. However, for many practical purposes, such as the applications to field theory, this axiom is often not essential.

A reader with little taste for formal mathematics could thus skip this chapter and the next and still be able to follow the rest of this book.

6.1 Factorization Algebras

A factorization algebra is a prefactorization algebra that satisfies a *local-to-global* axiom. This axiom is the analog of the gluing axiom for sheaves; it expresses how the values on big open sets are determined by the values on small open sets. We thus begin by reviewing the notion of a (co)sheaf, with more background and references available in Appendix A.5. Next, we introduce *Weiss covers*, which are the type of covers appropriate for factorization algebras. Finally, we give the definition of factorization algebras,

6.1.1 Sheaves and Cosheaves

In order to motivate our definition of factorization algebra, let us recall the sheaf axiom and then work out its dual, the cosheaf axiom.

Let M be a topological space and let Opens(M) denote the poset category of open subsets of a space M. A presheaf $\mathcal{A}$ on a topological space M with values in a category $\mathcal{C}$ is a functor from Opens$(M)^{op}$ to $\mathcal{C}$. Given an open cover $\{U_i \mid i \in I\}$ of an open set $U \subset M$, there is a canonical map

$$\mathcal{A}(U) \rightarrow \prod_i \mathcal{A}(U_i)$$

given by taking the product of the structure maps $\mathcal{A}(U) \rightarrow \mathcal{A}(U_i)$. A presheaf $\mathcal{A}$ is a sheaf if for every U and for every open cover of U, this map equalizes the pair of maps

$$\prod_i \mathcal{A}(U_i) \rightrightarrows \prod_{i,j} \mathcal{A}(U_i \cap U_j)$$

given by the two inclusions $U_i \cap U_j \rightarrow U_i$ and $U_i \cap U_j \rightarrow U_j$. That is, $\mathcal{A}(U)$ is the limit of that diagram. In words, this condition means that a section s of $\mathcal{A}$ on U can be described as sections s_i on each element of the cover that agree on the overlaps, i.e., $s_i|_{U_i \cap U_j} = s_j|_{U_i \cap U_j}$.

We now dualize this definition. A precosheaf Φ on M with values in a category $\mathcal{C}$ is a functor from Opens(M) to $\mathcal{C}$. It is a cosheaf if for every open cover of U, the canonical map given by taking the coproduct of the structure map $\Phi(U_k) \rightarrow \Phi(U)$,

$$\coprod_k \Phi(U_k) \rightarrow \Phi(U),$$

coequalizes the pair of maps

$$\coprod_{i,j} \Phi(U_i \cap U_j) \rightrightarrows \coprod_k \Phi(U_k)$$

given by the two inclusions $U_i \cap U_j \rightarrow U_i$ and $U_i \cap U_j \rightarrow U_j$. That is, $\Phi(U)$ is the colimit of that diagram. Thus, a cosheaf with values in $\mathcal{C}$ can be understood as a sheaf with values in $\mathcal{C}^{op}$.

In this book, our target categories are linear in nature, so that $\mathcal{C}$ is the category of vector spaces or of dg vector spaces or of cochain complexes with values in some additive category, such as differentiable vector spaces. In such a setting, we can write the cosheaf axiom in an equivalent but slightly different way. A precosheaf Φ on M with values in, say, vector spaces is a cosheaf if, for every open cover of any open set U, the sequence

$$\bigoplus_{i,j} \Phi(U_i \cap U_j) \to \bigoplus_k \Phi(U_k) \to \Phi(U) \to 0$$

is exact, where the first map is the difference of the structure maps for the two inclusions. In the remainder of this book, we phrase our gluing axioms in this style, as we always restrict to linear target categories.

6.1.2 Weiss Covers

For factorization algebras, we require our covers to be fine enough that they capture all the "multiplicative structure" – the structure maps – of the underlying prefactorization algebra. In fact, a factorization algebra will be a cosheaf with respect to this modified notion of cover.

Definition 6.1.1 Let U be an open set. A collection of open sets $\mathfrak{U} = \{U_i \mid i \in I\}$ is a *Weiss cover* of U if for any finite collection of points $\{x_1, \ldots, x_k\}$ in U, there is an open set $U_i \in \mathfrak{U}$ such that $\{x_1, \ldots, x_k\} \subset U_i$.

The Weiss covers define a Grothendieck topology on Opens(M), the poset category of open subsets of a space M. We call it the *Weiss topology* of M. (This notion was introduced in Weiss (1999) and it is explained nicely and further developed in Boavida de Brito and Weiss (2013).)

Remark: A Weiss cover is certainly a cover in the usual sense, but a Weiss cover typically contains an enormous number of opens. It is a kind of "exponentiation" of the usual notion of cover, because a Weiss cover is well suited to studying all configuration spaces of finitely many points in U. For instance, given a Weiss cover $\mathfrak{U}$ of U, the collection

$$\{U_i^n \mid i \in I\}$$

provides a cover in the usual sense of $U^n \subset M^n$ for every positive integer n. ◊

Example: For a smooth n-manifold M, there are several simple ways to construct a Weiss cover for M. One construction is simply to take the collection of open sets in M diffeomorphic to a disjoint union of finitely many copies of the open n-disc. Alternatively, the collection of opens $\{M \setminus q \mid q \in M\}$ is a Weiss cover of M. One can also produce a Weiss cover of a manifold with countably many elements. Fix a Riemannian metric on M. Pick a collection of points $\{q_n \in M \mid n \in \mathbb{N}\}$ such that the union of the unit "discs" $D_1(q_n) = \{q \in M \mid \mathrm{d}(q_n, q) < 1\}$ covers M. (Such an open may not be homeomorphic to a disc if the injectivity radius of q_n is less than 1.) Consider the collection of all "discs" $\mathfrak{D} = \{D_{1/m}(q_n) \mid m \in \mathbb{N}, n \in \mathbb{N}\}$. The collection of all finite, pairwise disjoint union of elements of $\mathfrak{D}$ is a countable Weiss cover. $\diamond$

The preceding examples suggest the following definition.

Definition 6.1.2 A cover $\mathfrak{U} = \{U_\alpha\}$ of M *generates* the Weiss cover $\mathfrak{V}$ if every open $V \in \mathfrak{V}$ is given by a finite disjoint union of opens U_α from $\mathfrak{U}$.

6.1.3 Strict Factorization Algebras

The value of a factorization algebra on U is determined by its behavior on a Weiss cover, just as the value of a cosheaf on an open set U is determined by its value on any cover of U. Throughout, the target multicategory $\mathcal{C}$ will be linear in nature (e.g., vector spaces or cochain complexes of differentiable vector spaces).

Definition 6.1.3 A prefactorization algebra on M with values in a multicategory $\mathcal{C}$ is a *factorization algebra* if it has the following property: for every open subset $U \subset M$ and every Weiss cover $\{U_i \mid i \in I\}$ of U, the sequence

$$\bigoplus_{i,j} \mathcal{F}(U_i \cap U_j) \to \bigoplus_k \mathcal{F}(U_k) \to \mathcal{F}(U) \to 0$$

is exact. That is, $\mathcal{F}$ is a factorization algebra if it is a cosheaf with respect to the Weiss topology.

A factorization algebra is *multiplicative* if, in addition, for every pair of disjoint open sets $U, V \in M$, the structure map

$$m_{U \sqcup V}^{U,V} : \mathcal{F}(U) \otimes \mathcal{F}(V) \to \mathcal{F}(U \sqcup V)$$

is an isomorphism.

To summarize, a multiplicative factorization algebra satisfies two conditions: an axiom of (co)descent and a multiplicativity axiom. The descent axiom says that it is a cosheaf with respect to the Weiss topology, and the multiplicativity axiom says that its value on finite collections of disjoint opens factors into a tensor product of the values on each open.

Remark: It is typically not easy to verify directly that a prefactorization algebra is a factorization algebra. In the locally constant case discussed in Section 6.4, we discuss a theorem that exhibits many more examples, including the prefactorization algebra on the real line arising from an associative algebra. Later in this chapter, we exhibit a large class of examples explicitly, arising from the enveloping construction for local Lie algebras. The next chapter provides analogs of methods familiar from sheaf theory, such as pushforwards and pullbacks and extension from a basis. ◊

6.1.4 The Čech Complex and Homotopy Factorization Algebras

If our target multicategory has a notion of weak equivalence – such as quasi-isomorphism between cochain complexes – then there is a natural modification of the notion of cosheaf. It arises from the fact that one should work with a corrected notion of colimit that takes into account that some objects are weakly equivalent, even if there is no strict equivalence between them. In other words, we need to work with the notion of colimit appropriate to ∞-categories. We will use the term *homotopy colimit* for this notion. (Appendix C.5 contains an extensive discussion in the context relevant to us: a category of cochain complexes in an additive category.) One then revises the cosheaf gluing axiom by using homotopy colimits, leading to the notion of a *homotopy cosheaf*. After describing these, we define homotopy factorization algebras.

Remark: Our target category will always be unbounded cochain complexes with values in some additive category C, and we want to treat quasi-isomorphic cochain complexes as equivalent. In other words, we want to treat $\mathrm{Ch}(C)$ as an ∞-category, and there are several approaches that are, in a precise sense, equivalent. When C is a Grothendieck Abelian category – as is the case for differentiable vector spaces DVS – this situation is well developed, and much of Appendix C.5 is devoted to understanding it. We discuss there a stable model category, a pretriangulated dg category, a simplicially enriched category, and a stable quasicategory associated to $\mathrm{Ch}(C)$, and how to transport constructions and statements between all these approaches. (We rely heavily on and summarize the approach in Lurie (2016).) In light of the results explained there, here we will write down explicit cochain complexes that describe the homotopy colimits of the diagrams we need. Readers familiar with higher categories will recognize how to generalize these definitions but we do not pursue such generality here. Work in that direction can be found in Ayala and Francis (2015) or chapter 5 of Lurie (2016). ◊

Let us provide the definition of a homotopy cosheaf. (See Appendix A.5.3 for more discussion.) Let Φ be a precosheaf on M with values in cochain complexes in an additive category C. Let $\mathfrak{U} = \{U_i \mid i \in I\}$ be a cover of some open subset U of M. Consider the cochain complex with values in cochain complexes whose $-k$th term is

$$\bigoplus_{\substack{j_0,\ldots,j_k \in I \\ j_i \text{ all distinct}}} \Phi(U_{j_0} \cap \cdots \cap U_{j_k}),$$

and note that this term has an "internal" differential arising from Φ. The "external" differential is the alternating sum of the structure maps arising from inclusion of $k + 1$-fold intersections into k-fold intersections. The *Čech complex of* $\mathfrak{U}$ *with coefficients in* Φ is the totalization of this double complex:

$$\check{C}(\mathfrak{U}, \Phi) = \bigoplus_{k=0}^{\infty} \left(\bigoplus_{\substack{j_0,\ldots,j_k \in I \\ j_i \text{ all distinct}}} \Phi(U_{j_0} \cap \cdots \cap U_{j_k})[k] \right).$$

There is a natural map from the 0th term (of the double complex) to $\Phi(U)$ given by the sum of the structure maps $\Phi(U_i) \to \Phi(U)$. Thus, the Čech complex possesses a natural map to $\Phi(U)$. We say that Φ is a *homotopy cosheaf* if the natural map from the Čech complex to $\Phi(U)$ is a quasi-isomorphism for every open $U \subset M$ and every open cover of U.

Remark: This Čech complex is simply the dual of the Čech complex for sheaves of cochain complexes. Note that this Čech complex is also the normalized cochain complex arising from a simplicial cochain complex, where we evaluate Φ on the simplicial space $\mathfrak{U}_\bullet$ associated to the cover $\mathfrak{U}$. In more sophisticated language, the Čech complex describes the homotopy colimit of Φ on this simplicial diagram. ◇.

We can now define our main object of interest. Let $\mathcal{C}$ be a multicategory whose underlying category is a Grothendieck abelian category. Then there is a natural multicategory whose underlying category is $\text{Ch}(\mathcal{C})$, the category of cochain complexes in $\mathcal{C}$ in which the weak equivalences are quasi-isomorphisms.

Definition 6.1.4 A *homotopy factorization algebra* is a prefactorization algebra $\mathcal{F}$ on X valued in $\text{Ch}(\mathcal{C})$, with the property that for every open set $U \subset X$ and Weiss cover $\mathfrak{U}$ of U, the natural map

$$\check{C}(\mathfrak{U}, \mathcal{F}) \to \mathcal{F}(U)$$

is a quasi-isomorphism. That is, $\mathcal{F}$ is a homotopy cosheaf with respect to the Weiss topology.

If $\mathcal{C}$ is a symmetric monoidal category, then we can formulate a multiplicativity condition, as follows.

Definition 6.1.5 A homotopy factorization algebra $\mathcal{F}$ is *multiplicative* if for every pair U, V of disjoint open subsets of X, the structure map

$$\mathcal{F}(U) \otimes \mathcal{F}(V) \to \mathcal{F}(U \sqcup V)$$

is a quasi-isomorphism.

Remark: In light of the weak equivalences, the strict notion of multiplicative factorization algebra is not appropriate for the world of cochain complexes. Whenever we refer to a factorization algebra in cochain complexes, we mean a homotopy factorization algebra. ◇

Because we will work primarily with differentiable cochain complexes, we introduce the following terminology.

Definition 6.1.6 A *differentiable factorization algebra* is a factorization algebra valued in the multicategory of differentiable cochain complexes.

Remark: Although the multiplicativity axiom does not directly extend to the setting of multicategories, we can examine the bilinear structure map

$$m_{U \sqcup V}^{U,V} \in \text{Ch(DVS)}(\mathcal{F}(U), \mathcal{F}(V) \mid \mathcal{F}(U \sqcup V))$$

for two disjoint opens U and V, and we can ask about the image of $m_{U \sqcup V}^{U,V}$. For almost all the examples constructed in this book, the differentiable vector spaces that appear are convenient vector spaces, which have a natural topology. In our constructions, one can typically see that the image of the structure map is dense with respect to this natural topology on $\mathcal{F}(U \sqcup V)$, which is a cousin of the multiplicativity axiom. In practice, though, it is the descent axiom that is especially important for our purposes. ◇

6.1.5 Weak Equivalences

The notion of weak equivalence, or quasi-isomorphism, on cochain complexes induces such a notion on homotopy factorization algebras.

Definition 6.1.7 Let F, G be homotopy factorization algebras valued in $\text{Ch}(\mathcal{C})$, as above. A map $\phi : F \to G$ of homotopy factorization algebras is a *weak equivalence* if, for all open subsets $U \subset M$, the map $F(U) \to G(U)$ is a quasi-isomorphism of cochain complexes.

We now provide an explicit criterion for checking weak equivalences, using the notion of a factorizing basis (see Definition 7.2.1).

Lemma 6.1.8 *A map $F \to G$ between differentiable factorization algebras is a weak equivalence if and only if, for every factorizing basis $\mathfrak{U}$ of X and every*

U in $\mathfrak{U}$*, the map*

$$F(U) \to G(U)$$

is a weak equivalence.

Proof For any open subset $V \subset X$, let $\mathfrak{U}_V$ denote the Weiss cover of V generated by all open subsets in $\mathfrak{U}$ that lie in V. By the descent axiom, the map

$$\check{C}(\mathfrak{U}_V, F) \to F(V)$$

is a weak equivalence, and similarly for G. Thus, it suffices to check that the map

$$\check{C}(\mathfrak{U}_V, F) \to \check{C}(\mathfrak{U}_V, G)$$

is a weak equivalence, for the following reason. Every Čech complex has a natural filtration by number of intersections. We thus obtain of spectral sequences from the map of Čech complexes. The fact that the maps

$$F(U_1 \cap \cdots \cap U_k) \to G(U_1 \cap \cdots \cap U_k)$$

are all weak equivalences implies that we have a quasi-isomorphism on the first page of the map of spectral sequences, so our original map is a quasi-isomorphism. $\square$

6.1.6 The (Multi)category Structure on Factorization Algebras

Every factorization algebra is itself a prefactorization algebra. Thus we define the category of factorization algebras $\mathrm{FA}(X, \mathcal{C})$ to be the full subcategory whose objects are the factorization algebras.

We have also described in Section 3.1.5 in Chapter 3 the symmetric monoidal category of prefactorization algebras. In brief, the tensor product of two prefactorization algebras is given by taking the tensor product of values at each open:

$$(F \otimes G)(U) = F(U) \otimes G(U),$$

and we simply define the structure maps as the tensor product of the structure maps. If the tensor product commutes (separately in each variable) with colimits over simplicial diagrams (i.e., geometric realizations), then the category of factorization algebras inherits this symmetric monoidal structure structure. The key point is to ensure that $F \otimes G$ satisfies the descent axiom, which follows quickly from the hypothesis on the tensor product. As we will not use this symmetric monoidal structure on factorization algebras, we do not develop this observation.

6.2 Factorization Algebras in Quantum Field Theory

We have seen in Section 1.3 in Chapter 1 how prefactorization algebras appear naturally when one thinks about the structure of observables of a quantum field theory. It is natural to ask whether the local-to-global axiom that distinguishes factorization algebras from prefactorization algebras also has a quantum field-theoretic interpretation.

The local-to-global axiom we posit states, roughly speaking, that all observables on an open set $U \subset M$ can be built up as sums of observables supported on arbitrarily small open subsets of M. To be concrete, let us consider a Weiss cover $\mathfrak{U}_\varepsilon$ of M, built out of all open discs in M of radius smaller than ε. Applied to this Weiss cover, our local-to-global axiom states that any observable $O \in \mathrm{Obs}(U)$ can be written as a sum of observables of the form $O_1 O_2 \cdots O_k$, where $O_i \in \mathrm{Obs}(D_{\delta_i}(x_i))$ and $x_1, \ldots, x_k \in M$.

By taking ε to be very small, we see that our local-to-global axiom implies that all observables can be written as sums of products of observables whose support arbitrarily close to being point-like in U.

This assumption is physically reasonable: most of the observables (or operators) considered in quantum field theory textbooks are supported at points, so it might make sense to restrict attention to observables built from these.

However, more global observables are also considered in the physics literature. For example, in a gauge theory, one might consider the observable which measures the monodromy of a connection around some loop in the space–time manifold. How would such observables fit into the factorization algebra picture?

The answer reveals a key limitation of our axioms: *the concept of factorization algebra is appropriate only for perturbative quantum field theories.*

Indeed, in a perturbative gauge theory, the gauge field (i.e., the connection) is taken to be an infinitesimally small perturbation $A_0 + \delta A$ of a fixed connection A_0, which is a solution to the equations of motion. There is a well-known formula (the time-ordered exponential) expressing the holonomy of $A_0 + \delta A$ as a power series in δA, where the coefficients of the power series are given as integrals over L^k, where L is the loop that we are considering. This expression shows that the holonomy of $A_0 + \delta A$ can be built up from observables supported at points (which happen to lie on the loop L). Thus, the holonomy observable will form part of our factorization algebra.

However, if we are not working in a perturbative setting, this formula does not apply, and we would not expect (in general) that the prefactorization algebra of observables satisfies the local-to-global axiom.

6.3 Variant Definitions of Factorization Algebras

In this section we sketch variations on the definition of (pre)factorization algebra used in this book. Our primary goal is to provide a bridge to the next section, where we discuss locally constant factorization algebras, which are well developed in the topology literature using an approach along the lines sketched here. The variations we describe undoubtedly admit improvements and modifications; they should not be taken as definitive. We hope and expect that approaches with this flavor may nonetheless play a role in future work.

The first observation is that many constructions of (pre)factorization algebras make sense on a large class of manifolds and not just on a fixed manifold. As an example, consider the factorization envelope of a dg Lie algebra $\mathfrak{g}$. Under a smooth open embedding $f : M \to N$, the natural map $f_* : \Omega_c^*(M) \to \Omega_c^*(N)$ of extending by zero preserves the wedge product. Hence, given a dg Lie algebra $\mathfrak{g}$, we obtain a natural map $f_* \otimes \mathrm{id}_{\mathfrak{g}} : \Omega_c^*(M) \otimes \mathfrak{g} \to \Omega_c^*(N) \otimes \mathfrak{g}$ of dg Lie algebras. Thus, the pushforward $f_* \mathbb{U}\mathfrak{g}$ of the factorization envelope on M is isomorphic to the factorization envelope on N, when restricted to the image of M. In other words, this factorization envelope construction naturally lives on a category of n-manifolds, not just on a fixed n-manifold.

We can formalize this kind of structure as follows.

Definition 6.3.1 Let Emb_n denote the category whose objects are smooth n-manifolds and whose morphisms are open embeddings. It possesses a symmetric monoidal structure under disjoint union.

Then we introduce the following variant of the notion of a prefactorization algebra. (In Section 6.3.4, we explain the appropriate local-to-global axiom.)

Definition 6.3.2 A *prefactorization algebra on n-manifolds* with values in a symmetric monoidal category $\mathcal{C}$ is a symmetric monoidal functor from Emb_n to $\mathcal{C}$.

Observe the following property of such a functor $\mathcal{F}$: the space of embeddings $\mathrm{Emb}_n(M, M)$ acts on the value $\mathcal{F}(M)$. In particular, the diffeomorphism group of M is a subset of $\mathrm{Emb}_n(M, M)$, and so $\mathcal{F}(M)$ has an action of the diffeomorphism group. Thus a prefactorization algebra $\mathcal{F}$ is sensitive to the topology of smooth manifolds; it is not a trivial generalization of the notion we've already developed.

As an example, for any dg Lie algebra $\mathfrak{g}$, the factorization envelope $\mathbb{U}_n \mathfrak{g} :$ $M \mapsto C_*(\Omega_c^*(M) \otimes \mathfrak{g})$ provides a prefactorization algebra on n-manifolds.

6.3.1 Varying in Smooth Families

It would be nice, for instance, to let the diffeomorphism group act smoothly, in some sense, and not just as a discrete group. Within the framework we've developed, there is a natural approach, parallel to our definition of a smoothly equivariant prefactorization algebra (see Section 3.7.3 in Chapter 3). Note that embeddings define a natural sheaf of sets on the site Mfld of smooth manifolds: given n-manifolds M, N, the sheaf $\mathcal{E}mb(M, N)$ assigns to the manifold X the subset

$$\{f \in C^\infty(X \times M, N) \mid \forall x \in X, f(x, -) \text{ is an embedding of } M \text{ in } N\}.$$

In other words, it gives smooth families over X of embeddings of M into N.

Definition 6.3.3 Let $\mathcal{E}mb_n$ denote the category enriched over Sh (Mfld) whose objects are smooth n-manifolds and in which $\mathcal{E}mb_n(M, N)$ is the sheaf $\mathcal{E}mb(M, N)$. It possesses a symmetric monoidal structure under disjoint union.

If we work with a multicategory enriched over Sh(Mfld), such as differentiable vector spaces DVS, we can define a *smooth* prefactorization algebra as an enriched multifunctor (or map of enriched colored operads). This definition captures a sense in which we have smooth families of structure maps. Our primary example – the factorization envelope – provides such a smooth prefactorization algebra because $\Omega^*(M)$ on a fixed manifold M defines a differentiable cochain complex.

6.3.2 Other Kinds of Geometry

This kind of construction works very generally. For instance, if we are interested in complex manifolds, we could work in the following setting.

Definition 6.3.4 Let Hol_n denote the category whose objects are complex n-manifolds and whose morphisms are open holomorphic embeddings. It possesses a symmetric monoidal structure under disjoint union.

Definition 6.3.5 A *prefactorization algebra on complex n-manifolds* with values in a symmetric monoidal category $\mathcal{C}$ is a symmetric monoidal functor from Hol_n to $\mathcal{C}$.

Again, there is a natural example arising from a factorization envelope construction. Compactly supported Dolbeault forms $\Omega_c^{0,*}$ provide a functor from Hol_n to nonunital commutative dg algebras. Hence, tensoring with a dg Lie algebra $\mathfrak{g}$ provides a symmetric monoidal functor from Hol_n to the category of dg Lie algebras, equipped with direct sum of underlying cochain complexes as a symmetric monoidal structure. Thus, the factorization envelope $\mathbb{U}_n^{\text{hol}}\mathfrak{g} : M \mapsto C_*(\Omega_c^{0,*}(M) \otimes \mathfrak{g})$ provides a prefactorization algebra on complex n-manifolds.

The Kac–Moody factorization algebras (see the example in Section 3.6.3 in Chapter 3) provide an example on Hol_1 (i.e., for Riemann surfaces). In Williams (2016), Williams constructs the Virasoro factorization algebras on Hol_1. These examples extend naturally to Hol_n for any n, although the choices of central extension become more interesting and complicated. (Indeed, one finds extensions as L_∞ algebras and not just dg Lie algebras.)

It should be clear that one can enrich this category Hol_n over Sh(Mfld) and hence to talk about smooth families of such holomorphic embeddings. (Or, indeed, one can enrich it over a site of all complex manifolds.) Thus, one can talk, for instance, about smooth prefactorization algebras on complex n-manifolds.

In general, let $\mathcal{G}$ denote some kind of local structure for n-manifolds, such a Riemannian metric or complex structure or orientation. In other words, $\mathcal{G}$ is a sheaf on Emb_n. A $\mathcal{G}$-structure on an n-manifold M is a section $G \in \mathcal{G}(M)$. There is a category $\text{Emb}_{\mathcal{G}}$ whose objects are n-manifolds with $\mathcal{G}$-structure (M, G_M) and whose morphisms are $\mathcal{G}$-structure-preserving embeddings, i.e., embeddings $f : M \hookrightarrow N$ such that $f^*G_N = G_M$. This category is fibered over $\text{Emb}_{\mathcal{G}}$. One can then talk about prefactorization algebras on $\mathcal{G}$-manifolds.

6.3.3 Application to Field Theory

This notion is quite useful in the context of field theory. One often studies a field theory that makes sense on a large class of manifolds. Indeed, physicists often search for action functionals that depend only on a particular geometric structure. For instance, a conformal field theory only depends on a conformal class of metric, and so the solutions to the equations of motion form a sheaf on the site of conformal manifolds. For example, the free $\beta\gamma$ system makes sense on all conformal 2-manifolds, indeed, its solutions form a sheaf on Hol_1. Similarly, the physicist's examples of a topological field theory depend only on the underlying smooth manifold (at least at the classical level). The solutions to the equations of motion for Abelian Chern–Simons theory (see Section 4.5 in Chapter 4) make sense on the site of oriented 3-manifolds. (More generally, Chern–Simons theory for gauge group G makes sense on the site of oriented 3-manifolds with a principal G-bundle.) Finally, the free scalar field theory with mass m makes sense on the site of Riemannian n-manifolds.

Since most of the constructions with field theories in this book have exploited the enveloping factorization algebra construction, the classical and quantum observables we have constructed are typically prefactorization algebras on $\mathcal{G}$-manifolds for some $\mathcal{G}$. This universality of these quantizations – that these prefactorization algebras of quantum observables simultaneously work for all manifolds with some geometric structure – illuminates why our constructions typically recover standard answers: these are the answers that work generally.

For interacting theories, the focus of Volume 2, the quantizations are rarely so easily obtained or described. On a fixed manifold, a quantization of the prefactorization algebra of classical observables might exist, but it might not exist for all manifolds with that structure. More explicitly, quantization proceeds via Feynman diagrammatics and renormalization, and hence it involves some explicit analysis (e.g., the introduction of counterterms). It is often difficult to quantize in a way that varies nicely over the collection of all $\mathcal{G}$-manifolds. There are situations in which quantizations exist over some category of $\mathcal{G}$-manifolds, however. For instance, Li and Li (2016) construct a version of the topological B-model in this formalism and show that a quantization exists for all smooth oriented 2-manifolds after choosing a holomorphic volume form on the target complex manifold X.

6.3.4 Descent Axioms

So far we have discussed only a variant of the notion of *pre*factorization algebra. We now turn to the local-to-global axiom in this context.

Definition 6.3.6 A *Weiss cover* of a $\mathcal{G}$-manifold M is a collection of $\mathcal{G}$-embeddings $\{\phi_i : U_i \to M\}_{i \in I}$ such that for any finite set of points $x_1, \ldots, x_n \in M$, there is some i such that $\{x_1, \ldots, x_n\} \subset \phi_i(U_i)$.

With this definition in hand, we can formulate the natural generalization of our earlier definition.

Definition 6.3.7 A *factorization algebra on $\mathcal{G}$-manifolds* is a symmetric monoidal functor $\mathcal{F} : \mathrm{Emb}_{\mathcal{G}} \to \mathrm{Ch}(\mathrm{DVS})$ that is a homotopy cosheaf in the Weiss topology.

Let's return to our running example. Our arguments in Section 6.6 show that the factorization envelope of a dg Lie algebra using the de Rham complex is a factorization algebra on Emb_n, and similarly the factorization enveloped using the Dolbeault complex is a factorization algebra on Hol_n.

Remark: These definitions, indeed the spirit of this variation on factorization algebras, are inspired by conversations with and the work of John Francis and

David Ayala. See Ayala and Francis (2015) for a wonderful overview of factorization algebras in the setting of topological manifolds, along with deep theorems about them. We note that they use the term "factorization homology," rather than factorization algebra, as they view the construction as a symmetric monoidal analog, for n-manifolds, of ordinary homology. In this picture, the value on a disc defines the coefficients for a homology theory on n-manifolds. (We discuss factorization homology in the next section.) In Ayala et al. (2016), they have developed a generalization of factorization techniques where the coefficients are (∞, n)-categories. One payoff of their results is a proof of the Cobordism Hypothesis, and hence an explanation of functorial topological field theories via factorization-theoretic thinking. (See Lurie (2009c) for the canonical reference on the Cobordism Hypothesis.) $\Diamond$

Remark: The quantizations mentioned in the preceding text produce factorization algebras and not just prefactorization algebras. Hence, their values on big manifolds can be computed from their values on smaller manifolds, via the local-to-global axiom. It is interesting to compare this approach to understanding the global behavior of a quantum field theory – by a kind of open cover – with the approach in functorial field theory, in the style of Atiyah–Segal–Lurie, where a manifold is soldered together from manifolds with corners. The Ayala–Francis work shows that the factorization approach undergirds the cobordism approach, for topological field theories; the Cobordism Hypothesis is a consequence of their approach. Moving in a geometric direction, Dwyer, Stolz, and Teichner have suggested a method of using a factorization algebra on $\mathcal{G}$-manifolds to build a functorial field theory on a category of $\mathcal{G}$-cobordisms, appropriately understood. $\Diamond$

6.4 Locally Constant Factorization Algebras

There is a family of operads that plays an important role in algebraic topology, known as the "little n-dimensional discs" operads, that were introduced by Boardman and Vogt (1973). The little n-discs operad is the operad in topological spaces whose k-ary operations is the space of embeddings of a disjoint union of k n-dimensional discs into a single n-dimensional disc via radial contraction and translation. (In other words, the embedding is completely specified by saying where the center of each disc goes and what its radius is.) We will use the term E_n operad for any topological operad weakly equivalent to the little n-discs, following a standard convention. An algebra over an E_n operad is called an E_n algebra. It has, by definition, families of k-fold multiplication maps parametrized by the configurations of k n-dimensional discs inside a single disc.

If M is an n-dimensional manifold, then prefactorization algebras on M locally bear a strong resemblance to E_n algebras. After all, a prefactorization algebra prescribes a way to combine an element sitting on each of k distinct discs into an element for a big disc containing all k discs. In fact, E_n algebras form a full subcategory of factorization algebras on $\mathbb{R}^n$, as we now indicate.

Definition 6.4.1 A factorization algebra $\mathcal{F}$ on an n-manifold M is *locally constant* if for each inclusion of open discs $D \subset D'$, then the map $\mathcal{F}(D) \to \mathcal{F}(D')$ is a quasi-isomorphism.

A central example is the locally constant factorization algebra A^{fact} on $\mathbb{R}$ given by an associative algebra A. (Recall the example in Section 3.1.1.)

Lurie has shown the following vast extension of this example. (See section 5.4.5 of Lurie (2016), particularly Theorem 5.4.5.9.)

Theorem 6.4.2 *There is an equivalence of $(\infty, 1)$-categories between E_n algebras and locally constant factorization algebras on $\mathbb{R}^n$.*

We remark that Lurie (and others) uses a different gluing axiom than we do. A careful comparison of the different axioms and a proof of their equivalence (for locally constant factorization algebras) can be found in Matsuoka (2014). We emphasize that higher categories are necessary for a rigorous development of these constructions and comparisons, although we will not emphasize that point in our discussion that follows, which is expository in nature.

Remark: This theorem, together with our work in these two books, builds an explicit bridge between topology and physics. Topological field theories in the physicist's sense, such as perturbative Chern–Simons theory, produce locally constant factorization algebras, which can then be viewed as E_n algebras and analyzed using the powerful machinery of modern homotopy theory. Conversely, intuition from physics about the behavior of topological field theories suggests novel constructions and examples to topology. Much of the amazing resonance between operads, homological algebra, and string-theoretic physics over the last few decades can be understood from this perspective. ◊

6.4.1 Motivation from Topology

Many examples of E_n algebras arise naturally from topology, such as labelled configuration spaces, as discussed in the work of Segal (1973), McDuff (1975), Bödigheimer (1987), Salvatore (2001), and Lurie (2016). We will discuss an important example, that of mapping spaces.

Recall that (in the appropriate category of spaces) there is an isomorphism $\text{Maps}(U \sqcup V, X) \cong \text{Maps}(U, X) \times \text{Maps}(V, X)$. This fact suggests that we might fix a target space X and define a prefactorization algebra by sending an open set U to $\text{Maps}(U, X)$. This construction almost works, but it is not clear how to "extend" a map $f : U \to X$ from U to a larger open set $V \supset U$. By working with "compactly supported" maps, we solve this issue.

Fix (X, p) a pointed space. Let F denote the prefactorization algebra on M sending an open set U to the space of compactly supported maps from U to (X, p). (Here, "f is compactly supported" means that the closure of $f^{-1}(X - p)$ is compact.) Then F is a prefactorization algebra in the category of pointed spaces. (Composing with the singular chains functor gives a prefactorization algebra in chain complexes of Abelian groups, but we will work at the level of spaces.)

Let's consider for a moment the case when the open set is an open disc $D \subset \mathbb{R}^n = M$. There is a weak homotopy equivalence

$$F(D) \simeq \Omega_p^n X$$

between the space of compactly supported maps $D \to X$ and the n-fold based loop space of X, based at the point p. (We are using the topologist's notation $\Omega_p^n X$ for the n-fold loop space; hopefully, this use does not confuse due to the standard notation for the space of n-forms.)

To see this equivalence, note that a compactly supported map $f : D \to X$ extends uniquely to a map from the closed $\overline{D}$, sending the boundary $\partial \overline{D}$ to the base point p of X. Since $\Omega_p^n X$ is defined to be the space of maps of pairs $(\overline{D}, \partial \overline{D}) \to (X, p)$, we have constructed the desired map from $F(D)$ to $\Omega_p^n X$. It is easily verified that this map is a homotopy equivalence.

Note that this prefactorization algebra is locally constant: if $D \hookrightarrow D'$ is an inclusion of open discs, then the map $F(D) \to F(D')$ is a weak homotopy equivalence.

Note also that there is a natural isomorphism

$$F(U_1 \sqcup U_2) = F(U_1) \times F(U_2)$$

if U_1, U_2 are disjoint opens. If D_1, D_2 are disjoint discs contained in a disc D_3, the prefactorization structure gives us a map

$$F(D_1) \times F(D_2) \to F(D_3).$$

These maps correspond to the standard E_n structure on the n-fold loop space.

For a particularly nice example, consider the case where the source manifold is $M = \mathbb{R}$. The structure maps of F then describe the standard product on the space $\Omega_p X$ of based loops in X. At the level of components, we recover the standard product on $\pi_0 \Omega_p X = \pi_1(X, p)$.

This prefactorization algebra F does not always satisfy the gluing axiom. However, Salvatore (2001) and Lurie (2016) have shown that if X is sufficiently connected, this prefactorization algebra is in fact a factorization algebra.

6.4.2 Relationship with Hochschild Homology

The direct relationship between E_n algebras and locally constant factorization algebras on $\mathbb{R}^n$ raises the question of what the local-to-global axiom means from an algebraic point of view. Evaluating a locally constant factorization algebra on a manifold M is known as *factorization homology* or *topological chiral homology*.

The following result, for $n = 1$, is striking and helpful.

Theorem 6.4.3 *For A an E_1 algebra (e.g., an associative algebra), there is a weak equivalence*

$$A^{fact}(S^1) \simeq HH_*(A),$$

where A^{fact} denotes the locally constant factorization algebra on $\mathbb{R}$ associated to A and $HH_*(A)$ denotes the Hochschild homology of A.

Here $HH_*(A) \simeq A \otimes^{\mathbb{L}}_{A \otimes A} A$ means any cochain complex quasi-isomorphic to the usual bar complex (i.e., we are interested in more than the mere cohomology groups). In particular, this theorem says that the cohomology of the global sections on S^1 of A^{fact} equals the Hochschild homology of the E_1 algebra A.

This result is one of the primary motivations for the higher dimensional generalizations of factorization homology. It has several proofs in the literature, depending on choice of gluing axiom and level of generality (for instance, one can work with algebra objects in more general ∞-categories). See Theorem 3.9 of Ayala and Francis (2015) or Theorem 5.5.3.11 of Lurie (2016); a bare-handed proof for dg algebras can be found in section 4.3 of Gwilliam (2012).

There is a generalization of this result even in dimension 1. Note that there are more general ways to extend A^{fact} from the real line to the circle, by allowing "monodromy." Let σ denote an automorphism of A. Pick an orientation of S^1 and fix a point p in S^1. Let $(A^{fact})^\sigma$ denote the prefactorization algebra on S^1 such that on $S^1 - \{p\}$ it agrees with A^{fact} but where the structure maps across p use the automorphism σ. For instance, if L is a small interval to the left of p, R is a small interval to the right of p, and M is an interval containing both L and

R, then the structure map is

$$
\begin{array}{ccc}
A \otimes A & \to & A \\
a \otimes b & \mapsto & a \otimes \sigma(b)
\end{array}
$$

where the leftmost copy of A corresponds to L and so on. It is natural to view the copy of A associated to an interval containing p as the $A - A$ bimodule A^σ where A acts as the left by multiplication and the right by σ-twisted multiplication.

Theorem 6.4.4 *There is a weak equivalence* $(A^{fact})^\sigma(S^1) \simeq HH_*(A, A^\sigma)$.

Remark: In Section 8.1.2 in Chapter 8, we use this theorem to analyze the factorization homology of the quantum observables of the massive free scalar field on the circle.   ◊

Beyond the one-dimensional setting, some of the most useful insights into the meaning of factorization homology arise from its connection to the cobordism hypothesis and fully extended topological field theories. See section 4 of Lurie (2009c) for an overview of these ideas, and Scheimbauer (n.d.) for development and detailed proofs. As mentioned in the remarks at the end of Section 6.3, Ayala et al. (2016) extends factorization homology by allowing coefficients in (∞, n)-categories, which leads to richer invariants. There is closely related work by Morrison and Walker (2012) using the formalism of blob homology.

6.4.3 Factorization Envelopes

The connection between locally constant factorization algebras and E_n algebras also provides a beautiful universal property to the factorization envelope of a dg Lie algebra $\mathfrak{g}$,

$$
\mathbb{U}_n \mathfrak{g} : M \mapsto C_*(\Omega_c^*(M) \otimes \mathfrak{g}),
$$

where M runs over the category Emb_n of smooth n-manifolds and embeddings. This enveloping construction provides a functor from dg Lie algebras to E_n algebras in the category of cochain complexes over a characteristic zero field. (Here we view $\mathbb{U}_n \mathfrak{g}$ as an E_n algebra, by Lurie's theorem.) On the other hand, there is a "forgetful" functor from this category of E_n algebras in cochain complexes to the category of dg Lie algebras. For $n = 1$, this functor is essentially the familiar functor from dg associative algebras to dg Lie algebras given by remembering only the commutator bracket and not the full multiplication. For $n > 1$, such a functor exists owing to the formality of the E_n operad: as an operad in cochain complexes, $E_n \simeq H^*(E_n) = P_n$, and there is clearly a forgetful functor from P_n algebras to dg Lie algebras.

The factorization envelope provides the left adjoint for this forgetful functor: for any map of Lie algebras $\mathfrak{g} \to A$, where A is an E_n algebra viewed as simply a Lie algebra, then there is a unique map $\mathbb{U}_n \mathfrak{g} \to A$ of E_n algebras, and conversely. To actually state this kind of result, we need to use ∞-categories, of course. Knudsen (2014) has provided a proof of this universal property using factorization techniques.

Theorem 6.4.5 *Let k be a field of characteristic zero. Let $\mathcal{C}$ be a stable, k-linear, presentably symmetric monoidal ∞-category. There is an adjunction*

$$\mathbb{U}_n : \mathrm{Alg}_{\mathrm{Lie}}(\mathcal{C}) \leftrightarrows \mathrm{Alg}_{E_n}(\mathcal{C}) : F$$

such that for any $X \in \mathcal{C}$, $\mathrm{Free}_{E_n} X \simeq \mathbb{U}_n \mathrm{Free}_{\mathrm{Lie}}(\Sigma^{n-1} X)$.

Here $\mathrm{Free}_{\mathrm{Lie}} : \mathcal{C} \to \mathrm{Alg}_{\mathrm{Lie}}(\mathcal{C})$ denotes the free Lie algebra functor adjoint to the forgetful functor $\mathrm{Alg}_{\mathrm{Lie}}(\mathcal{C}) \to \mathcal{C}$, and likewise $\mathrm{Free}_{E_n} : \mathcal{C} \to \mathrm{Alg}_{\mathrm{Lie}}(\mathcal{C})$ denotes the free E_n algebra functor adjoint to the forgetful functor $\mathrm{Alg}_{E_n}(\mathcal{C}) \to \mathcal{C}$. As is conventional in topology, and hence in much work on stable ∞-categories, $\Sigma : \mathcal{C} \to \mathcal{C}$ denotes the "suspension functor" sending X to the pushout $0 \coprod_X 0$. In the setting of cochain complexes, suspension is given by a shift.

6.4.4 References

For a deeper discussion of factorization homology than we've given, we recommend Ayala and Francis (2015) to start, as it combines a clear overview with a wealth of applications. A lovely expository account is Ginot (2015). Locally constant factorization algebras already possess a substantial literature, as they sit at the nexus of manifold topology and higher algebra. See, for instance, Ayala et al. (2015, 2016), Francis (2013), Ginot et al. (2012, 2014), Horel (2015), Lurie (2016), Matsuoka (2014), and Morrison and Walker (2012). Some striking applications to representation theory, particularly quantum character varieties, can be found in Ben Zvi et al. (2016).

6.5 Factorization Algebras from Cosheaves

The goal of this section is to describe a natural class of factorization algebras. The factorization algebras that we construct from classical and quantum field theory will be closely related to the factorization algebras discussed here.

The main result of this section is that, given a nice cosheaf of vector spaces or cochain complexes F on a manifold M, the functor

$$\operatorname{Sym} F : U \mapsto \operatorname{Sym}(F(U))$$

is a factorization algebra. It is clear how this functor is a prefactorization algebra (see the example in Section 3.1.1 in Chapter 3); the hard part is verifying that it satisfies the local-to-global axiom. The examples in which we are ultimately interested arise from cosheaves F that are compactly supported sections of a vector bundle, so we focus on cosheaves of this form.

We begin by providing the definitions necessary to state the main result of this section. We then state the main result and explain its role for the rest of the book. Finally, we prove the lemmas that culminate in the proof of the main result.

6.5.1 Preliminary Definitions

Definition 6.5.1 A *local cochain complex* on M is a graded vector bundle E on M (with finite rank), whose smooth sections will be denoted by $\mathscr{E}$, equipped with a differential operator $d : \mathscr{E} \to \mathscr{E}$ of cohomological degree 1 satisfying $d^2 = 0$.

Recall the notation from Section 3.5 in Chapter 3. For E be a local cochain complex on M and U an open subset of M, we use $\mathscr{E}(U)$ to denote the cochain complex of smooth sections of E on U, and we use $\mathscr{E}_c(U)$ to denote the cochain complex of compactly supported sections of E on U. Similarly, let $\overline{\mathscr{E}}(U)$ denote the distributional sections on U and let $\overline{\mathscr{E}}_c(U)$ denote the compactly supported distributional sections of E on U. In Appendix B, it is shown that these four cochain complexes are differentiable cochain complexes in a natural way.

We use $E^! = E \otimes \operatorname{Dens}_M$ to denote the appropriate dual object. We give $E^!$ the differential that is the formal adjoint to d on E. Note that, ignoring the differential, $\overline{\mathscr{E}}_c(U)$ is the continuous dual to $\mathscr{E}^!(U)$ and that $\mathscr{E}_c(U)$ is the continuous dual to $\overline{\mathscr{E}}^!(U)$.

Lemma 6.5.2 *A local cochain complex* $(\mathscr{E}, d)$ *is a sheaf of differentiable cochain complexes. In fact, it is also a homotopy sheaf. Similarly, compactly supported sections* $(\mathscr{E}_c, d)$ *is a cosheaf of differentiable cochain complexes, as well as a homotopy cosheaf.*

Proof We describe the cosheaf case as the sheaf case is parallel.

In Section B.7.2 in Appendix B, we showed that precosheaf $\mathscr{E}_c^k$ given the degree k vector space is a cosheaf with values in DVS. Thus we know that

$\mathcal{E}_c$, without the differential, is a cosheaf of graded differentiable vector spaces. As colimits of cochain complexes (note: not homotopy hocolimits) are computed degreewise, we see that $(\mathcal{E}_c, Q)$ is a cosheaf of differentiable cochain complexes. Identical arguments apply to $\overline{\mathcal{E}}_c$.

The homotopy cosheaf condition is only marginally more difficult. Suppose we are interested in an open U and a cover $\mathfrak{U}$ for U. Recall that the Čech complex is the totalization of a double complex; we will view d as the vertical differential and the Čech differential as horizontal, so that the double complex is concentrated in the left half plane. We will work with the "augmented" double complex in which $\mathcal{E}_c(U)$ is added as the degree 1 column. The contracting cochain homotopy constructed in Section B.7.2 in Appendix B applies to the cosheaf $\mathcal{E}_c^k$ of degree k elements. Hence this augmented double complex is acyclic in the horizontal direction. The usual "staircase" or "zigzag" then ensures that the whole double complex is quasi-isomorphic to the rightmost column. □

The factorization algebras we will discuss are constructed from the symmetric algebra on the vector spaces $\mathcal{E}_c(U)$ and $\overline{\mathcal{E}}^{\,!}_c(U)$. Note that, since $\overline{\mathcal{E}}^{\,!}_c(U)$ is dual to $\mathcal{E}(U)$, we can view $\widehat{\mathrm{Sym}}\, \overline{\mathcal{E}}^{\,!}_c(U)$ as the algebra of formal power series on $\mathcal{E}(U)$. Thus, we often write

$$\widehat{\mathrm{Sym}}\, \overline{\mathcal{E}}^{\,!}_c(U) = \mathcal{O}(\mathcal{E}(U)),$$

because we view this algebra as the space of "functions on $\mathcal{E}(U)$."

Note that if $U \to V$ is an inclusion of open sets in M, then there are natural maps of commutative dg algebras

$$\mathrm{Sym}\, \mathcal{E}_c(U) \to \mathrm{Sym}\, \mathcal{E}_c(V),$$
$$\widehat{\mathrm{Sym}}\, \mathcal{E}_c(U) \to \widehat{\mathrm{Sym}}\, \mathcal{E}_c(V),$$
$$\mathrm{Sym}\, \overline{\mathcal{E}}_c(U) \to \mathrm{Sym}\, \overline{\mathcal{E}}_c(V),$$
$$\widehat{\mathrm{Sym}}\, \overline{\mathcal{E}}_c(U) \to \widehat{\mathrm{Sym}}\, \overline{\mathcal{E}}_c(V).$$

Thus, each of these symmetric algebras forms a precosheaf of commutative algebras, and thus a prefactorization algebra. We denote these prefactorization algebras by $\mathrm{Sym}\, \mathcal{E}_c$ and so on.

6.5.2 The Main Theorem

The main result of this section is the following.

Theorem 6.5.3 *We have the following parallel results for vector bundles and local cochain complexes.*

(i) *Let E be a vector bundle on M. Then*

   (a) $\mathrm{Sym}\,\mathscr{E}_c$ and $\mathrm{Sym}\,\overline{\mathscr{E}}_c$ *are strict (nonhomotopical) factorization algebras valued in the category of differentiable vector spaces, and*
   (b) $\widehat{\mathrm{Sym}}\,\mathscr{E}_c$ and $\widehat{\mathrm{Sym}}\,\overline{\mathscr{E}}_c$ *are strict (nonhomotopical) factorization algebras valued in the category of differentiable pro-vector spaces.*

(ii) *Let E be a local cochain complex on M. Then*

   (a) $\mathrm{Sym}\,\mathscr{E}_c$ and $\mathrm{Sym}\,\overline{\mathscr{E}}_c$ *are homotopy factorization algebras valued in the category of differentiable cochain complexes, and*
   (b) $\widehat{\mathrm{Sym}}\,\mathscr{E}_c$ and $\widehat{\mathrm{Sym}}\,\overline{\mathscr{E}}_c$ *are homotopy factorization algebras valued in the category of differentiable pro-cochain complexes.*

Proof Let us first prove the strict version of the result. To start with, consider the case of $\mathrm{Sym}\,\mathscr{E}_c$. We need to verify the local-to-global axiom.

Let U be an open set in M and let $\mathfrak{U} = \{U_i \mid i \in I\}$ be a Weiss cover of U. We need to prove that $\mathrm{Sym}^* \mathscr{E}_c(U)$ is the cokernel of the map

$$\bigoplus_{i,j\in I} \mathrm{Sym}(\mathscr{E}_c(U_i \cap U_j)) \to \bigoplus_{i\in I} \mathrm{Sym}(\mathscr{E}_c(U_i))$$

that sends a section on $U_i \cap U_j$ to its extension to a section of U_i minus its extension to U_j. This map preserves the decomposition of $\mathrm{Sym}\,\mathscr{E}_c(U)$ into symmetric powers: a section of $\mathrm{Sym}^k(\mathscr{E}_c(U_i \cap U_j))$ extends to a section of $\mathrm{Sym}^k(\mathscr{E}_c(U_i))$. Thus, it suffices to show that

$$\mathrm{Sym}^m \mathscr{E}_c(U) = \mathrm{coker}\left(\oplus_{i,j\in I} \mathrm{Sym}^m(\mathscr{E}_c(U_i \cap U_j)) \to \oplus_{i\in I} \mathrm{Sym}^m \mathscr{E}_c(U_i)\right),$$

for all m.

Now, observe that

$$\mathscr{E}_c(U)^{\widehat{\otimes}_\pi m} = \mathscr{E}_c^{\boxtimes m}(U^m)$$

where $\mathscr{E}_c^{\boxtimes m}$ is the cosheaf on U^m of compactly supported smooth sections of the vector bundle $E^{\boxtimes m}$, which denotes the external product of E with itself m times.

Thus it is enough to show that

$$\mathscr{E}_c^{\boxtimes m}(U^m) = \mathrm{coker}\left(\bigoplus_{i,j\in I} \mathscr{E}_c^{\boxtimes m}\left((U_i \cap U_j)^m\right) \to \bigoplus_{i\in I} \mathscr{E}_c^{\boxtimes m}\left(U_i^m\right)\right).$$

Our cover $\mathfrak{U}$ is a Weiss cover. This means that, for every finite set of points $x_1, \ldots, x_m \in U$, we can find an open U_i in the cover $\mathfrak{U}$ containing every x_j.

This implies that the subsets of U^m of the form $(U_i)^m$, where $i \in I$, cover U^m. Further,

$$(U_i)^m \cap (U_j)^m = (U_i \cap U_j)^m.$$

The desired isomorphism now follows from the fact that $\mathscr{E}_c^{\boxtimes m}$ is a cosheaf on M^m.

The same argument applies to show that $\operatorname{Sym} \mathscr{E}_c^!$ is a factorization algebra. In the completed case, essentially the same argument applies, with the subtlety (see Section C.4 in Appendix C) that, when working with pro-cochain complexes, the direct sum is completed.

For the homotopy case, the argument is similar. Let $\mathfrak{U} = \{U_i \mid i \in I\}$ be a Weiss cover of an open subset U of M. Let $\mathcal{F} = \operatorname{Sym} \mathscr{E}_c$ denote the prefactorization algebras we are considering (the argument that follows will apply when we use the completed symmetric product or use $\overline{\mathscr{E}}_c$ instead of $\mathscr{E}_c$). We need to show that the map

$$\check{C}(\mathfrak{U}, \mathcal{F}) \to \mathcal{F}(U)$$

is an equivalence, where the left-hand side is equipped with the standard Čech differential.

Let $\mathcal{F}^m(U) = \operatorname{Sym}^m \mathscr{E}_c$. Both sides above split as a direct sum over m, and the map is compatible with this splitting. (If we use the completed symmetric product, this decomposition is as a product rather than a sum.)

We thus need to show that the map

$$\bigoplus_{i_1,\dots,i_n} \operatorname{Sym}^m \mathscr{E}_c \left(U_{i_1} \cap \cdots \cap U_{i_n} \right) [n-1] \to \operatorname{Sym}^m \left(\mathscr{E}_c(U) \right)$$

is a weak equivalence.

For $i \in I$, we get an open subset $U_i^m \subset U^m$. Since $\mathfrak{U}$ is a Weiss cover of U, these open subsets form a cover of U^m. Note that

$$U_{i_1}^m \cap \cdots \cap U_{i_n}^m = \left(U_{i_1} \cap \cdots \cap U_{i_n} \right)^m.$$

Note that $\mathscr{E}_c(U)^{\otimes m}$ can be naturally identified with $\Gamma_c(U^m, E^{\boxtimes m})$ (where the tensor product is the completed projective tensor product).

Thus, to show that the Čech descent axiom holds, we need to verify that the map

$$\bigoplus_{i_1,\dots,i_n \in I} \Gamma_c \left(U_{i_1}^m \cap \cdots \cap U_{i_n}^m, E^{\boxtimes m} \right) [n-1] \to \Gamma_c(V^m, E^{\boxtimes m})$$

is a quasi-isomorphism. The left-hand side above is the Čech complex for the cosheaf of compactly supported sections of $E^{\boxtimes m}$ on V^m. Standard partition of unity arguments show that this map is a weak equivalence. (See Section A.5.4 in Appendix A for discussion and references.)   □

6.6 Factorization Algebras from Local Lie Algebras

We just showed that for a local cochain complex E, the prefactorization algebra $\widehat{\text{Sym}}\, \mathscr{E}_c$ is, in fact, a factorization algebra. What this construction says is that the "functions on $\mathscr{E}$," viewed as a space, satisfy a locality condition on the manifold M over which $\mathscr{E}$ lives. We can reconstruct functions on $\mathscr{E}(M)$ from knowing about functions on $\mathscr{E}$ with very small support on M. But $\mathscr{E}$ is a simple kind of space, as it is linear in nature. (We should remark that $\mathscr{E}$ is a simple kind of *derived* space because it is a cochain complex.) We now extend to a certain type of nonlinear situation.

In Section 3.6 in Chapter 3, we introduced the notion of a local dg Lie algebra. Since a local cochain complex is an Abelian local dg Lie algebra, we might hope that the Chevalley–Eilenberg cochain complex $C^*\mathcal{L}$ of a local Lie algebra $\mathcal{L}$ also forms a factorization algebra. As we explain in Volume 2, a local dg Lie algebra can be interpreted as a derived space that is nonlinear in nature. In this setting, the Chevalley–Eilenberg cochain complex are the "functions" on this space. Hence, if $C^*\mathcal{L}$ is a factorization algebra, we would know that functions on this nonlinear space $\mathcal{L}$ can also be reconstructed from functions localized on the manifold M.

In Section 3.6 in Chapter 3, we also constructed an important class of prefactorization algebras: the factorization envelope of a local dg Lie algebra. Again, in the preceding section, we showed $\widehat{\text{Sym}}\, \mathscr{E}_c$ is a factorization algebra, which we can view as the factorization envelope of the Abelian local Lie algebra $\mathscr{E}[-1]$. Thus, we might expect that the factorization envelope of a local Lie algebra satisfies the local-to-global axiom.

We will now demonstrate that both these Lie-theoretic constructions are factorization algebras.

Theorem 6.6.1 *Let L be a local dg Lie algebra on a manifold M. Then the prefactorization algebras*

$$\mathbb{U}\mathcal{L} : U \mapsto C_*(\mathcal{L}_c(U))$$
$$\mathbb{O}\mathcal{L} : U \mapsto C^*(\mathcal{L}(U))$$

are factorization algebras.

Remark: The argument that follows applies, with very minor changes, to a local L_∞ algebra, a modest generalization we introduce later.   ◊

Proof The proof is a spectral sequence argument, and we will reuse this idea throughout the book (notably in proving that quantum observables form a factorization algebra).

We start with the factorization envelope. Note that for any dg Lie algebra $\mathfrak{g}$, the Chevalley–Eilenberg chains $C_*\mathfrak{g}$ have a natural filtration $F^n = \mathrm{Sym}^{\leq n}(\mathfrak{g}[1])$ compatible with the differential. The first page of the associated spectral sequence is simply the cohomology of $\mathrm{Sym}(\mathfrak{g}[1])$, where $\mathfrak{g}$ is viewed as a cochain complex rather than a Lie algebra (i.e., we extend the differential on $\mathfrak{g}[1]$ as a coderivation to the cocommutative coalgebra $\mathrm{Sym}(\mathfrak{g}[1])$).

Consider the Čech complex of $\mathbb{U}\mathcal{L}$ with respect to some Weiss cover $\mathfrak{U}$ for an open U in M. Applying the filtration above to each side of the map

$$\check{C}(\mathfrak{U}, \mathbb{U}\mathcal{L}) \to \mathbb{U}\mathcal{L}(U),$$

we get a map of spectral sequences. On the first page, this map is a quasi-isomorphism by Theorem 6.5.3. Hence the original map of filtered complexes is a quasi-isomorphism.

Now we provide the analogous argument for $\mathbb{O}\mathcal{L}$. For any dg Lie algebra $\mathfrak{g}$, the Chevalley–Eilenberg cochains $C^*\mathfrak{g}$ have a natural filtration $F^n = \widehat{\mathrm{Sym}}^{\geq n}(\mathfrak{g}^\vee[-1])$ compatible with the differential. The first page of the associated spectral sequence is simply the cohomology of $\widehat{\mathrm{Sym}}(\mathfrak{g}^\vee[-1])$, where we view $\mathfrak{g}^\vee[-1]$ as a cochain complex and extend its differential as a derivation to the completed symmetric algebra.

Consider the Čech complex of $\mathbb{O}\mathcal{L}$ with respect to some Weiss cover $\mathfrak{U}$ for an open U in M. Applying the filtration above to each side of the map

$$\check{C}(\mathfrak{U}, \mathbb{O}\mathcal{L}) \to \mathbb{O}\mathcal{L}(U),$$

we get a map of spectral sequences. On the first page, this map is a quasi-isomorphism by Theorem 6.5.3. Hence the original map of filtered complexes is a quasi-isomorphism. $\qquad\square$

7

Formal Aspects of Factorization Algebras

In this chapter, we describe some natural constructions with factorization algebras. For example, we show how factorization algebras, like sheaves, push forward along maps of spaces. We also describe ways in which a factorization algebra can be reconstructed from local data.

7.1 Pushing Forward Factorization Algebras

A crucial feature of factorization algebras is that they push forward nicely. Let M and N be topological spaces admitting Weiss covers and let $f : M \to N$ be a continuous map. Given a Weiss cover $\mathfrak{U} = \{U_\alpha\}$ of an open $U \subset N$, let $f^{-1}\mathfrak{U} = \{f^{-1}U_\alpha\}$ denote the preimage cover of $f^{-1}U \subset M$. Observe that $f^{-1}\mathfrak{U}$ is Weiss: given a finite collection of points $\{x_1, \ldots, x_n\}$ in $f^{-1}U$, the image points $\{f(x_1), \ldots, f(x_n)\}$ are contained in some U_α in $\mathfrak{U}$ and hence $f^{-1}U_\alpha$ contains the x_j.

Definition 7.1.1 Given a factorization algebra $\mathcal{F}$ on a space M and a continuous map $f : M \to N$, the *pushforward factorization algebra* $f_*\mathcal{F}$ on N is defined by

$$f_*\mathcal{F}(U) = \mathcal{F}(f^{-1}(U)).$$

Note that for the map to a point $f : M \to pt$, the pushforward factorization algebra $f_*\mathcal{F}$ is simply the global sections of $\mathcal{F}$. We also call this the *factorization homology* of $\mathcal{F}$ on M.

7.2 Extension from a Basis

Let X be a topological space, and let $\mathfrak{U}$ be a basis for X, that is closed under taking finite intersections. It is well-known that there is an equivalence of categories between sheaves on X and sheaves that are defined only for open sets in the basis $\mathfrak{U}$. In this section we prove a similar statement for factorization algebras.

In this section, we require topological spaces to be Hausdorff.

7.2.1 Preliminary Definitions

We begin with a paired set of definitions.

Definition 7.2.1 A *factorizing basis* $\mathfrak{U} = \{U_i\}_{i \in I}$ for a space X is a basis for the (usual) topology of X with the following properties:

(i) For every finite set $\{x_1, \ldots, x_n\} \subset X$, there exists $i \in I$ such that $\{x_1, \ldots, x_n\} \subset U_i$.
(ii) If U_i and U_j are disjoint, then $U_i \cup U_j$ is in $\mathfrak{U}$.
(iii) $U_i \cap U_j \in \mathfrak{U}$ for every U_i and U_j in $\mathfrak{U}$.

As an example, for $X = \mathbb{R}^n$, consider the collection of all opens that are a disjoint union of finitely many convex open sets.

Note that a factorizing basis provides a Weiss cover for every open U. Given any finite set $\{x_1, \ldots, x_n\} \subset U$, fix a collection of pairwise disjoint opens $V_i \subset U$ such that $x_i \in V_i$. (Here we use that X is Hausdorff.) Then there is an element of the basis $U_i' \subset V_i$ with $x_i \in U_i'$. The disjoint union $U_1' \sqcup \cdots \sqcup U_n' \subset U$ is in $\mathfrak{U}$ and contains $\{x_1, \ldots, x_n\} \subset U$. Thus, the collection of opens in $\mathfrak{U}$ contained in U form a Weiss cover for U.

Let $\mathcal{C}$ denote a symmetric monoidal Grothendieck Abelian category.

Definition 7.2.2 Given a factorizing basis $\mathfrak{U}$, a $\mathfrak{U}$-*prefactorization algebra* consists of the following:

(i) For every $U_i \in \mathfrak{U}$, an object $\mathcal{F}(U_i) \in \mathcal{C}$.
(ii) For every finite tuple of pairwise disjoint opens $U_{i_0}, \ldots, U_{i_n}$ all contained in U_j – with all these opens from $\mathfrak{U}$ – a structure map

$$\mathcal{F}(U_{i_0}) \otimes \cdots \otimes \mathcal{F}(U_{i_n}) \to \mathcal{F}(U_j)).$$

(iii) The structure maps are equivariant and associative.

In other words, a $\mathfrak{U}$-prefactorization algebra $\mathcal{F}$ is like a prefactorization algebra, except that $\mathcal{F}(U)$ is defined only for sets U in $\mathfrak{U}$.

A $\mathfrak{U}$-*factorization algebra* is a $\mathfrak{U}$-prefactorization algebra with the property that, for every U in $\mathfrak{U}$ and every Weiss cover $\mathfrak{V}$ of U consisting of open sets in $\mathfrak{U}$, the natural map

$$\check{C}(\mathfrak{V}, \mathcal{F}) \overset{\simeq}{\Rightarrow} \mathcal{F}(U)$$

is a weak equivalence, where $\check{C}(\mathfrak{V}, \mathcal{F})$ denotes the Čech complex described earlier (Section 6.1 in Chapter 6).

Note that we have *not* required that $\mathcal{F}(U) \otimes \mathcal{F}(V) \to \mathcal{F}(U \sqcup V)$ is an equivalence, so that we are focused here on factorization algebras without the multiplicativity condition. The extension construction will rely on having a lax symmetric monoidal functor from simplicial cochain complexes to cochain complexes via an Eilenberg–Zilber-type shuffle map. That is, we want the classical formula to hold in this setting. (Those familiar with higher categories will recognize that a similar proof works with any symmetric monoidal ∞-category whose symmetric monoidal product preserves sifted colimits.)

7.2.2 The Statement

In this section we show that any $\mathfrak{U}$-factorization algebra on X extends to a factorization algebra on X. This extension is unique up to quasi-isomorphism.

For an open $V \subset X$, we use $\mathfrak{U}_V$ to denote the collection of all the opens in $\mathfrak{U}$ contained inside V.

Let $\mathcal{F}$ be a $\mathfrak{U}$-factorization algebra. Let us define a prefactorization algebra $\mathrm{ext}(\mathcal{F})$ on X by

$$\mathrm{ext}(\mathcal{F})(V) = \check{C}(\mathfrak{U}_V, \mathcal{F}),$$

for each open $V \subset X$. Conversely, if $\mathcal{G}$ is a factorization algebra on X, let $\mathrm{res}(\mathcal{G})$ denote its restriction to $\mathfrak{U}$-factorization algebra. Our main result here is then the following.

Proposition 7.2.3 *For any $\mathfrak{U}$-factorization algebra $\mathcal{F}$, its extension $\mathrm{ext}(\mathcal{F})$ is a factorization algebra on X such that $\mathrm{res}(\mathrm{ext}(\mathcal{F}))$ is naturally isomorphic to $\mathcal{F}$. Conversely, if $\mathcal{G}$ is a factorization algebra on X, then $\mathrm{ext}(\mathrm{res}(\mathcal{G}))$ is naturally quasi-isomorphic to $\mathcal{G}$.*

Thus, we will prove that there is a kind of equivalence of categories

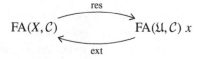

between the category $\mathrm{PreFA}(X\mathcal{C})$ of factorization algebras on X and the category $\mathrm{PreFA}(\mathfrak{U}, \mathcal{C})$ of $\mathfrak{U}$-factorization algebras. The subtlety here is that $\mathrm{ext}(\mathrm{res}(\mathcal{F}))$ will not be isomorphic to $\mathcal{F}$ but just weakly equivalent.

Remark: With these definitions in hand, it should be clear why we can recover a factorization algebra on X from just a factorization algebra on a factorizing basis $\mathfrak{U}$. The first property of $\mathfrak{U}$ ensures that the "atomic" structure maps factor through the basis. (Compare with how an ordinary cover means that the "atomic" restriction maps of a sheaf factor through the cover: every sufficiently small neighborhood of any point is contained in some open in the cover.) The second property ensures that we know how to multiply within the factorizing basis. In particular, we have the maps

$$\mathcal{F}(U_{i_0}) \otimes \cdots \otimes \mathcal{F}(U_{i_n}) \to \mathcal{F}(U_{i_0} \cup \cdots \cup U_{i_n})$$

and

$$\mathcal{F}(U_{i_0} \cup \cdots \cup U_{i_n}) \to \mathcal{F}(U_k),$$

for any opens in our basis. Associativity thus determines the multiplication map

$$\mathcal{F}(U_{i_0}) \otimes \cdots \otimes \mathcal{F}(U_{i_n}) \to \mathcal{F}(U_k).$$

In other words, any multiplication map factors through a unary structure map and a multiplication map arising from disjoint union. Finally, the third property ensures that we know $\mathcal{F}$ on the intersections that appear in the gluing condition. $\Diamond$

7.2.3 The Proof

Let $\mathfrak{U}$ be a factorizing basis and $\mathcal{F}$ a $\mathfrak{U}$-factorization algebra. We will construct a factorization algebra $\mathrm{ext}(\mathcal{F})$ on X in several stages:

- We give the value of $\mathrm{ext}(\mathcal{F})$ on every open V in X.
- We construct the structure maps and verify associativity and equivariance under permutation of labels.
- We verify that $\mathrm{ext}(\mathcal{F})$ satisfies the gluing axiom.

Finally, it will be manifest from our construction how to extend maps of $\mathfrak{U}$-factorization algebras to maps of their extensions.

We use the following notations. Given a simplicial cochain complex $A_\bullet$ (so each A_n is a cochain complex), let $C(A_\bullet)$ denote the totalization of the double complex obtained by taking the *un*normalized cochains. Let $C_N(A_\bullet)$ denote the totalization of the double complex by taking the *normalized* cochains. Finally, let

$$sh_{AB} : C_N(A_\bullet) \otimes C_N(B_\bullet) \to C_N(A_\bullet \otimes B_\bullet)$$

denote the Eilenberg–Zilber shuffle map, which is a lax symmetric monoidal functor from simplicial cochain complexes to cochain complexes. See Theorem A.2.8 in Appendix A for some discussion.

Extending Values

For an open $V \subset X$, recall $\mathfrak{U}_V$ denotes the Weiss cover of V given by all the opens in $\mathfrak{U}$ contained inside V. We then define

$$\mathrm{ext}(\mathcal{F})(V) := \check{C}(\mathfrak{U}_V, \mathcal{F}).$$

This construction provides a precosheaf on X.

Extending Structure Maps

The Čech complex for the factorization gluing axiom arises as the normalized cochain complex of a simplicial cochain complex, as should be clear from its construction. We use $\check{C}(\mathfrak{V}, \mathcal{F})_\bullet$ to denote this simplicial cochain complex, so that

$$\check{C}(\mathfrak{V}, \mathcal{F}) = C_N(\check{C}(\mathfrak{V}, \mathcal{F})_\bullet),$$

with $\mathcal{F}$ a factorization algebra and $\mathfrak{V}$ a Weiss cover.

We construct the structure maps by using the simplicial cochain complexes $\check{C}(\mathfrak{V}, \mathcal{F})_\bullet$, as many properties are manifest at that level. For instance, the unary maps $\mathrm{ext}(\mathcal{F})(V) \to \mathrm{ext}(\mathcal{F})(W)$ arising from inclusions $V \hookrightarrow W$ are easy to understand: $\mathfrak{U}_V$ is a subset of $\mathfrak{U}_W$ and thus we get a map between every piece of the simplicial cochain complex.

We now explain in detail the map

$$m_{VV'} : \mathrm{ext}(\mathcal{F})(V) \otimes \mathrm{ext}(\mathcal{F})(V') \to \mathrm{ext}(\mathcal{F})(V \cup V'),$$

where $V \cap V' = \emptyset$. Note that knowing this map, we recover every other multiplication map

by postcomposing with a unary map.

The n-simplices $\check{C}(\mathfrak{U}_V, \mathcal{F})_n$ is the direct sum of the complexes

$$\mathcal{F}(U_{i_0} \cap \cdots \cap U_{i_n}),$$

with each U_{i_k} in V. The n-simplices of the tensor product of simplicial cochain complexes $\check{C}(\mathfrak{U}_V, \mathcal{F})_\bullet \otimes \check{C}(\mathfrak{U}_{V'}, \mathcal{F})_\bullet$ are precisely the levelwise tensor product

$\check{C}(\mathfrak{U}_V, \mathcal{F})_n \otimes \check{C}(\mathfrak{U}_{V'}, \mathcal{F})_n$, which breaks down into a direct sum of terms

$$\mathcal{F}(U_{i_0} \cap \cdots \cap U_{i_n}) \otimes \mathcal{F}(U_{j_0} \cap \cdots \cap U_{j_n})$$

with the U_{i_k}'s in V and the U_{j_k}'s in V'. We need to define a map

$$m_{VV',n} : \check{C}(\mathfrak{U}_V, \mathcal{F})_n \otimes \check{C}(\mathfrak{U}_{V'}, \mathcal{F})_n \to \check{C}(\mathfrak{U}_{V \cup V'}, \mathcal{F})_n$$

for every n, and we will express it in terms of these direct summands.
Now

$$(U_{i_0} \cap \cdots \cap U_{i_n}) \cup (U_{j_0} \cap \cdots \cap U_{j_n}) = (U_{i_0} \cup U_{j_0}) \cap \cdots \cap (U_{i_n} \cup U_{j_n}).$$

Thus we have a given map

$$\mathcal{F}(U_{i_0} \cap \cdots \cap U_{i_n}) \otimes \mathcal{F}(U_{j_0} \cap \cdots \cap U_{j_n}) \to \mathcal{F}\Big((U_{i_0} \cup U_{j_0}) \cap \cdots \cap (U_{i_n} \cup U_{j_n})\Big),$$

because $\mathcal{F}$ is defined on a factorizing basis. The right-hand term is one of the direct summands for $\check{C}(\mathfrak{U}_{V \cup V'}, \mathcal{F})_n$. Summing over the direct summands on the left, we obtain the desired levelwise map.
Finally, the composition

$$C_N(\check{C}(\mathfrak{U} \cap V, \mathcal{F})_\bullet) \otimes C_N(\check{C}(\mathfrak{U}_{V'}, \mathcal{F})_\bullet)$$

$$\Big\downarrow {}^{sh}$$

$$C_N\Big(\check{C}(\mathfrak{U} \cap V, \mathcal{F})_\bullet \otimes \check{C}(\mathfrak{U}_{V'}, \mathcal{F})_\bullet\Big)$$

$$\Big\downarrow {}^{C_N(m_{VV',\bullet})}$$

$$C_N(\check{C}(\mathfrak{U}_{V \cup V'}, \mathcal{F})_\bullet)$$

gives us $m_{VV'}$.
A parallel argument works to construct the multiplication maps from n disjoint opens to a bigger open.
The desired associativity and equivariance are clear at the level of the simplicial cochain complexes $\check{C}(\mathfrak{U}_V, \mathcal{F})$, since they are inherited from $\mathcal{F}$ itself.

Verifying Gluing

We have constructed $\text{ext}(\mathcal{F})$ as a prefactorization algebra, but it remains to verify that it is a factorization algebra. Thus, our goal is the following.

Proposition 7.2.4 *The extension* $\text{ext}(\mathcal{F})$ *is a cosheaf with respect to the Weiss topology. In particular, for every open subset W and every Weiss cover $\mathfrak{W}$, the complex $\check{C}(\mathfrak{W}, \text{ext}(\mathcal{F}))$ is quasi-isomorphic to $\text{ext}(\mathcal{F})(W)$.*

As our gluing axiom is simply the axiom for cosheaf – but using a funny class of covers – the standard refinement arguments about Čech homology apply. We now spell this out.

For W an open subset of X and $\mathfrak{W} = \{W_j\}_{j \in J}$ a Weiss cover of W, there are two associated covers that we will use:

(a) $\mathfrak{U}_W = \{U_i \subset W \mid i \in I\}$ and
(b) $\mathfrak{U}_\mathfrak{W} = \{U_i \mid \exists j \text{ such that } U_i \subset W_j\}$.

The first is just the factorizing basis of W induced by $\mathfrak{U}$, but the second consists of the opens in $\mathfrak{U}$ subordinate to the cover $\mathfrak{W}$. Both are Weiss covers of W.

To prove the proposition, we break the argument into two steps and exploit the intermediary Weiss cover $\mathfrak{U}_\mathfrak{W}$.

Lemma 7.2.5 *There is a natural quasi-isomorphism*

$$f : \check{C}(\mathfrak{W}, \mathrm{ext}(\mathcal{F})) \xrightarrow{\simeq} \check{C}(\mathfrak{U}_\mathfrak{W}, \mathcal{F}),$$

for every Weiss cover $\mathfrak{W}$ of an open W.

Proof This argument boils down to combinatorics with the covers. Some extra notation will clarify what's going on. In the Čech complex for the cover $\mathfrak{W}$, for instance, we run over $n + 1$-fold intersections $W_{j_0} \cap \cdots \cap W_{j_n}$. We will denote this open by $W_{\vec{j}}$, where $\vec{j} = (j_0, \ldots, j_n) \in J^{n+1}$, with J the index set for $\mathfrak{W}$. Since we are in intersections for which all the indices are pairwise distinct, we let $\widehat{J^{n+1}}$ denote this subset of J^{n+1}.

First, we must exhibit the desired map f of cochain complexes. The source complex $\check{C}(\mathfrak{W}, \mathrm{ext}(\mathcal{F}))$ is constructed out of $\mathcal{F}$'s behavior on the opens U_i. Explicitly, we have

$$\check{C}(\mathfrak{W}, \mathrm{ext}(\mathcal{F})) = \bigoplus_{n \geq 0} \bigoplus_{\vec{j} \in \widehat{J^{n+1}}} \check{C}(\mathfrak{U}_{W_{\vec{j}}}, \mathcal{F})[n].$$

Note that each term $\mathcal{F}(U_{\vec{i}})$ appearing in this source complex appears *only once* in the target complex $\check{C}(\mathfrak{U}_\mathfrak{W}, \mathcal{F})$. Let f send each such term $\mathcal{F}(U_{\vec{i}})$ to its unique image in the target complex via the identity. This map f is a cochain map: it clearly respects the internal differential of each term $\mathcal{F}(U_{\vec{i}})$, and it is compatible with the Čech differential by construction.

Second, we need to show f is a quasi-isomorphism. We will show this by imposing a filtration on the map and showing the induced spectral sequence is a quasi-isomorphism on the first page.

We filter the target complex $\check{C}(\mathfrak{U}_\mathfrak{W}, \mathcal{F})$ by

$$F^n \check{C}(\mathfrak{U}_\mathfrak{W}, \mathcal{F}) := \bigoplus_{k \leq n} \bigoplus_{\vec{i} \in \widehat{I^{k+1}}} \mathcal{F}(U_{\vec{i}})[k].$$

Equip the source complex $\check{C}(\mathfrak{W}, \text{ext}(\mathcal{F}))$ with a filtration by pulling this filtration back along f. In particular, for any $U_{\vec{i}} = U_{i_0} \cap \cdots \cap U_{i_n}$, the preimage under f consists of a direct sum over all tuples $W_{\vec{j}} = W_{j_0} \cap \cdots \cap W_{j_m}$ such that every U_{i_k} is a subset of $W_{\vec{j}}$. Here, m can be any nonnegative integer (in particular, it can be bigger than n).

Consider the associated graded complexes with respect to these filtrations. The source complex has

$$\text{Gr}\,\check{C}(\mathfrak{U}_{\mathfrak{W}}, \mathcal{F}) = \bigoplus_{n \geq 0} \bigoplus_{\vec{i} \in \widehat{I^{n+1}}} \mathcal{F}(U_{\vec{i}})[n] \otimes \left(\bigoplus_{\vec{j} \in \widehat{J^{m+1}} \text{ such that } U_{i_k} \subset W_{\vec{j}} \, \forall k} \mathbb{C}[m] \right).$$

The rightmost term (after the tensor product) corresponds to the chain complex for a simplex – here, the simplex is infinite-dimensional – and hence is contractible. In consequence, the map of spectral sequences is a quasi-isomorphism. $\square$

We now wish to relate the Čech complex on the intermediary $\mathfrak{U}_{\mathfrak{W}}$ to that on $\mathfrak{U}_W$.

Lemma 7.2.6 *The complexes $\check{C}(\mathfrak{U}_W, \mathcal{F})$ and $\check{C}(\mathfrak{U}_{\mathfrak{W}}, \mathcal{F})$ are quasi-isomorphic.*

Proof We will produce a proof

$$\check{C}(\mathfrak{U}_W, \mathcal{F}) \overset{\simeq}{\leftarrow} \check{C}(\mathfrak{U}_W, \text{ext}_{\mathfrak{W}}(\mathcal{F})) \overset{\simeq}{\to} \check{C}(\mathfrak{U}_{\mathfrak{W}}, \mathcal{F}).$$

Recall that $\mathfrak{U}_W$ is a factorizing basis for W. Then $\mathcal{F}$, restricted to W, is a $\mathfrak{U}_W$-factorizing basis. Hence, for any $V \in \mathfrak{U}_W$ and any Weiss cover $\mathfrak{V} \subset \mathfrak{U}_W$ of V, we have

$$\check{C}(\mathfrak{V}, \mathcal{F}) \overset{\simeq}{\to} \mathcal{F}(V).$$

For any $V \in \mathfrak{U}_W$, let $\mathfrak{U}_{\mathfrak{W}}|_V = \{U_i \in \mathfrak{U} \mid U_i \subset V \cap W_j \text{ for some } j \in J\}$. Note that this is a Weiss cover for V. We define

$$\text{ext}_{\mathfrak{W}}(\mathcal{F})(V) := \check{C}(\mathfrak{U}_{\mathfrak{W}}|_V, \mathcal{F}).$$

By construction, the natural map

$$\text{ext}_{\mathfrak{W}}(\mathcal{F})(V) \to \mathcal{F}(V) \tag{7.2.3.1}$$

is a quasi-isomorphism.

Thus we have a quasi-isomorphism

$$\check{C}(\mathfrak{U}_W, \text{ext}_{\mathfrak{W}}(\mathcal{F})) \overset{\simeq}{\to} \check{C}(\mathfrak{U}_W, \mathcal{F}),$$

by using the natural map from line (7.2.3.1) on each open in the Čech complex for $\mathfrak{U}_W$.

The map

$$\check{C}(\mathfrak{U}_W, \text{ext}_{\mathfrak{W}}(\mathcal{F})) \xrightarrow{\simeq} \check{C}(\mathfrak{U}_{\mathfrak{W}}, \mathcal{F})$$

arises by mimicking the construction in the preceding lemma. $\qquad\square$

7.3 Pulling Back Along an Open Immersion

Factorization algebras do not pull back along an arbitrary continuous map, at least not in a simple way. On the other hand, when U is an open subset of a space X, a factorization algebra $\mathcal{F}$ on X clearly restricts to a factorization algebra on U. We would like to generalize this observation to open immersions $f : Y \to X$ and to have a pullback operation producing a factorization algebra $f^*\mathcal{F}$ on Y from a factorization algebra on X. In particular, we would like to have pullbacks to covering spaces.

For simplicity (and because it encompasses all the examples we care about), let Y and X be smooth manifolds of the same definition and let $f : Y \to X$ be a local diffeomorphism, i.e., for any $y \in Y$ there is some open neighborhood U_y such that $f|_{U_y}$ is a diffeomorphism from U_y to $f(U_y)$.

We also restrict to multiplicative factorization algebras, a restriction that will be visible in the construction of the pullback construction f^*.

Equip X with a factorizing basis as follows. Fix a Riemannian metric on X and let $\mathfrak{C} = \{C_i\}_{i \in I}$ be the collection of all strongly convex open sets in X. (A set $U \subset M$ is strongly convex if for any two points in U, the unique minimizing geodesic in M between the points is wholly contained in U and no other geodesic connecting the points in is U.) As every sufficiently small ball is strongly convex, $\mathfrak{C}$ forms a basis for the ordinary topology of X. Note that the intersection of two strongly convex opens is another strongly convex open. Let $\overline{\mathfrak{C}}$ denote the Weiss cover generated by $\mathfrak{C}$, which is thus a factorizing basis for X.

There is a closely associated basis on Y. Let $\mathfrak{C}'$ denote the collection of all connected opens U in Y such that $f(U)$ is in $\mathfrak{C}$. Let $\overline{\mathfrak{C}'}$ be the factorizing basis of Y generated by $\mathfrak{C}'$.

Given $\mathcal{F}$ a multiplicative factorization algebra on X, let $\mathcal{F}_{\mathfrak{C}}$ denote the associated multiplicative factorization algebra on the basis $\overline{\mathfrak{C}}$. We now define a multiplicative factorization algebra $f^*\mathcal{F}_{\mathfrak{C}'}$ on the basis $\overline{\mathfrak{C}'}$ as follows. To each $U \in \mathfrak{C}'$, let $f^*\mathcal{F}_{\mathfrak{C}'}(U) = \mathcal{F}(f(U))$. To a finite collection of disjoint opens

$V_1, \ldots, V_n$ in $\mathfrak{C}'$, we assign

$$f^* \mathcal{F}_{\mathfrak{C}'}(V_1 \sqcup \cdots \sqcup V_n) = f^* \mathcal{F}_{\mathfrak{C}'}(V_1) \otimes \cdots \otimes f^* \mathcal{F}_{\mathfrak{C}'}(V_n),$$

the tensor product of the values on those components, by multiplicativity. For U in $\mathfrak{C}'$ (and thus convex) and V in $\overline{\mathfrak{C}'}$ such that $V \subset U$, equip $f^* \mathcal{F}_{\mathfrak{C}'}$ with the obvious structure map

$$f^* \mathcal{F}_{\mathfrak{C}'}(V) = \mathcal{F}(V) \to \mathcal{F}(U) = f^* \mathcal{F}_{\mathfrak{C}'}(U).$$

By multiplicativity, this determines the structure map for any inclusion in the factorizing basis $\overline{\mathfrak{C}'}$

$$\sqcup_j V_j \subset \sqcup_i U_i,$$

where each $U_i \in \mathfrak{C}'$. This inclusion is the disjoint union over i of the inclusions

$$U_i \cap (\sqcup_j V_j) \subset U_i,$$

and we have already specified the structure map for each i. Let $f^* \mathcal{F}$ denote the factorization algebra on Y obtained by extending $f^* \mathcal{F}_{\mathfrak{C}'}$ from the basis $\overline{\mathfrak{C}'}$.

Note that although we used a choice of Riemannian metric in making the construction, the factorization algebra $f^* \mathcal{F}$ is unique up to isomorphism. It is enough to exhibit a map locally and the pullbacks are locally isomorphic on Y.

7.4 Descent Along a Torsor

Let G be a discrete group acting on a space X. Recall the definition of discretely G-equivariant prefactorization algebra (see Definition 3.7.1 in Chapter 3).

Proposition 7.4.1 *Let G be a discrete group acting properly discontinuously on X, so that $X \to X/G$ is a principal G-bundle. Then there is an equivalence of categories between G-equivariant multiplicative factorization algebras on X and multiplicative factorization algebras on X/G.*

Proof If $\mathcal{F}$ is a factorization algebra on X/G, then $f^* \mathcal{F}$ is a G-equivariant factorization algebra on X.

Conversely, let $\mathcal{F}$ be a G-equivariant factorization algebra on $\mathcal{F}$. Let $\mathfrak{U}_{con}$ be the open cover of X/G consisting of those connected sets where the G-bundle $X \to X/G$ admits a section. Let $\mathfrak{U}$ denote the factorizing basis for X/G

generated by $\mathfrak{U}_{con}$. We will define a $\mathfrak{U}$-factorization algebra $\mathcal{F}^G$ by defining

$$\mathcal{F}^G(U) = \mathcal{F}(\sigma(U)),$$

where σ is any section of the G-bundle $\pi^{-1}(U) \to U$.

Because $\mathcal{F}$ is G-equivariant, $\mathcal{F}(\sigma(U))$ is independent of the section σ chosen. Since $\mathfrak{U}$ is a factorizing basis, $\mathcal{F}^G$ extends canonically to a factorization algebra on X/G. $\qquad\qquad\square$

8

Factorization Algebras: Examples

8.1 Some Examples of Computations

The examples at the heart of this book have an appealing aspect: for these examples, it is straightforward to compute global sections – the factorization homology – because we do not need to use the gluing axiom directly. Instead, we can use the theorems in the preceding sections to compute global sections via analysis. By analogy, consider how de Rham cohomology, which exploits partitions of unity and analysis, compares to the derived functor of global sections for the constant sheaf.

In this section, we compute the global sections of several examples we've already studied in the preceding chapters.

8.1.1 Enveloping Algebras

Let $\mathfrak{g}$ be a Lie algebra and let $\mathfrak{g}^{\mathbb{R}}$ denote the local Lie algebra $\Omega^*_{\mathbb{R}} \otimes \mathfrak{g}$ on the real line $\mathbb{R}$. Recall Proposition 3.4.1, which showed that the factorization envelope $\mathbb{U}\mathfrak{g}^{\mathbb{R}}$ recovers the universal enveloping algebra $U\mathfrak{g}$.

This factorization algebra $\mathbb{U}\mathfrak{g}^{\mathbb{R}}$ is also defined on the circle.

Proposition 8.1.1 *There is a weak equivalence*

$$\mathbb{U}\mathfrak{g}^{\mathbb{R}}(S^1) \simeq C_*(\mathfrak{g}, U\mathfrak{g}^{\mathrm{ad}}) \simeq HH_*(U\mathfrak{g}),$$

where in the middle we mean the Lie algebra homology complex for $U\mathfrak{g}$ as a $\mathfrak{g}$-module via the adjoint action

$$v \cdot a = va - av,$$

where $v \in \mathfrak{g}$ and $a \in U\mathfrak{g}$ and va denotes multiplication in $U\mathfrak{g}$.

243

The second equivalence is a standard fact about Hochschild homology (see, e.g., Loday (1998)). It arises by constructing the natural map from the Chevalley–Eilenberg complex to the Hochschild complex, which is a quasi-isomorphism because the filtration (inherited from the tensor algebra) induces a map of spectral sequences that is an isomorphism on the E_1 page.

Note one interesting consequence of this theorem: the structure map for an interval mapping into the whole circle corresponds to the trace map $U\mathfrak{g} \to U\mathfrak{g}/[U\mathfrak{g}, U\mathfrak{g}]$.

Proof The wonderful fact here is that we do not need to pick a Weiss cover and work with the Čech complex. Instead, we simply need to examine the cochain complex $C_*(\Omega^*(S^1) \otimes \mathfrak{g})$.

One approach is to use the natural spectral sequence arising from the filtration $F^k = \mathrm{Sym}^{\leq k}$. The E_1 page is $C_*(\mathfrak{g}, U\mathfrak{g})$, and one must verify that there are no further pages.

Alternatively, recall that the circle is formal, so $\Omega^*(S^1)$ is quasi-isomorphic to its cohomology $H^*(S^1)$ as a dg algebra. Thus, we get a homotopy equivalence of dg Lie algebras

$$\Omega^*(S^1) \otimes \mathfrak{g} \simeq \mathfrak{g} \oplus \mathfrak{g}[-1],$$

where, on the right, $\mathfrak{g}$ acts by the (shifted) adjoint action on $\mathfrak{g}[-1]$. Now,

$$C_*(\mathfrak{g} \oplus \mathfrak{g}[-1]) \cong C_*(\mathfrak{g}, \mathrm{Sym}\, \mathfrak{g}),$$

where $\mathrm{Sym}\, \mathfrak{g}$ obtains a $\mathfrak{g}$-module structure from the Chevalley–Eilenberg homology complex. Direct computation verifies this action is precisely the adjoint action of $\mathfrak{g}$ on $U\mathfrak{g}$ (we use, of course, that $U\mathfrak{g}$ and $\mathrm{Sym}\, \mathfrak{g}$ are isomorphic as vector spaces). $\qquad\Box$

8.1.2 The Free Scalar Field in Dimension 1

Recall from Section 4.3 in Chapter 4 that the Weyl algebra is recovered from the factorization algebra of quantum observables Obs^q for the free scalar field on $\mathbb{R}$. We know that the global sections of Obs^q on a circle should thus have some relationship with the Hochschild homology of the Weyl algebra, but we will see that the relationship depends on the ratio of the mass to the radius of the circle.

For simplicity, we restrict attention to the case where S^1 has a rotation-invariant metric, radius r, and total length $L = 2\pi r$. Let S_L^1 denote this circle $\mathbb{R}/\mathbb{Z}L$. We also work with $\mathbb{C}$-linear observables, rather than $\mathbb{R}$-linear.

Proposition 8.1.2 *If the mass $m = 0$, then $H^k \operatorname{Obs}^q(S_L^1)$ is $\mathbb{C}[\hbar]$ for $k = 0, -1$ and vanishes for all other k.*

If the mass m satisfies $mL = in$ for some integer n, then $H^k \operatorname{Obs}^q(S_L^1)$ is $\mathbb{C}[\hbar]$ for $k = 0, -2$ and vanishes for all other k.

For all other values of mass m, $H^0 \operatorname{Obs}^q(S_L^1) \cong \mathbb{C}[\hbar]$ and all other cohomology groups vanish.

Hochschild homology with monodromy (recall Theorem 6.4.4) provides an explanation for this result. The equations of motion for the free scalar field locally have a two-dimensional space of solutions, but on a circle the space of solutions depends on the relationship between the mass and the length of the circle. In the massless case, a constant function is always a solution, no matter the length. In the massive case, there is either a two-dimensional space of solutions (for certain imaginary masses, because our conventions) or a zero-dimensional space. Viewing the space of local solutions as a local system, we have monodromy around the circle, determined by the Hamiltonian.

When the monodromy is trivial (so $mL \in i\mathbb{Z}$), we are simply computing the Hochschild homology of the Weyl algebra. Otherwise, we have a nontrivial automorphism of the Weyl algebra.

Note that this proposition demonstrates, in a very primitive example, that the factorization homology can detect spectral properties of the Hamiltonian. The proof will exhibit how the existence of solutions to the equations of motion (in this case, a very simple ordinary differential equation) affects the factorization homology. It would be interesting to develop further this kind of relationship.

Proof We directly compute the global sections, in terms of the analysis of the local Lie algebra

$$\mathcal{L} = \left(C^\infty(S_L^1) \xrightarrow{\Delta + m^2} C^\infty(S_L^1)[-1] \right) \oplus \mathbb{C}\hbar$$

where $\hbar$ has cohomological degree 1 and the bracket is

$$[\phi^0, \phi^1] = \hbar \int_{S_L^1} \phi^0 \phi^1$$

with ϕ^k a smooth function with cohomological degree $k = 0$ or 1. We need to compute $C_*(\mathcal{L})$.

Fourier analysis allows us to make this computation easily. We know that the exponentials $e^{ikx/L}$, with $k \in \mathbb{Z}$, form a topological basis for smooth functions on S^1 and that

$$(\Delta + m^2)e^{ikx/L} = \left(\frac{k^2}{L^2} + m^2 \right) e^{ikx/L}.$$

Hence, for instance, when $m^2 L^2$ is not a square integer, it is easy to verify that the complex

$$C^\infty(S_L^1) \xrightarrow{\Delta+m^2} C^\infty(S_L^1)[-1]$$

has trivial cohomology. Thus, $H^* \mathcal{L} = \mathbb{C}\hbar$ and so $H^*(C_* \mathcal{L}) = \mathbb{C}[\hbar]$, where in the Chevalley–Eilenberg complex $\hbar$ is shifted into degree 0.

When $m = 0$, we have

$$H^* \mathcal{L} = \mathbb{R}x \oplus \mathbb{R}\xi \oplus \mathbb{R}\hbar,$$

where x represents the constant function in degree 0 and ξ represents the constant function in degree 1. The bracket on this Lie algebra is

$$[x, \xi] = \hbar L.$$

Hence $H^0(C_* \mathcal{L}) \cong \mathbb{C}[\hbar]$ and $H^{-1}(C_* \mathcal{L}) \cong \mathbb{C}[\hbar]$, and the remaining cohomology groups are zero.

When $m = in/L$ for some integer n, we see that the operator $\Delta + m^2$ has two-dimensional kernel and cokernel, both spanned by $e^{\pm inx/L}$. By a parallel argument to the case with $m = 0$, we see $H^0(C_* \mathcal{L}) \cong \mathbb{C}[\hbar] \cong H^{-2}(C_* \mathcal{L})$ and all other groups are zero. $\qquad\square$

8.1.3 The Kac–Moody Factorization Algebras

Recall from the example in Section 3.6.3 in Chapter 3 and Section 5.5 in Chapter 5 that there is a factorization algebra on a Riemann surface Σ associated to every affine Kac–Moody Lie algebra. For $\mathfrak{g}$ a simple Lie algebra with symmetric invariant pairing $\langle -, - \rangle_\mathfrak{g}$, we have a shifted central extension of the local Lie algebra $\Omega_\Sigma^{0,*} \otimes \mathfrak{g}$ with shift

$$\omega(\alpha, \beta) = \int_\Sigma \langle \alpha, \partial\beta \rangle_\mathfrak{g}.$$

Let $\mathbb{U}_\omega \mathfrak{g}^\Sigma$ denote the twisted factorization envelope for this extended local Lie algebra.

Proposition 8.1.3 *Let g denote the genus of a closed Riemann surface Σ. Then*

$$H^*(\mathbb{U}_\omega^\sigma \mathfrak{g}(\Sigma)) \cong H^*(\mathfrak{g}, U\mathfrak{g}^{\otimes g})[c]$$

where c denotes a parameter of cohomological degree 0.

Proof We need to understand the Lie algebra homology

$$C_*(\Omega^{0,*}(\Sigma) \otimes \mathfrak{g} \oplus \mathbb{C}c),$$

whose differential has the form $\overline{\partial} + d_{\text{Lie}}$. (Note that in the Lie algebra, c has cohomological degree 1.)

Consider the filtration $F^k = \text{Sym}^{\leq k}$ by polynomial degree. The purely analytic piece $\overline{\partial}$ preserves polynomial degree while the Lie part d_{Lie} lowers polynomial degree by 1. Thus the E_1 page of the associated spectral sequence has $\text{Sym}(H^*(\Sigma, \mathcal{O}) \otimes \mathfrak{g} \oplus \mathbb{C}c)$ as its underlying graded vector space. The differential now depends purely on the Lie-theoretic aspects. Moreover, the differential on further pages of the spectral sequence are zero.

In the untwisted case (where there is no extension), we find that the E_1 page is precisely $H^*(\mathfrak{g}, U\mathfrak{g}^{\otimes g})$, by computations directly analogous to those for Proposition 8.1.1.

In the twisted case, the only subtlety is to understand what happens to the extension. Note that $H^0(\Omega^{0,*}(\Sigma)$ is spanned by the constant functions. The pairing ω vanishes if a constant function is an input, so we know that the central extension does not contribute to the differential on the E_1 page.

An alternative proof is to use the fact the $\Omega^{0,*}(\Sigma)$ is homotopy equivalent to its cohomology as a dg algebra. Hence we can compute $C_*(H^*(\Sigma, \mathcal{O}) \otimes \mathfrak{g} \oplus \mathbb{C}c)$ instead. □

We can use the ideas of Sections 4.6 and 4.7 in Chapter 4 to understand the "correlation functions" of this factorization algebra $\mathbb{U}_\omega \mathfrak{g}^\Sigma$. More precisely, given a structure map

$$\mathbb{U}_\omega \mathfrak{g}^\Sigma(V_1) \otimes \cdots \otimes \mathbb{U}_\omega \mathfrak{g}^\Sigma(V_n) \to \mathbb{U}_\omega \mathfrak{g}^\Sigma(\Sigma)$$

for some collection of disjoint opens $V_1, \ldots, V_n \subset \Sigma$, we will provide a method for describing the image of this structure map. It has the flavor of a Wick's formula.

Let $\lfloor \mathcal{O} \rfloor$ denote the cohomology class of an element $\mathcal{O}$ of $\mathbb{U}_\omega \mathfrak{g}^\Sigma$. We want to encode relations between cohomology classes.

As Σ is closed, Hodge theory lets us construct an operator $\overline{\partial}^{-1}$, which vanishes on the harmonic functions and $(0,1)$-forms but provides an inverse to $\overline{\partial}$ on the complementary spaces. (We must make a choice of Riemannian metric, of course, to do this.) Now consider an element $a_1 \cdots a_k$ of cohomological degree 0 in $\mathbb{U}_\omega \mathfrak{g}^\Sigma(\Sigma)$, where each a_j lives in $\Omega^{0,1}(\Sigma) \otimes \mathfrak{g} \oplus \mathbb{C}c$.

Then we have

$$(\bar{\partial} + \mathrm{d}_{\mathrm{Lie}}) \left((\bar{\partial}^{-1} a_1) a_2 \cdots a_k \right) = \bar{\partial}(\bar{\partial}^{-1} a_1) a_2 \cdots a_k + \mathrm{d}_{\mathrm{Lie}} \left((\bar{\partial}^{-1} a_1) a_2 \cdots a_k \right)$$

$$= \bar{\partial}(\bar{\partial}^{-1} a_1) a_2 \cdots a_k$$

$$+ \sum_{j=2}^{k} [\bar{\partial}^{-1} a_1, a_j] a_2 \cdots \widehat{a_j} \cdots a_k$$

$$+ \sum_{j=2}^{k} \int_\Sigma \left\langle \bar{\partial}^{-1} a_1, \partial a_j \right\rangle_{\mathfrak{g}} c\, a_2 \cdots \widehat{a_j} \cdots a_k.$$

When a_1 is in the complementary space to the harmonic forms, we know $\bar{\partial}\bar{\partial}^{-1} a_1 = a_1$ so we have the relation

$$0 = \lfloor a_1 \cdots a_k \rfloor + \sum_{j=2}^{k} \lfloor [\bar{\partial}^{-1} a_1, a_j] a_2 \cdots \widehat{a_j} \cdots a_k \rfloor$$

$$+ \sum_{j=2}^{k} \int_\Sigma \left\langle \bar{\partial}^{-1} a_1, \partial a_j \right\rangle_{\mathfrak{g}} \lfloor c\, a_2 \cdots \widehat{a_j} \cdots a_k \rfloor,$$

which allows us to iteratively reduce the "polynomial" degree of the original term $\lfloor a_1 \cdots a_k \rfloor$ (from k to $k-1$ or less, in this case).

This relation looks somewhat complicated but it is easy to understand for k small. For instance, in $k = 1$, we see that

$$0 = \lfloor a_1 \rfloor$$

whenever a_1 is in the space orthogonal to the harmonic forms. We get

$$0 = \lfloor a_1 a_2 \rfloor + \lfloor [\bar{\partial}^{-1} a_1, a_2] \rfloor + \int_\Sigma \left\langle \bar{\partial}^{-1} a_1, \partial a_2 \right\rangle_{\mathfrak{g}} \lfloor c \rfloor$$

for $k = 2$.

8.1.4 The Free $\beta\gamma$ System

In Section 5.4 of Chapter 5 we studied the local structure of the free $\beta\gamma$ theory; in other words, we carefully examined the simplest structure maps for the factorization algebra Obs^q of quantum observables on the plane $\mathbb{C}$. It is natural to ask about the global sections on a closed Riemann surface.

There are many ways, however, to extend this theory to a Riemann surface Σ. Let $\mathcal{L}$ be a holomorphic line bundle on Σ. Then the $\mathcal{L}$-twisted free $\beta\gamma$ system has fields

$$\mathscr{E} = \Omega^{0,*}(\Sigma, \mathcal{L}) \oplus \Omega^{1,*}(\Sigma, \mathcal{L}^\vee),$$

where $\mathcal{L}^\vee$ denotes the dual line bundle. The -1-symplectic pairing on fields is to apply pointwise the evaluation pairing between $\mathcal{L}$ and $\mathcal{L}^\vee$ and then to integrate the resulting density. The differential is the $\bar{\partial}$ operator for these holomorphic line bundles.

Let $\mathrm{Obs}^q_{\mathcal{L}}$ denote the factorization algebra of quantum observables for the $\mathcal{L}$-twisted $\beta\gamma$ system. Locally on Σ, we know how to understand the structure maps: pick a trivialization of $\mathcal{L}$ and employ our work from Section 5.4 in Chapter 5.

Proposition 8.1.4 *Let $b_k = \dim H^k(\Sigma, \mathcal{L})$. Then $H^* \mathrm{Obs}^q_{\mathcal{L}}(\Sigma)$ is a rank-one free module over $\mathbb{C}[\hbar]$ and concentrated in degree $-b_0 - b_1$.*

Proof As usual, we use the spectral sequence arising from the filtration $F^k = \mathrm{Sym}^{\leq k}$ on Obs^q. The first page just depends on the cohomology with respect to $\bar{\partial}$, so the underlying graded vector space is

$$\mathrm{Sym}\left(H^*(\Sigma, \mathcal{L})[1] \oplus H^*(\Sigma, \mathcal{L}^\vee \otimes \Omega^1_{hol})[1]\right)[\hbar],$$

where Ω^1_{hol} denotes the holomorphic cotangent bundle on Σ. By Serre duality, we know this is the symmetric algebra on a $+1$-symplectic vector space, concentrated in degrees 0 and -1 and with dimension $b_0 + b_1$ in each degree. The remaining differential in the spectral sequence is the BV Laplacian for the pairing, and so the cohomology is spanned by the maximal purely odd element, which has degree $-b_0 - b_1$. □

8.2 Abelian Chern–Simons Theory and Quantum Groups

In this section, we analyze Abelian Chern–Simons theory from a point of view complementary to our analysis in Section 4.5 in Chapter 4. It is standard in mathematics to interpret Chern–Simons theory in terms of a quantum group: line operators in Chern–Simons theory are representations of the quantum group, and expected values of line operators (which provide invariants of knots in 3-manifolds) can be constructed using the representation theory of the quantum group.

In the language of physics, the connection between Chern0-Simons theory and the quantum group is explained as follows. In any topological field theory, one expects to be able to define a category whose objects are line operators and whose morphisms are interfaces between line operators. More explicitly, we have in mind fixing a line in space and every line operator in this category is supported on a line parallel to the fixed line. In a three-dimensional topological field theory (TFT), we thus have two directions orthogonal to the fixed direction, and we can imagine taking the operator product of line operators in this plane. This operator product expansion (OPE) is expected to give the category of line operators the structure of a braided monoidal category. For Chern–Simons theory, the expectation from theoretical physics is that the braided monoidal category of line operators is the category of representations of a quantum group.

Our primary goal in this section is to implement this idea in the language of factorization algebras, but developed in the particular example of Abelian Chern–Simons theory. This example demonstrates, in a substantially simpler context, many of the ideas and techniques used to construct the Yangian from a non-Abelian four-dimensional gauge theory in Costello (2013b). (The gauge theory yielding the Yangian should be viewed as a variant of non-Abelian Chern–Simons theory that lives $\mathbb{C} \times \mathbb{R}^2$. The theory is holomorphic in the $\mathbb{C}$ direction and topological along the $\mathbb{R}^2$ directions.)

The construction here uses deep theorems about E_n algebras, Koszul duality, and Tannakian reconstruction, which take us outside the central focus of this text. So we will explain the structure of the argument and provide references to the relevant literature, but we will not give a complete proof. Interested readers encouraged to consult Costello (2013b) for more.

Overview of this Section

Before we can explain the argument, we need to formulate the statement. Thus, we begin by explaining how every E_n algebra – and in particular $\mathrm{Obs}_{\mathrm{CS}}^q$, which is an E_3 algebra – has an associated ∞-category of modules with a natural E_{n-1}-monoidal structure. This terminology means that in this category of modules, there is a *space* of ways to tensor together modules, and this space is homotopy equivalent to the configuration space of points in $\mathbb{R}^{n-1}$. In our situation, $n = 3$ so the ∞-category of modules for $\mathrm{Obs}_{\mathrm{CS}}^q$ will be E_2-monoidal, which is the ∞-categorical version of being braided monoidal.

With these ideas about modules in hand, we can give a precise version of our claim and provide a strategy of proof. In particular, we explain how these modules for $\mathrm{Obs}_{\mathrm{CS}}^q$ are equivalent to the category of representations of a quantum group.

We finish this section by connecting the concept of module for an E_3 algebra to that of a line operator defined in a more physical sense. This will justify our interpretation of the E_2-monoidal category of left modules of $\mathrm{Obs}_{\mathrm{CS}}^q$ as being the category of line operators for Abelian Chern–Simons theory.

Remark: The kinds of assertions and constructions we want to make here rely on methods and terminology from higher category theory and homotopical algebra. Formalizing our ideas here would be a nontrivial endeavor, and orthogonal to our primary goals in this book. Hence we will use the language of higher categories here quite loosely, although our expressions can be made precise. In particular, in this section only, we will sometimes simply use "category" in place of ∞-category, except where the distinction becomes important. We will also use terminology like "functor" as if working with ordinary categories. $\Diamond$

8.2.1 Modules for E_n Algebras

Even in classical algebra, there are different notions of module for an associative algebra: left modules, right modules, and bimodules. If we view associative algebras as one-dimensional in nature, it makes us suspect that for an n-dimensional algebra – such as an E_n algebra – there might be a proliferation of distinct notions reasonably called "modules," and that suspicion is true. (See Ayala et al. (2015) for examples.) Nevertheless, in the context here, we will only need the notion of left (equivalently, right) modules.

The reason is that every E_n algebra A has an "underlying" E_1 algebra A, which is unique up to equivalence (we explain this "forgetful functor" momentarily). Thus we can associate to A its category of left modules, which we will denote LMod_A to emphasize that we mean *left* modules (and not bimodules). As we will explain, this category LMod_A is naturally equipped with an E_{n-1}-monoidal structure.

More generally, every E_n algebra has an underlying E_k algebra for all $0 \leq k \leq n$. This fact can be seen in two distinct ways: via the operadic definition or via the factorization algebra picture. They are both geometric in nature.

We start with the operadic approach. Fix a linear inclusion of $\iota : \mathbb{R}^k \hookrightarrow \mathbb{R}^n$. We will use $\iota : (x_1, \ldots, x_k) \mapsto (x_1, \ldots, x_k, 0, \ldots, 0)$. Then if $D_r^k(p)$ denotes the k-disc of radius r in $\mathbb{R}^k$ centered at p, then $D_r^n(\iota(p)) \cap \iota(\mathbb{R}^k) = \iota(D_r^k(p))$, where $D_r^n(\iota(p))$ denotes the n-disc of radius r in $\mathbb{R}^n$ centered at p. In this way, ι induces a map $\iota(m) : E_k(m) \to E_n(m)$ from the configuration space of m disjoint little k-discs to the configuration space of m disjoint little n-discs. Altogether, we get a map $\iota : E_k \to E_n$ of topological operads. Algebras over operads pull back along maps of operads, so for $A \in \mathrm{Alg}_{E_n}$, we obtain $\iota^* A \in \mathrm{Alg}_{E_k}$.

One can rephrase this from the point of view of factorization algebras. We will need a construction of factorization algebras that we have not used before. If M is a manifold, $\mathcal{F}$ a factorization algebra on M, and $X \subset M$ a submanifold (possibly with boundary), we can define a factorization algebra $i_X^* \mathcal{F}$ by setting

$$i_X^* \mathcal{F}(U) = \lim_{V \supset U} \mathcal{F}(V)$$

where the limit is taken over those opens V in X which contain U. The factorization product on $\mathcal{F}$ defines one on $i_X^* \mathcal{F}$.

If $\mathcal{F}$ is locally constant, this construction is particularly nice: in that case, the limit is eventually constant so that $i_X^* \mathcal{F}(U)$ is $\mathcal{F}(V)$ for some sufficiently small open V containing U.

If $\mathbb{R} \subset \mathbb{R}^n$ is a line, we can restrict a locally constant factorization algebra on $\mathbb{R}^n$ to one on $\mathbb{R}$ in this way, to produce an associative algebra. This restriction procedure is how one describes the E_1 algebra underlying an E_k algebra in the language of factorization algebras.

Modules via Geometry

We now want to examine the type of monoidal structure on LMod_A arising from an E_n algebra structure on A. Again, one can use results about operads and categories or one can use factorization algebras. Here we begin with the factorization approach.

Let

$$\mathbb{H}^n = \{(x_1, \ldots, x_n) \in \mathbb{R}^n \mid x_n \geq 0\}$$

denote the "upper half space." Let $\partial \mathbb{H}^n$ denote the wall $\{x_n = 0\}$. By "disc in $\mathbb{H}$" we will mean the intersection $D \cap \mathbb{H}^n$ for any open disc $D \subset \mathbb{R}^n$. Since we are always working in $\mathbb{H}^n$ for the moment, we will use an open U to denote its intersection with $\mathbb{H}^n$.

Let $\mathcal{A}$ be a locally constant factorization algebra on $\mathbb{R}^n$ associated to the E_n algebra A. Given $M \in \mathrm{LMod}_A$ and a point $p \in \partial \mathbb{H}^n$ living on the wall $\{x_n = 0\}$, we construct a factorization algebra $\mathcal{M}_p$ on $\mathbb{H}^n$ by extending from the following factorization algebra on a basis. For an open U not containing p, we set $\mathcal{M}_p(U) = \mathcal{A}(U)$, and for any disc D in $\mathbb{H}^n$ containing p, we set $\mathcal{M}_p(D) = M$. For any disjoint union $U \sqcup D$ with $p \in D$, we set $\mathcal{M}_p(U \sqcup D) = \mathcal{A}(U) \otimes M$. The structure maps are easy to specify. If $U \sqcup D \subset D'$, then $U \subset D' \setminus \overline{D}$, and on this hemispherical shell, $\mathcal{A}(D' \setminus \overline{D}) \simeq A$. Since we know how A acts on M, we have the composition

$$\mathcal{M}_p(U) \otimes \mathcal{M}_p(D) = \mathcal{A}(U) \otimes M \to \mathcal{A}(D' \setminus \overline{D}) \otimes M \to A \otimes M \to M.$$

Now extend $\mathcal{M}_p$ from this basis to all opens.

To help visualize this factorization algebra, here is a picture in dimension 2.

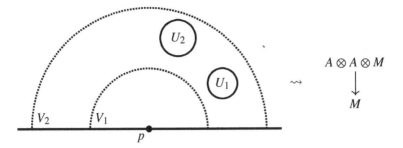

To each of the opens U_1 and U_2, $\mathcal{M}_p$ assigns a copy of A. To the half-discs V_1 and V_2, $\mathcal{M}_p$ assigns the left A-module M.

Remark: It can help to use the following pushforward to understand the structure maps. The distance-from-p function $d_p : x \mapsto d(p, x)$ sends $\mathbb{H}^n$ to $\mathbb{R}_{\geq 0}$. Pushing forward $\mathcal{M}_p$ along d_p produces the factorization algebra encoding A on $\mathbb{R}_{>0}$ with M supported at 0, as discussed in Section 3.3 in Chapter 3. This pushforward makes clear how the action on M of $\mathcal{A}$, evaluated on complicated opens in $\mathbb{H}^n$, factors through the action of A. This pushforward encodes precisely the action of A on M with which we began. $\Diamond$

The same type of procedure allows us to assign a factorization algebra behaving like $\mathcal{A}$ in the bulk but with modules "inserted" at points on the boundary. Let $p_1, \ldots, p_k$ be distinct points on $\mathbb{H}^n$. Let $M_1, \ldots, M_k$ be objects of LMod$_A$. Abusively, let $\vec{p}$ denote this collection $\{(M_i, p_i)\}$. We construct a factorization algebra $\mathcal{M}_{\vec{p}}$ as follows. For any open U disjoint from the points $\{p_i\}$, set $\mathcal{M}_{\vec{p}}(U) = \mathcal{A}(U)$. For any disc D containing exactly one distinguished point p_i, set $\mathcal{M}_{\vec{p}}(D) = M_i$. For any disc D' containing $U \sqcup D$, with D and D' containing p_i but not other distinguished point p_j, we provide the structure map as above. In other words, in a sufficiently small neighborhood of p_i, $\mathcal{M}_{\vec{p}}$ becomes $\mathcal{M}_{p_i}$ as described earlier. Extend to all opens to produce $\mathcal{M}_{\vec{p}}$.

In the picture that follows, we use M_p to denote the left A-module associated with the point p and M_q to denote the left A-module associated with the point q.

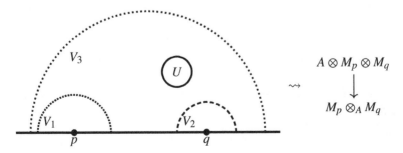

To the half-disc V_1, $\mathcal{M}_{p,q}$ assigns M_p; to the half-disc V_2, $\mathcal{M}_{p,q}$ assigns M_q; and to the open U, $\mathcal{M}_{pq}$ assigns A. The value on the big half-disc V_3 must receive maps from all these values separately and all intermediary opens (e.g., through a half-disc containing V_2 and U but not V_1), so its value must be some version of "$M_p \otimes_A M_q$."

From such an $\mathcal{M}_{\vec{p}}$, we can recover a left A-module in a very simple way. Consider the projection map $\pi : \mathbb{H}^n \to \mathbb{R}_{\geq 0}$ sending $(x_1, \ldots, x_n)$ to x_n. Then $\pi_* \mathcal{M}_{\vec{p}}$ returns A on any interval not containing the boundary point 0, but it has interesting value on an interval containing 0.

In this way, we produce a left A-module for any configuration of distinct points $\{p_1, \ldots, p_k\}$ in $\partial \mathbb{H}^n \cong \mathbb{R}^{n-1}$ and any corresponding k-tuple of left A-modules $\{M_1, \ldots, M_k\}$. In other words, we have sketched a process that can lead to making LMod_A an E_{n-1} algebra in some category of categories.

Modules via Operads

Developing the picture sketched earlier into a formal theorem would be somewhat nontrivial. Fortunately, Lurie has proved the theorem we need in Lurie (2016), using the operadic approach to thinking about E_n algebras. We begin by stating the relevant result from his work and then gloss the essential argument (although we hide a huge amount of technical machinery developed to make such an argument possible).

We need some notation to state the result. If $\mathcal{C}$ is a symmetric monoidal ∞-category, then $\mathcal{C}$ is itself an E_∞ algebra in the ∞-category Cat_∞ of ∞-categories, equipped with the Cartesian symmetric monoidal structure (i.e., the product of categories). Hence, it makes sense to talk about left module categories over $\mathcal{C}$, which we denote $\mathrm{LMod}_\mathcal{C}(\mathrm{Cat}_\infty)$. Let $\mathrm{Cat}_\infty(\Delta)$ denote the subcategory of ∞-categories possessing geometric realizations and whose functors preserve geometric realizations.

Theorem 8.2.1 (Theorem 4.8.5.5 and Corollary 5.1.2.6 of Lurie (2016))
Let $\mathcal{C}$ denote a symmetric monoidal ∞-category possessing geometric realizations and whose tensor product preserves colimits separately in each variable. Then there is a fully faithful symmetric monoidal functor

$$\mathrm{LMod}_- : \mathrm{Alg}_{E_1}(\mathcal{C}) \to \mathrm{LMod}_\mathcal{C}(\mathrm{Cat}_\infty(\Delta))$$

sending an E_1 algebra A to LMod_A. This functor determines a fully faithful functor

$$\mathrm{Alg}_{E_n}(\mathcal{C}) \to \mathrm{Alg}_{E_{n-1}}(\mathrm{LMod}_\mathcal{C}(\mathrm{Cat}_\infty(\Delta))).$$

This result might seem quite technical but it is a natural generalization of more familiar constructions from ordinary algebra. One might interpret the first part as saying that

(i) There is a big category whose objects are categories with an action of the symmetric monoidal category $\mathcal{C}$.
(ii) The category of left modules over an algebra in $\mathcal{C}$ gives an object in this big category.
(iii) There is a way of "tensoring" categories of modules such that

$$\text{LMod}_A \boxtimes \text{LMod}_B \simeq \text{LMod}_{A \otimes B} \, .$$

Tensoring categories might not be familiar, but it is suggested by generalizing the construction of the tensor product of ordinary vector spaces: given a product of categories $\mathcal{D} \times \mathcal{D}'$, we might hope there is a category $\mathcal{D} \boxtimes \mathcal{D}'$ receiving a bifunctor from $\mathcal{D} \times \mathcal{D}'$ such that all other bifunctors from $\mathcal{D} \times \mathcal{D}'$ factor through it. Note that we have a natural bifunctor $\text{LMod}_A \times \text{LMod}_B \to \text{LMod}_{A \otimes B}$.

To understand the second part of the theorem, it helps to know about *Dunn additivity*: it is equivalent to give an object $A \in \mathcal{C}$ the structure of an E_n algebra as it is to give A first the structure of an E_1 algebra and then the structure of an E_{n-1} algebra in the category of E_1 algebras. In other words, there is an equivalence of categories

$$\text{Alg}_{E_{n-1}}(\text{Alg}_{E_1}(\mathcal{C})) \simeq \text{Alg}_{E_n}(\mathcal{C}).$$

(See Theorem 5.1.2.2 of Lurie (2016).) By induction, to give A an E_n algebra is to specify n compatible E_1 structures on A.

Geometrically, this assertion should seem plausible. If one has an E_n structure on A, then each way of stacking discs along a coordinate axes in $\mathbb{R}^n$ provides an E_1 multiplications. Hence, we obtain n different E_1 multiplications on A. In the other direction, knowing how to multiply along these axes, one should be able to reconstruct a full E_n structure.

The theorem then says that we get a natural functor

$$\text{Alg}_{E_n}(\mathcal{C}) \simeq \text{Alg}_{E_{n-1}}(\text{Alg}_{E_1}(\mathcal{C})) \xrightarrow{\text{LMod}} \text{Alg}_{E_{n-1}}(\text{LMod}_\mathcal{C}(\text{Cat}_\infty(\Delta))),$$

where the first equivalence is Dunn additivity and the second functor exists because LMod_- is symmetric monoidal. This composed arrow is what allows one to turn an E_n algebra into an E_{n-1} monoidal category.

8.2.2 The Main Statement and the Strategy of Proof

Let $\mathcal{A}_{\text{CS}}$ denote the E_3 algebra associated to the locally constant factorization algebra Obs^q_{CS} of quantum observables for Abelian Chern–Simons theory with the rank 1 Abelian Lie algebra. Since we will do everything over the field of complex numbers, we will identify this Lie algebra with $\mathbb{C}$. By the theory described earlier, the category $\text{LMod}_{\mathcal{A}_{CS}}$ is an E_2-monoidal category. Recall that E_2-monoidal categories are the ∞-categorical version of braided monoidal categories. There is another braided monoidal category intimately connected with Chern–Simons theory: representations of the quantum group. We want to relate these two categories.

In the case of interest, the quantum group is a quantization of the universal enveloping algebra of the Abelian Lie algebra $\mathbb{C}$. Typically, a quantum group depends on a parameter often denoted q or $\hbar$. The size of this parameter is related to the choice of an invariant pairing on the Lie algebra. In the Abelian case, multiplication by scalars is a symmetry of the Lie algebra that also scales the invariant pairing. Thus, all values of the quantum parameter are equivalent, and we find only a single quantum group.

It will be convenient to consider a completed version of the universal enveloping algebra. Thus, our Abelian quantum group will be $\widehat{U}(\mathbb{C}) = \mathbb{C}[[t]]$. The Hopf algebra structure will be that on the universal enveloping algebra. Thus, it is commutative and co-commutative and the variable t is primitive. The only interesting structure we find is the R-matrix, given by the formula

$$R = e^{t \otimes t} \in \widehat{U}(\mathbb{C}) \widehat{\otimes} \widehat{U}(\mathbb{C}).$$

This R-matrix gives us a quasi-triangular Hopf algebra (in a completed sense).

It is a standard result in the theory of quantum groups that the category of modules over a quasi-triangular Hopf algebra is a braided monoidal category. We will let $\text{Rep}(\widehat{U}(\mathbb{C}))$ denote the category of *finite-dimensional* representations of our quantum group. Such a finite-dimensional representation is the same as a vector space V with a nilpotent endomorphism.

We would like to compare $\text{Rep}(\widehat{U}(\mathbb{C}))$ with a braided monoidal category arising from our E_3-algebra $\mathcal{A}_{CS}$. The category of left modules over this E_3 algebra is an E_2-monoidal ∞-category. To construct a braided monoidal category, we need to turn this ∞-category into an ordinary category, and we need this ordinary category to be equivalent as a *monoidal* category to $\text{Rep}(\widehat{U}(\mathbb{C}))$.

Identifying the Category of Interest

To isolate the classical category we need, the following technical proposition will be useful.

Proposition 8.2.2 *Let us view $\mathcal{A}_{CS}$ as an E_2 algebra via the map of operads from E_2 to E_3. Then there is a canonical equivalence of E_2 algebras from $\mathcal{A}_{CS}$*

to the commutative algebra $\mathbb{C}[\varepsilon]$, *where the parameter* ε *is of degree* 1. *We view* $\mathbb{C}[\varepsilon]$ *as an* E_2 *algebra via the map from* E_2 *to* E_∞.

Since the proof is both technical and unilluminating, we will defer it to the end of this section.

With this proposition in hand, we can understand the category of modules over $\mathcal{A}_{CS}$ much better. Since $\mathcal{A}_{CS}$ is equivalent as an E_2 algebra to $\mathbb{C}[\varepsilon]$, the monoidal category LMod$_{\mathcal{A}_{CS}}$ is equivalent to the monoidal category LMod$_{\mathbb{C}[\varepsilon]}$. We will isolate in this monoidal ∞-category, a subcategory whose homotopy category is equivalent to Rep$(\widehat{U}(\mathbb{C}))$.

Let us start by indicating why there is any reason to expect a relationship between such categories. A first observation is that Koszul duality of associative algebras interchanges exterior and symmetric algebras. Indeed, $\mathbb{C}[\varepsilon]$ is an exterior algebra on one variable, and $\mathbb{C}[[t]]$ is a (completed) symmetric algebra on one variable, and they are Koszul dual to one another. Alternatively, $\mathbb{C}[[t]]$ is the (completed) universal enveloping algebra of the abelian Lie algebra $\mathbb{C}$, $\mathbb{C}[\varepsilon]$ is the Chevalley–Eilenberg cochains of the same Lie algebra, and Koszul duality interchanges the universal enveloping algebras and the Chevalley–Eilenberg cochains of a given Lie algebra.

It is then well known that Koszul duality induces an equivalence between (appropriate) categories of modules. In the case at hand, this relationship works as follows. If V is a finite-dimensional representation of $\widehat{U}(\mathbb{C})$, in which the generator t acts by a nilpotent endomorphism $\rho(t)$ of V, we define a dg module $\Phi(V)$ for $\mathbb{C}[\varepsilon]$ by setting

$$\Phi(V) = V \otimes \mathbb{C}[\varepsilon], \quad d = \varepsilon \rho(t).$$

More formally, since V is a module for the Lie algebra $\mathbb{C}$, we can take the Chevalley–Eilenberg cochains $C^*(\mathbb{C}, V)$ of $\mathbb{C}$ with coefficients in V. This complex is a dg module for $C^*(\mathbb{C}) = \mathbb{C}[\varepsilon]$. Let LMod$^0(\mathbb{C}[\varepsilon])$ be the full subcategory of the ∞-category of $\mathbb{C}[\varepsilon]$-modules given by the essential image of the functor Φ.

One can describe this category in a different way. There is a functor

$$\text{Aug}: \quad \text{LMod}(\mathbb{C}[\varepsilon]) \quad \to \quad \text{dgVect}$$
$$M \quad \mapsto \quad M \otimes^{\mathbb{L}}_{\mathbb{C}[\varepsilon]} \mathbb{C}.$$

We call this functor Aug because it is constructed from the augmentation map $\mathbb{C}[\varepsilon] \to \mathbb{C}$ sending $\varepsilon \to 0$. Notice that Aug is a monoidal functor, since the map $\mathbb{C}[\varepsilon] \to \mathbb{C}$ is a map of E_2 algebras. The category LMod$^0(\mathbb{C}[\varepsilon])$ then consists of those modules M for which Aug(M) has finite-dimensional cohomology concentrated in degree 0. (In other words, it is the homotopy fiber product of Aug and the inclusion of finite-dimensional vector space Vect^{fd} into dgVect.) Since, as we have seen, $\mathcal{A}_{CS}$ is equivalent to $\mathbb{C}[\varepsilon]$ as an E_2 algebra, and so also as an E_1 algebra, we can view this category as being a subcategory of LMod($\mathcal{A}_{CS}$).

The Main Result

The classical category we want to compare with $\text{Rep}(\widehat{U}(\mathbb{C}))$ is the homotopy category $\text{Ho}(\text{LMod}^0(\mathcal{A}_{CS}))$. Note that, since $\text{LMod}^0(\mathcal{A}_{CS})$ is an E_2-monoidal category, its homotopy category is a braided monoidal category.

Our main theorem is the following.

Theorem 8.2.3 *Equip* $\text{Rep}(\widehat{U}(\mathbb{C}))$ *with the structure of braided monoidal category by the R-matrix* $R = e^{t \otimes t}$ *for the Abelian quantum group. The functor*

$$\Phi : \text{Rep}(\widehat{U}(\mathbb{C})) \to \text{Ho}(\text{LMod}^0(\mathcal{A}_{CS}))$$

is an equivalence of braided monoidal categories, where

The interesting part of this theorem is that it replaces the somewhat mysterious E_3 algebra $\mathcal{A}_{CS}$ with the much more explicit and concrete quasi-triangular Hopf algebra $\mathbb{C}[[t]]$ with R-matrix $e^{t \otimes t}$.

The Argument

We will break the proof up into several lemmas.

Lemma 8.2.4 *The functor*

$$\Phi : \text{Rep}(\widehat{U}(\mathbb{C})) \to \text{Ho}(\text{LMod}^0(\mathcal{A}_{CS}))$$

is an equivalence of monoidal categories.

This equivalence is as ordinary categories, not ∞-categories.

Proof Since $\mathcal{A}_{CS}$ is equivalent as an E_2 algebra to $\mathbb{C}[\varepsilon]$, the monoidal product on the category of $\mathbb{C}[\varepsilon]$-modules is given by tensoring over $\mathbb{C}[\varepsilon]$. Let us check that Φ is monoidal. If V, W are representations of $\widehat{U}(\mathbb{C}) = \mathbb{C}[[t]]$, on which t acts by $\rho_V(t), \rho_W(t)$, respectively, then t acts on $V \otimes W$ by $\rho_V(t) \otimes 1 + 1 \otimes \rho_W(t)$. There is an isomorphism

$$\Phi(V) \otimes_{\mathbb{C}[\varepsilon]} \Phi(W) = (\mathbb{C}[\varepsilon] \otimes V) \otimes_{\mathbb{C}[\varepsilon]} (\mathbb{C}[\varepsilon] \otimes W) \to \mathbb{C}[\varepsilon] \otimes (V \otimes W) = \Phi(V \otimes W)$$

that takes the differential on the left-hand side, coming from the tensor product of the differentials on $\Phi(V)$ and $\Phi(W)$, to the differential on $\Phi(V \otimes W)$. This natural isomorphism makes Φ monoidal. Next, let us check that Φ is an equivalence of categories. By the definition of $\text{LMod}^0(\mathcal{A}_{\text{CS}})$, the functor Φ is essentially surjective. It is easy to verify that Φ is full and faithful.   □

Lemma 8.2.5 *The equivalence of the previous lemma transfers the braided monoidal category structure on* $\text{Ho}(\text{LMod}^0(\mathcal{A}_{\text{CS}}))$ *to a braided monoidal category on* $\text{Rep}(\widehat{U}(\mathbb{C}))$. *This braided monoidal category structure is realized by some R-matrix*

$$R \in \widehat{U}(\mathbb{C}) \widehat{\otimes} \widehat{U}(\mathbb{C}),$$

giving $\widehat{U}(\mathbb{C})$ *the structure of quasi-triangular Hopf algebra.*

Proof Let us sketch how one can see this using the Tannakian formalism, before giving a more concrete approach to constructing the quasi-triangular Hopf algebra structure.

The category $\text{Ho}(\text{LMod}^0(\mathcal{A}_{\text{CS}}))$ is a braided monoidal category. There is a functor

$$\text{Aug} : \text{Ho}(\text{LMod}^0(\mathcal{A}_{\text{CS}})) \to \text{Vect}$$

sending a module M to $M \otimes_{\mathbb{C}[\varepsilon]} \mathbb{C}$, using the equivalence of E_2 algebras between $\mathcal{A}_{\text{CS}}$ and $\mathbb{C}[\varepsilon]$. The augmentation map $\mathbb{C}[\varepsilon] \to \mathbb{C}$ is a map of E_2 algebras, so this augmentation functor is monoidal. Under the equivalence

$$\text{Rep}(\widehat{U}(\mathbb{C})) \simeq \text{Ho}(\text{LMod}^0(\mathcal{A}_{\text{CS}}))$$

the functor Aug becomes the forgetful functor, which takes a representation of $\widehat{U}(\mathbb{C})$ and forgets the $\widehat{U}(\mathbb{C})$ action.

Let us recall how the Tannakian formalism works for braided monoidal categories. (See, e.g., Etingof et al. (2015).) Suppose we have a braided monoidal category $\mathcal{C}$ with a monoidal (but not necessarily braided monoidal) functor $F : \mathcal{C} \to \text{Vect}$. Then, by considering the automorphisms of the fibre functor F, one can construct a quasi-triangular Hopf algebra whose representations will be our original braided monoidal category $\mathcal{C}$, and where the functor F is the forgetful functor. (One needs some additional technical hypothesis on $\mathcal{C}$ and F for this construction to work.)

We are in precisely the situation where this formalism applies. The braided monoidal category is $\text{Ho}(\text{LMod}^0(\mathcal{A}_{\text{CS}}))$, and the fiber monoidal functor is the functor Aug given by the augmentation of $\mathcal{A}_{\text{CS}}$ as an E_2 algebra. Since we already know that $\text{Ho}(\text{LMod}^0(\mathcal{A}_{\text{CS}}))$ is equivalent to representations of the Hopf algebra $\widehat{U}(\mathbb{C})$, in such a way that the fiber functor becomes the forgetful functor, the Hopf algebra produced by the Tannakian story is $\widehat{U}(\mathbb{C})$. We therefore find that the braiding is described by some R-matrix on this Hopf algebra.

Since this approach is so abstract, let us describe how to see concretely the R-matrix on $\widehat{U}(\mathbb{C})$.

The equivalence with $\text{Ho}(\text{LMod}^0(\mathcal{A}_{CS}))$ gives us a braided monoidal structure on representations of $\widehat{U}(\mathbb{C})$. For any two representations V, W of $\widehat{U}(\mathbb{C})$, this braiding gives an isomorphism $V \otimes W \cong W \otimes V$. There is already such an isomorphism, just of vector spaces, since vector spaces are a symmetric monoidal category. Composing these two isomorphisms gives, for all representations V, W of $\widehat{U}(\mathbb{C})$, an isomorphism $R_{V,W} : V \otimes W \xrightarrow{\cong} V \otimes W$. It is an isomorphism of vector spaces and not necessarily of $\widehat{U}(\mathbb{C})$-modules, and it is natural in V and W.

The functor

$$\text{Rep}(\widehat{U}(\mathbb{C})) \times \text{Rep}(\widehat{U}(\mathbb{C})) \quad \to \quad \text{Vect}$$
$$(V, W) \quad \mapsto \quad V \otimes W$$

is represented by the right $\widehat{U}(\mathbb{C})^{\widehat{\otimes}2}$ module $\widehat{U}(\mathbb{C}) \widehat{\otimes} \widehat{U}(\mathbb{C})$, i.e., the functor is given by tensoring with this right module. The natural automorphism $R_{V,W}$ of this functor is thus represented by an automorphism of this right module, which must be given by left multiplication with some element

$$R \in \widehat{U}(\mathbb{C}) \widehat{\otimes} \widehat{U}(\mathbb{C}).$$

This element is the universal R-matrix.

Now, the axioms of a braided monoidal category translate directly into the properties that R must satisfy to define a quasi-triangular Hopf algebra structure on $\widehat{U}(\mathbb{C})$. □

So far, we have seen that we can identify the braided monoidal category of representations of $\widehat{U}(\mathbb{C})$ with the braided monoidal category $\text{Ho} \, \text{LMod}^0(\mathcal{A}_{CS})$, where we equip $\widehat{U}(\mathbb{C})$ with *some* R-matrix. It remains to show that there is essentially only one possible R-matrix.

Proposition 8.2.6 *Every R-matrix for $\widehat{U}(\mathbb{C})$ satisfying the axioms defining a quasi-triangular Hopf algebra is of the form*

$$R = \exp(c(t \otimes t)) \in \widehat{U}(\mathbb{C}) \widehat{\otimes} \widehat{U}(\mathbb{C})$$

where c is a complex number.

Proof Recall that an R-matrix of a Hopf algebra A is an invertible element R of $A \otimes A$ such that

$$\Delta^{\mathrm{op}}(x) = R\Delta(x)R^{-1} \text{ for all } x \in A,$$
$$\Delta \otimes \mathrm{id}_A(R) = R_{13}R_{23},$$
$$\mathrm{id}_A \otimes \Delta(R) = R_{13}R_{12},$$

where Δ denotes the coproduct on A, Δ^{op} denotes Δ postcomposed with the flip operation, and R_{13}, for example, denotes

$$\sum_i a_i \otimes 1 \otimes b_i \in A \otimes A \otimes A$$

where $R = \sum_i a_i \otimes b_i \in A \otimes A$.

For us, owing to the completion of the tensor product, an element R will live in $\mathbb{C}[[t_1]] \widehat{\otimes} \mathbb{C}[[t_2]]$. As $\mathbb{C}[[t]]$ is cocommutative, we see that the first equation will be trivially satisfied by any invertible element. The only conditions are then the remaining equations. As we are working with a deformation of the usual symmetric braiding on $\mathbb{C}[[t]]$-modules, an R-matrix is of the form

$$1 \otimes 1 + c_1 t_1 \otimes 1 + c_2 1 \otimes t_2 + \cdots.$$

To simplify the problem further, note that, because $\mathbb{C}[[t_1, t_2]]$ is a commutative algebra, we can take the logarithm of R. If we let $\overline{R}$ denote this logarithm, then the equations then become additive:

$$\Delta \otimes \mathrm{id}_A(\overline{R}) = \overline{R}_{13} + \overline{R}_{23},$$
$$\mathrm{id}_A \otimes \Delta(\overline{R}) = \overline{R}_{13} + \overline{R}_{12}.$$

Writing $\overline{R} = \sum_i x_{(i)} \otimes y_{(i)}$ with $x_{(i)} \in \mathbb{C}[[t_1]]$ and $y_{(i)} \in \mathbb{C}[[t_2]]$, we see that the first equation becomes

$$\sum_i \Delta(x_{(i)}) \otimes y_{(i)} = \sum_i (x_{(i)} \otimes 1 + 1 \otimes x_{(i)}) \otimes y_{(i)},$$

which forces each $x_{(i)}$ to be primitive. The other equation likewise forces each $y_{(i)}$ to be primitive. Hence $\overline{R} = c t_1 \otimes t_2$ for some constant c and so $R = \exp(c t_1 \otimes t_2)$. The constant c is the only freedom available in choosing an R-matrix. $\square$

Thus, we have shown the desired equivalence of braided monoidal categories, where $\widehat{U}(\mathbb{C})$ is equipped with an R-matrix of the form $\exp(c(t \otimes t))$

for some value of the constant c. This constant cannot be zero, as this would force $\mathcal{A}_{CS}$ to be a commutative E_3 algebra, and $\mathcal{A}_{CS}$ is not commutative. Any two nonzero values of the constant c are related by scaling t, and so all the braided monoidal categories are equivalent. This completes the proof of Theorem 8.2.3.

8.2.3 Line Operators

In this section, we explain a general definition of a Wilson line operator in the language of factorization algebras, and we will attempt to match our definition with ideas prevalent in the physics literature. We then show that, in the case of Abelian Chern–Simons theory, the category of Wilson line operators is the category we considered earlier.

The Basic Idea of Defects

Let us first explain how line operators (and more general defects) are understood from the physics point of view. Suppose we have a field theory on a manifold M, and $X \subset M$ is a submanifold. A *defect* for our field theory on X is a way of changing the field theory on X while leaving it the same outside X. Often, one can realize a defect by coupling a theory living on X to our original theory living on M. In the case that X is a one-dimensional submanifold, we call such a defect a line operator (or line defect).

For example, if $K : S^1 \to \mathbb{R}^3$ is a knot, we could couple a free fermion on S^1 to Abelian Chern–Simons theory on $\mathbb{R}^3$. If A is the field of Chern–Simons theory and $\psi : S^1 \to \mathbb{C}$ the fermionic field on S^1, we can couple the theories with the action

$$S_K(A, \psi) = \int_{S^1} \psi(\theta)(\mathrm{d} + K^*A)\psi(\theta) + \int_{\mathbb{R}^3} A \wedge \mathrm{d}A.$$

The quantization of this field theory, which we will not develop here, should produce a factorization algebra of observables $\mathrm{Obs}^q_{CS,K}$ that agrees with Obs^q_{CS} on opens disjoint from K but that on opens intersecting K also includes observables on ψ. A small tubular neighborhood $\widehat{K}$ of K inside $\mathbb{R}^3$ is diffeomorphic to $K \times \mathbb{R}^2$, so we can pushforward $\mathrm{Obs}^q_{CS,K}$ from this neighborhood to S^1. The underlying E_1 algebra $\mathcal{A}_K$ should look something like $\mathcal{A}_{CS}$ tensored with a Clifford algebra (the observables for the uncoupled free fermion). Thus the line defect for Abelian Chern–Simons theory produces an interesting modification $\mathrm{Obs}^q_{CS,K}$ of Obs^q_{CS} near K; we will view this factorization algebra as a model for what a line defect should be in the setting of factorization algebras.

By picking different field theories supported on K and coupling them to Chern–Simons theory in the bulk, we obtain other factorization algebras. One can do this, for instance, by varying the coupling constants in the example above or adding more interactions terms. One can similarly imagine picking a two-dimensional submanifold and coupling a field theory on it to the three-dimensional bulk.

Defects via Factorization Algebras

Let us now explain in the language of factorization algebras how to couple a field theory to a quantum mechanical system living on a line. Suppose we have a factorization algebra $\mathcal{F}$ on $\mathbb{R}^n$, which we assume here is locally constant. Let us fix a copy of $\mathbb{R} \subset \mathbb{R}^n$. Let us place a locally constant factorization algebra $\mathcal{G}$ on $\mathbb{R}$, which we think of as the observables of a quantum mechanical system. We will view $\mathcal{G}$ as a homotopy-associative algebra or E_1 algebra or A_∞ algebra, but we will tend to write formulas as if everything were strictly associative. (One can always replace an A_∞ algebra by a weakly equivalent but strictly associative dg algebra.)

Let $i_*\mathcal{G}$ denote the pushforward of $\mathcal{G}$ to a factorization algebra on $\mathbb{R}^n$. Note that $i_*\mathcal{G}$ takes value $\mathbb{C}$ on any open that does not intersect $\mathbb{R}$, and it takes value $\mathcal{G}(i^{-1}(U))$ on any open that does intersect $\mathbb{R}$.

Before we define a coupled factorization algebra, we need to understand the situation where the field theory on $\mathbb{R}^n$ and the field theory on $\mathbb{R}$ are uncoupled. In this case, the factorization algebra of observables for the system is simply $\mathcal{F} \otimes i_*\mathcal{G}$. We are interested in deforming this uncoupled situation.

Definition 8.2.7 A *coupling* of our factorization algebra $\mathcal{F}$ to the factorization algebra $\mathcal{G}$ living on $\mathbb{R}$ is an element $\alpha \in \mathcal{F}|_{\mathbb{R}} \otimes \mathcal{G}$ of cohomological degree 1 satisfying the Maurer–Cartan equation

$$d\alpha + \tfrac{1}{2}[\alpha, \alpha] = d\alpha + \alpha \star \alpha = 0,$$

where $\star$ is the product in $\mathcal{F}|_{\mathbb{R}} \otimes \mathcal{G}$ viewed as an associative algebra.

Here, by $\mathcal{F}|_{\mathbb{R}}$, we mean the factorization algebra on $\mathbb{R}$ obtained by restricting $\mathcal{F}$ (e.g., by taking a small tubular neighborhood of this copy of $\mathbb{R}$ and then pushing forward to $\mathbb{R}$). We can view $\mathcal{F}|_{\mathbb{R}}$ as a homotopy-associative algebra. The tensor product $\mathcal{F}|_{\mathbb{R}} \otimes \mathcal{G}$ is then also a homotopy-associative algebra.

Remark: Bear in mind the following:

(i) It is common to model homotopy associative algebras by A_∞ algebras. If one takes this route, then one should use the L_∞ version of the Maurer–Cartan equation.

(ii) We are always imagining a situation in which our line defect is a small deformation of a decoupled system. One can capture this requirement by including a formal parameter, viewed as a coupling constant, next to α in this definition, but we will not do so. ◊

Given such a coupling α, we can deform the factorization algebra $\mathcal{F}|_{\mathbb{R}} \otimes \mathcal{G}$ by simply adding $[\alpha, -]$ to the differential of $\mathcal{F}|_{\mathbb{R}} \otimes \mathcal{G}$. In other words, the twisted differential d^α is given by

$$d^\alpha(x) = dx + [\alpha, x] = dx + \alpha \star x + x \star \alpha.$$

We denote this factorization algebra on $\mathbb{R}$ by $\mathcal{F}|_{\mathbb{R}} \otimes_\alpha \mathcal{G}$.

We can also deform $\mathcal{F} \otimes i_* \mathcal{G}$ using this data. On $\mathbb{R}$, it is simply $\mathcal{F}|_{\mathbb{R}} \otimes_\alpha \mathcal{G}$. Away from $\mathbb{R}$, we change nothing and the factorization algebra is just given by $\mathcal{F}$. It is not completely obvious that this construction makes sense, but we will see in the following lemma that it does.

Lemma 8.2.8 *This construction defines a factorization algebra $\mathcal{F} \otimes_\alpha i_* \mathcal{G}$ on $\mathbb{R}^n$ deforming $\mathcal{F} \otimes i_* \mathcal{G}$.*

We call this factorization algebra a *line operator*, and in the text that follows we will relate it explicitly to the standard example of a Wilson observable.

Proof We will sketch a proof using technology developed in the literature on E_n algebras. For simplicity, let us suppose that $\mathbb{R}$ is a straight line in $\mathbb{R}^n$.

Let us identify $\mathbb{R}^n = \mathbb{R} \times \mathbb{R}^{n-1}$ and use coordinates (x, v) where $x \in \mathbb{R}$ and $v \in \mathbb{R}^{n-1}$. Our chosen copy of $\mathbb{R}$ is $\{v = 0\}$. Define a projection

$$\pi : \mathbb{R}^n \setminus \mathbb{R} \to \mathbb{R} \times \mathbb{R}_{>0}$$
$$(x, v) \mapsto (x, \|v\|).$$

The pushforward $\pi_* \mathcal{F}$ will be a locally constant factorization algebra on the two-dimensional manifold $\mathbb{R} \times \mathbb{R}_{>0}$, and so an E_2 algebra.

We are interested in extensions of $\pi_* \mathcal{F}$ into a factorization algebra on the half-space $\mathbb{R} \times \mathbb{R}_{\geq 0}$, which is constructible with respect to the stratification where the two strata are $\mathbb{R} \times 0$ and $\mathbb{R} \times \mathbb{R}_{>0}$. Such an extension will be an E_1 algebra on $\mathbb{R} \times 0$, together with some compatibility between this E_1 algebra and the E_2 algebra $\pi_* \mathcal{F}$. We would like the E_1 algebra on $\mathbb{R} \times 0$ to be $\mathcal{F}|_{\mathbb{R}} \otimes_\alpha \mathcal{G}$.

The work of Ayala, Francis, and Tanaka (2016) tell us that such an extension of $\pi_* \mathcal{F}$ is the same as a constructible factorization algebra on $\mathbb{R}^n$ that is constructible with respect to the stratification where the strata are $\mathbb{R}^n \setminus \mathbb{R}$ and $\mathbb{R}$, and which on $\mathbb{R}^n \setminus \mathbb{R}$ is $\mathcal{F}$ and on $\mathbb{R}$ is $\mathcal{F}|_{\mathbb{R}} \otimes_\alpha \mathcal{G}$.

One can ask, what compatibility is required to glue the E_2 algebra $\pi_* \mathcal{F}$ on $\mathbb{R} \times \mathbb{R}_{>0}$ to the E_1 algebra $\mathcal{F}|_{\mathbb{R}} \otimes_\alpha \mathcal{G}$? The answer is that we need to give a morphism of E_2 algebras from $\pi_* \mathcal{F}$ to the Hochschild cochain complex of $\mathcal{F}|_{\mathbb{R}} \otimes_\alpha \mathcal{G}$. Now, since the deformation of $\mathcal{F}|_{\mathbb{R}} \otimes \mathcal{G}$ given by $[\alpha, -]$ is an *inner* deformation, it does not change the Hochschild cochain complex (up to weak equivalence). Thus, the gluing data we need is the same as in the case when $\alpha = 0$. But when $\alpha = 0$, we already have the desired factorization algebra on $\mathbb{R}^n$, namely $\mathcal{F} \otimes i_* \mathcal{G}$. $\qquad\square$

Remark: It might seem restrictive to consider only deformations of $\mathcal{F}|_{\mathbb{R}} \otimes \mathcal{G}$ that are inner (i.e., given by an element α) rather than arbitrary deformations of the algebra structure. The last paragraph of the proof indicates one reason why we define "coupling" via inner deformations, and we provide some further motivations in the next section. $\qquad\Diamond$

Motivating this Construction

There are two natural questions remaining from our discussion of line operators. First, in the case of Chern–Simons theory, how do we match up line operators with representations of the quantum group? And second, why does our notion of "coupling," Definition 8.2.7, match up with the physical idea of coupling a quantum mechanical system to our field theory?

Let us address the first point. In the factorization algebra approach, a one-dimensional topological field theory is specified by an associative algebra, viewed as the algebra of operators (or observables) for the field theory. In other approaches to (topological) field theory Lurie (2009c), however, one is required to have a Hilbert space as well. A one-dimensional topological field theory is then specified by a finite-dimensional vector space V, which is the Hilbert space attached to a point. The algebra of operators is the matrix algebra $\mathrm{End}(V)$. It is thus natural to consider coupling a field theory to a topological quantum mechanical system of Atiyah–Segal–Lurie type, where the algebra of operators is $\mathrm{End}(V)$.

If we do this, we find that a coupling is a Maurer–Cartan element in $\mathcal{F}|_{\mathbb{R}} \otimes \mathrm{End}(V)$. But such an element also provides a deformation of $\mathcal{F}|_{\mathbb{R}} \otimes V$ as a free left $\mathcal{F}|_{\mathbb{R}}$-module into a nontrivial projective left module. In this way, we have sketched why the following lemma is true.

Lemma 8.2.9 *The following are equivalent:*

(i) *Specifying a projective rank k left module for the E_n algebra $\mathcal{F}$.*

(ii) *Coupling a locally constant factorization algebra $\mathcal{F}$ on $\mathbb{R}^n$ to the trivial quantum mechanical system on $\mathbb{R} \subset \mathbb{R}^n$ whose algebra of operators is the matrix algebra $\mathfrak{gl}_k$.*

For Chern–Simons theory, this class of line operators matches up nicely with the usual Wilson observables. As a particularly simple example, let's return to Abelian Chern–Simons theory. In that case, we know that $\mathcal{F}|_{\mathbb{R}}$ is equivalent to the algebra $\mathbb{C}[\varepsilon]$, where ε has cohomological degree one. Every degree one element in $\mathbb{C}[\varepsilon] \otimes \mathfrak{gl}_k$ is of the form εT, with $T \in \mathfrak{gl}_k$, and every εT is a Maurer–Cartan element since the differential d here is trivial and $\varepsilon^2 = 0$. A deformation of the action of $\mathbb{C}[\varepsilon]$ on $\mathbb{C}[\varepsilon] \otimes V$, of course, amounts to changing how ε acts, and we are free to change its action by left multiplication from the free action

$$m_\varepsilon : M + \varepsilon N \mapsto \varepsilon M$$

to any linear map of the form

$$\widetilde{m}_\varepsilon : M + \varepsilon N \mapsto \varepsilon A M,$$

where $A \in \mathfrak{gl}_k$. Now set $T = A - 1$.

Next, let us explain how to interpret our concept of "coupling" from the point of view of physics. To do this, we need to say a little bit about local operators (that is, operators supported at a point) from the point of view of factorization algebras. This topic is treated in more detail in Volume 2. Here, we will be brief.

In any field theory on $\mathbb{R}^n$ whose observables are given by a factorization algebra $\mathcal{F}$, there is a *co-stalk* $\mathcal{F}_x$ of $\mathcal{F}$ for each point $x \in \mathbb{R}^n$. This co-stalk is the limit $\lim_{x \in U} \mathcal{F}(U)$ of the spaces of observables on neighborhoods U of x. The co-stalk $\mathcal{F}_x$ is a way to define local observables at x for the field theory: by definition an element of $\mathcal{F}_x$ is an observable contained in $\mathcal{F}(U)$ for every neighborhood U of x. For a general field theory, this limit might produce something hard to describe (and one might also worry about the difference between the limit and the homotopy limit). For a topological field theory, however, the limit is essentially constant, so that we can identify $\mathcal{F}_x$ with $\mathcal{F}(D)$ for any disc D around x.

As x varies in $\mathbb{R}^n$, the space $\mathcal{F}_x$ forms a vector bundle, typically of infinite rank. This vector bundle has a flat connection, reflecting the fact that we can differentiate local operators. Because $\mathcal{F}_x$ may be infinite-dimensional, one cannot in general solve the parallel transport equation from the flat connection. Thus, for a general factorization algebra, $\mathcal{F}_x$ should be thought of as a D-module. For a topological field theory, however, the bundle $\mathcal{F}_x$ is a local system, meaning one can solve the parallel transport equation. The equivalence $\mathcal{F}_x \simeq \mathcal{F}_y$ given by parallel transport arises by the quasi-isomorphisms $\mathcal{F}_x \xrightarrow{\sim} \mathcal{F}(D) \xleftarrow{\sim} \mathcal{F}_y$ for any disc D containing both x and y.

Let us consider, as before, a line $\mathbb{R} \subset \mathbb{R}^n$, with a field theory on $\mathbb{R}^n$ whose observables are described by a factorization algebra $\mathcal{F}$ and with a field theory on $\mathbb{R}$ whose observables are described by a factorization algebra $\mathcal{G}$. Let us assume that both factorization algebras are locally constant. For $x \in \mathbb{R}$, we can consider the space $\mathcal{F}_x \otimes \mathcal{G}_x$, which is the tensor product of local operators of our two field theories at $x \in \mathbb{R} \subset \mathbb{R}^n$. As we have seen, the space $\mathcal{F}_x \otimes \mathcal{G}_x$ forms, as x varies, a vector bundle with flat connection on $\mathbb{R}$. Let us denote this vector bundle by $\mathcal{F}_{\mathrm{loc}} \otimes \mathcal{G}_{\mathrm{loc}}$. Since $\mathcal{F}_{\mathrm{loc}} \otimes \mathcal{G}_{\mathrm{loc}}$ is a local system, there is a quasi-isomorphism

$$\mathcal{F}|_{\mathbb{R}}(D) \otimes \mathcal{G}(D) \simeq \Omega^*(\mathbb{R}, \mathcal{F}_{\mathrm{loc}} \otimes \mathcal{G}_{\mathrm{loc}})$$

for any disc $D \subset \mathbb{R}$.

From a physics point of view, we might expect that to couple the system specified by $\mathcal{F}$ to that specified by $\mathcal{G}$, we need to give a Lagrangian density on the real line $\mathbb{R}$. In this case, such a Lagrangian would be a 1-form on $\mathbb{R}$ with values in the bundle of local observables $\mathcal{F}_{\mathrm{loc}} \otimes \mathcal{G}_{\mathrm{loc}}$, viewed as observables on the combined fields for the two field theories.

Such a Lagrangian density does appear naturally in our earlier mathematical definition. Under the isomorphism

$$\mathcal{F}|_{\mathbb{R}}(D) \otimes \mathcal{G}(D) \simeq \Omega^*(\mathbb{R}, \mathcal{F}_{\mathrm{loc}} \otimes \mathcal{G}_{\mathrm{loc}}),$$

a Maurer–Cartan element α on the left-hand side corresponds to a sum

$$S_\alpha = S_\alpha^{(0)} + S_\alpha^{(1)},$$

where $S_\alpha^{(0)}$ is a function along $\mathbb{R}$ with values in the degree 1 component of the vector bundle of local observables and $S_\alpha^{(1)}$ is a 1-form with values in the degree 0 component of the local observables. The term $S_\alpha^{(1)}$ is thus a Lagrangian density. To understand better what it means, we need to go a bit further.

Under the preceding correspondence, the Maurer–Cartan equation for α becomes the equation

$$d_{dR} S_\alpha^{(0)} + d_{\mathcal{F} \otimes \mathcal{G}} S_\alpha^{(0)} + d_{\mathcal{F} \otimes \mathcal{G}} S_\alpha^{(1)} + \left[S_\alpha^{(0)}, S_\alpha^{(1)} \right] = 0,$$

where $d_{\mathcal{F} \otimes \mathcal{G}}$ is the differential on $\mathcal{F}_{\mathrm{loc}} \otimes \mathcal{G}_{\mathrm{loc}}$. This equation is essentially a version of the master equation in the Batalin–Vilkovisky (BV) formalism, but this relationship is manifest only with factorization algebras produced by BV quantization, which is the subject of Volume 2. Solving the equation perturbatively, however, we can get the flavor of the relationship.

Consider the situation where we try to solve for S_α iteratively in some formal variable c:

$$S_\alpha = c \left(S_{\alpha,1}^{(0)} + S_{\alpha,1}^{(1)} \right) + c^2 \left(S_{\alpha,2}^{(0)} + S_{\alpha,2}^{(1)} \right) + \cdots.$$

At the first stage, we work modulo c^2. We then want to solve the equation

$$d_{dR} S_{\alpha,1}^{(0)} + d_{\mathcal{F} \otimes \mathcal{G}} S_{\alpha,1}^{(1)} = 0.$$

(Note that for grading reasons, we always need $d_{\mathcal{F} \otimes \mathcal{G}} S_\alpha^{(0)} = 0$.) And we are interested in solutions modulo the image of the operator $d_{dR} + d_{\mathcal{F} \otimes \mathcal{G}}$ on elements in $\Omega^*(\mathbb{R}, \mathcal{F}_{\mathrm{loc}} \otimes \mathcal{G}_{\mathrm{loc}})$ of cohomological degree 0; this space is the first-order version of gauge equivalence between solutions of the Maurer–Cartan equation. Unraveling these conditions, we see that we care about a Lagrangian density in $\mathcal{F}_{\mathrm{loc}} \otimes \mathcal{G}_{\mathrm{loc}}$ up to a total derivative. These data are what one expects from physics. Continuing onto higher powers of c, one finds that our algebraic definition for coupling the quantum mechanical system $\mathcal{G}$ to the field theory $\mathcal{F}$ matches with what one expects from physics.

In sum, for any topological field theory whose observables are given by a factorization algebra $\mathcal{F}$ on $\mathbb{R}^n$, this discussion indicates that the category of projective left modules for $\mathcal{F}$ captures the category of Wilson line operators for the n-dimensional field theory described by $\mathcal{F}$. These categories both admit interesting ways to "combine" or "fuse" objects: the left $\mathcal{F}$-modules have an E_{n-1}-monoidal structure, by Lurie's work, and the line defects can be fused, equipping them with an E_{n-1}-monoidal structure as well. (For a physically oriented introduction to the relationship between defects and higher categories, see Kapustin (2010).) We assert that these agree: when one examines the physical constructions, they behave as the homotopical algebra suggests. For some discussion of these issues, see Costello (2014).

The Take-away Message

Let us apply this discussion to Abelian Chern–Simons theory, where we have an E_3 algebra $\mathcal{A}_{\mathrm{CS}}$. The left modules that we find in our discussion above – projective modules of rank n – match exactly with the category $\mathrm{LMod}^0(\mathcal{A}_{\mathrm{CS}})$ we introduced earlier. If we define the E_2-monoidal category of line operators of Abelian Chern–Simons theory to be the E_2-monoidal category $\mathrm{LMod}^0(\mathcal{A}_{\mathrm{CS}})$, we find that our theorem relating Abelian Chern–Simons theory to the quantum group takes the following satisfying form.

Theorem 8.2.10 *The following are equivalent as braided monoidal categories:*

(i) *The homotopy category of the category of Wilson line operators of Abelian Chern–Simons theory.*
(ii) *The category of representations of the quantum group for the Abelian Lie algebra* $\mathbb{C}$.

In this case, the identification between braided monoidal structures can be seen explicitly. We have already shown that the factorization algebra encodes the Wilson loop observables and recovers the Gauss linking number, but it is these relations that are used in providing the braiding on the line operators. (See, e.g., the discussion in Kapustin and Saulina (2011).)

8.2.4 Proof of a Technical Proposition

Let us now give the proof of the following proposition, whose proof we skipped earlier.

Proposition 8.2.11 *View* $\mathcal{A}_{CS}$ *as an* E_2 *algebra via the map of operads from* E_2 *to* E_3. *There is a canonical equivalence of* E_2 *algebras from* $\mathcal{A}_{CS}$ *to the commutative algebra* $\mathbb{C}[\varepsilon]$, *where the parameter* ε *is of degree 1. We view* $\mathbb{C}[\varepsilon]$ *as an* E_2 *algebra via the map of operads from* E_2 *to* E_∞.

Proof Recall that $\mathcal{A}_{CS}$ is a twisted factorization envelope. Explicitly, it assigns to an open subset $U \subset \mathbb{R}^3$, the cochain complex

$$C_*(\Omega_c^*(U)[1] \oplus \mathbb{C} \cdot \hbar[-1]) \otimes_{\mathbb{C}[\hbar]} \mathbb{C}_{\hbar=1}.$$

The Lie bracket on the dg Lie algebra $\Omega_c^*(U)[1] \oplus \mathbb{C} \cdot \hbar[-1]$ is given by the shifted central extension of the Abelian Lie algebra $\Omega_c^*(U)[1]$ for the cocycle

$$\alpha \otimes \beta \mapsto \int \alpha \wedge \beta.$$

We are interested in the restriction of this twisted enveloping factorization algebra to a plane $\mathbb{R}^2 \subset \mathbb{R}^3$. Recall that

$$\left(i_{\mathbb{R}^2}^* \mathcal{A}_{CS}\right)(U) = \lim_{V \subset U} \mathcal{A}_{CS}(V),$$

where the limit is taken over those opens V in $\mathbb{R}^3$ that contain our chosen open $U \subset \mathbb{R}^2$. Note that since $\mathcal{A}_{CS}$ is locally constant, this limit is eventually constant. If we identify $\mathbb{R}^3 = \mathbb{R}^2 \times \mathbb{R}$, we can identify

$$\left(i_{\mathbb{R}^2}^* \mathcal{A}_{CS}\right)(U) = \mathcal{A}_{CS}(U \times \mathbb{R}).$$

In other words, we can identify $i_{\mathbb{R}^2}^* \mathcal{A}_{CS}$ with the pushforward $\pi_* \mathcal{A}_{CS}$ where $\pi : \mathbb{R}^3 \to \mathbb{R}^2$ is the projection. This identification is as factorization algebras.

There is a quasi-isomorphism, for all $U \subset \mathbb{R}^2$,

$$\Omega_c^*(U)[-1] \simeq \Omega_c^*(U \times \mathbb{R}).$$

Concretely, we can define this cochain map as follows. Choose a function f on $\mathbb{R}$ with compact supprt such that $\int f(t) \, dt = 1$. The quasi-isomorphism sends a form $\omega \in \Omega_c^*(U)$ to $\omega \wedge f(t) \, dt$, where t indicates the coordinate on $\mathbb{R}$.

The factorization algebra $\pi_* \mathcal{A}_{CS}$ is the twisted factorization algebra of the precosheaf of Abelian dg Lie algebras $\pi_* \Omega_c^*[1]$. To an open subset $U \subset \mathbb{R}^2$, this Abelian Lie algebra assigns $\Omega_c^*(U \times \mathbb{R})[1]$. The cocycle defining the central extension is given by the formula

$$[\alpha, \beta] = \hbar \int_{U \times \mathbb{R}} \alpha \wedge \beta.$$

We can restrict this cocycle to one on the sub-cosheaf $\Omega_c^*(U)$, where we use the map $\Omega_c^*(U) \to \Omega_c^*(U \times \mathbb{R})[1]$ discussed earlier. We find that it is zero, since for $\omega_1, \omega_2 \in \Omega_c^*(U)[1]$, we have

$$\int_{U \times \mathbb{R}} \omega_1 f(t) dt \wedge \omega_2 f(t) \, dt = 0.$$

It follows that we have a quasi-isomorphism of precosheaves of Lie algebras on $\mathbb{R}^2$:

$$\mathbb{C} \cdot \hbar[-1] \oplus \Omega_c^*(U) \simeq \mathbb{C} \cdot \hbar[-1] \oplus \Omega_c^*(U \times \mathbb{R})[1].$$

On the right hand side, the central extension is the one defined above, and on the left-hand side it is zero.

By applying the twisted factorization envelope construction to both sides, we find that that $\pi_* \mathcal{A}_{CS}$ is quasi-isomorphic to the un-twisted factorization envelope of the precosheaf of Abelian dg Lie algebras Ω_c^* on $\mathbb{R}^2$. Since this Lie algebra is Abelian, and there is no central extension, the result is a commutative factorization algebra. The corresponding E_2 algebra must therefore come from a commutative algebra.

Let us now identify this E_2 algebra explicitly. Note that, for every open $U \subset \mathbb{R}^2$, there is a cochain map

$$\Omega_c^*(U) \to \mathbb{C}[-2]$$

given by integration over U. This map is a quasi-isomorphism if U is a disc. It also respects the cosheaf operations, and so gives a map of cosheaves on $\mathbb{R}^2$

$$\Omega_c^* \to \underline{\mathbb{C}}[-2]$$

to the constant cosheaf $\underline{\mathbb{C}}[-2]$. The constant cosheaf $\underline{\mathbb{C}}$ assigns to any open $U \subset \mathbb{R}^2$ the space $H_0(U, \mathbb{C})$, which is a coproduct of a copy of $\mathbb{C}$ for each connected component of U.

Applying the factorization envelope construction, we find a map of prefactorization algebras

$$C_*(\Omega_c^*) = \operatorname{Sym} \Omega_c^*[1] \to \operatorname{Sym} \underline{\mathbb{C}}[-1].$$

This map is a quasi-isomorphism when restricted to disjoint unions of discs. It therefore identifies the E_2 algebra with the commutative algebra $\operatorname{Sym} \mathbb{C}[-1] = \mathbb{C}[\varepsilon]$. $\qquad\square$

Remark: There is another approach to this lemma that is less direct but applicable more broadly, by using the machinery of deformation theory.

There is a Hochschild complex for E_n algebras (in cochain complexes over a field of characteristic zero) akin to the Hochschild cochain complex for associative algebras. Furthermore, given an E_n algebra A, there is an exact sequence of the form

$$\operatorname{Def}(A) \to HH_{E_n}^*(A) \to A$$

relating the E_n Hochschild complex of A and the complex describing E_n algebra deformations of A (see Francis (2013)). First-order deformations of the E_n algebra structure are given by the $n + 1$st cohomology group of the complex $\operatorname{Def}(A)$. The funny shift is due to the fact that the E_n deformation complex is an E_{n+1} algebra and the underlying dg Lie algebra is obtained by shifting to turn the Poisson bracket into an unshifted Lie bracket. Compare, for instance, the fact that $HH^2(A)$ classifies first-order deformations of associative algebras.

Moreover, there is an analog of the Hochschild–Kostant–Rosenberg theorem for E_n algebras: for A a commutative dg algebra A, it identifies the E_n-deformation complex for A with the shifted polyvector fields $\operatorname{Sym}_A(\mathbb{T}_A[-n])$. Here $\mathbb{T}_A$ denotes the tangent complex of A. (See Calaque and Willwacher (2015) for the full statement and a proof.)

For us, the classical observables of Abelian Chern–Simons are equivalent to the commutative dg algebra $\mathbb{C}[\varepsilon]$, where ε has cohomological degree 1, because it is the factorization envelope of an Abelian Lie algebra. The E_2 HKR theorem then tells us that

$$HH_{E_2}^*(\mathbb{C}[\varepsilon]) \simeq \mathbb{C}\left[\varepsilon, \partial/\partial\varepsilon\right],$$

where ε and $\partial/\partial\varepsilon$ have cohomological degree 1. There are no degree 3 elements of our deformation complex, so the E_2 deformation must be trivial. Hence $\mathcal{A}_{CS}$ is a trivial deformation as an E_2 algebra. Note, by contrast, that there is a one-dimensional space of first-order deformations of $\mathbb{C}[\varepsilon]$ as an E_3 algebra because $HH^*_{E_3}(\mathbb{C}[\varepsilon])$ is $\mathbb{C}[\varepsilon, \partial/\partial\varepsilon]$, with $\partial/\partial\varepsilon$ now in degree 2. $\Diamond$

Appendix A

Background

We use techniques from disparate areas of mathematics throughout this book and not all of these techniques are standard knowledge, so here we provide a terse introduction to

- Simplicial sets and simplicial techniques
- Operads, colored operads (or multicategories), and algebras over colored operads
- Differential graded (dg) Lie algebras and their (co)homology
- Sheaves, cosheaves, and their homotopical generalizations
- Elliptic complexes, formal Hodge theory, and parametrices

along with pointers to more thorough treatments. By no means do readers need to be expert in all these areas to use our results or follow our arguments. They just need a working knowledge of this background machinery, and this appendix aims to provide the basic definitions, to state the results relevant for us, and to explain the essential intuition.

We do assume that readers are familiar with basic homological algebra and basic category theory. For homological algebra, there are numerous excellent sources, in books and online, among which we recommend the complementary texts by Weibel (1994) and Gelfand and Manin (2003). For category theory, the standard reference Mac Lane (1998) is adequate for our needs; we also recommend the series Borceux (1994a).

Remark: Our references are not meant to be complete, and we apologize in advance for the omission of many important works. We simply point out sources that we found pedagogically oriented or particularly accessible. ◊

A.1 Reminders and Notation

We overview some terminology and notations before embarking on our expositions.

For $\mathcal{C}$ a category, we often use $x \in \mathcal{C}$ to indicate that x is an object of $\mathcal{C}$. We typically write $\mathcal{C}(x, y)$ to the denote the set of morphisms between the objects x and y, although occasionally we use $\mathrm{Hom}_{\mathcal{C}}(x, y)$. The opposite category $\mathcal{C}^{op}$ has the same objects but $\mathcal{C}^{op}(x, y) = \mathcal{C}(y, x)$.

Given a collection of morphisms S in $\mathcal{C}$, a *localization of $\mathcal{C}$ with respect to* S is a category $\mathcal{C}[S^{-1}]$ and a functor $q : \mathcal{C} \to \mathcal{C}[S^{-1}]$ satisfying the following conditions.

(i) For every morphism $s \in S$, its image $q(s)$ is an isomorphism.
(ii) For any functor $F : \mathcal{C} \to \mathcal{D}$ such that $F(s)$ is an isomorphism for every $s \in S$, there is a functor $\overline{F} : \mathcal{C}[S^{-1}] \to \mathcal{C}$ and a natural isomorphism $\tau : \overline{F} \circ F \Rightarrow F$.
(iii) For every category $\mathcal{D}$, the functor $q^* : Fun(\mathcal{C}[S^{-1}], \mathcal{D}) \to Fun(\mathcal{C}, \mathcal{D})$ sending G to $G \circ q$ is full and faithful.

These ensure that the localization is unique up to equivalence of categories, if it exists. (We do not discuss size issues in this book.) The localization is the category in which every $s \in S$ becomes invertible. A morphism $f : \mathcal{C}[S^{-1}](x, y)$ can be concretely understood as a "zig zag"

$$x \xleftarrow{\simeq} z_1 \to z_2 \xleftarrow{\simeq} \cdots \to y$$

where the arrows going to the left (the "wrong way") are all in S and the arrows going to the right are arbitrary morphisms from $\mathcal{C}$.

We typically apply localization in the setting of a category $\mathcal{C}$ with a class W of *weak equivalences*. We call W a class of weak equivalences if W contains all isomorphisms in $\mathcal{C}$ and satisfies the two-out-of-three property, which states that given morphisms $f \in \mathcal{C}(x, y)$ and $g \in \mathcal{C}(y, z)$ such that two morphisms in the set $\{f, g, g \circ f\}$ are in W, then all three are in W. As examples, consider

(i) Any category with isomorphisms as weak equivalences.
(ii) The category of cochain complexes with cochain homotopy equivalences as weak equivalences.
(iii) The category of cochain complexes with quasi-isomorphisms as weak equivalences.
(iv) The category of topological spaces with weak homotopy equivalences as weak equivalences.

We often denote the localization $\mathcal{C}[W^{-1}]$ by $\mathrm{Ho}(\mathcal{C}, W)$, or just $\mathrm{Ho}(\mathcal{C})$, and call it the "homotopy category."

A.2 Simplicial Techniques

Simplicial sets are a combinatorial substitute for topological spaces, so it should be no surprise that they can be quite useful. On the one hand, we can borrow intuition for them from algebraic topology; on the other, simplicial sets are extremely concrete to work with because of their combinatorial nature. In fact, many constructions in homological algebra are best understood via their simplicial origins. Analogs of simplicial sets (i.e., simplicial objects in other categories) are useful as well.

In this book, we use simplicial sets in two ways:

• When we want to talk about a family of quantum field theories (QFTs) or space of parametrices, we will usually construct a simplicial set of such objects instead.
• We accomplish some homological constructions by passing through simplicial sets (e.g., in constructing the extension of a factorization algebra from a basis).

After giving the essential definitions, we state the main theorems we use.

Definition A.2.1 Let Δ denote the category whose objects are totally ordered finite sets and whose morphisms are nondecreasing maps between ordered sets. We usually work with the skeletal subcategory whose objects are

$$[n] = \{0 < 1 < \cdots < n\}.$$

A morphism $f : [m] \to [n]$ then satisfies $f(i) \leq f(j)$ if $i < j$.

We will relate these objects to geometry in the text that follows, but it helps to bear in mind the following picture. The set $[n]$ corresponds to the n-simplex Δ^n equipped with an ordering of its vertices, as follows. View the n-simplex Δ^n as living in $\mathbb{R}^{n+1}$ as the set

$$\left\{ (x_0, x_1, \ldots, x_n) \;\middle|\; \sum_j x_j = 1 \text{ and } 0 \leq x_j \leq 1 \; \forall j \right\}.$$

Identify the element $0 \in [n]$ with the 0th basis vector $e_0 = (1, 0, \ldots, 0)$, $1 \in [n]$ with $e_1 = (0, 1, 0, \ldots, 0)$, and $k \in [n]$ with the kth basis vector e_k. The ordering on $[n]$ then prescribes a path along the edges of Δ^n, starting at e_0, then going to e_1, and continuing till the path ends at e_n.

Every map $f : [m] \to [n]$ induces a linear map $f_* : \mathbb{R}^{m+1} \to \mathbb{R}^{n+1}$ by setting $f_*(e_k) = e_{f(k)}$. This linear map induces a map of simplices $f_* : \Delta^m \to \Delta^n$.

There are particularly simple maps that play an important role throughout the subject. Note that every map f factors into a surjection followed by an injection. We can then factor every injection into a sequence of *coface maps*, namely maps of the form

$$f_k : \quad [n] \quad \to \quad\quad\quad [n+1]$$
$$i \quad \mapsto \quad \begin{cases} i, & i \le k \\ i+1, & i > k \end{cases} \quad .$$

Similarly, we can factor every surjection into a sequence of *codegeneracy maps*, namely maps of the form

$$d_k : \quad [n] \quad \to \quad\quad\quad [n-1]$$
$$i \quad \mapsto \quad \begin{cases} i, & i \le k \\ i-1, & i > k \end{cases} \quad .$$

The names *face* and *degeneracy* fit nicely with the picture of the geometric simplices: a coface map corresponds to a choice of n-simplex in the boundary of the $n+1$-simplex, and a codegeneracy map corresponds to "collapsing" an edge of the n-simplex to project the n-simplex onto an $n-1$-simplex.

We now introduce the main character.

Definition A.2.2 A *simplicial set* is a functor $X : \Delta^{op} \to Set$, often denoted $X_\bullet$. The set $X([n])$, often denoted X_n, is called the "set of n-simplices of X." A *map of simplicial sets* $F : X \to Y$ is a natural transformation of functors. We denote this category of simplicial sets by sSet.

Let's quickly examine what the coface maps tell us about a simplicial set X. For example, by definition, a map $f_k : [n] \to [n+1]$ in Δ goes to $X(f_k) : X_{n+1} \to X_n$. We interpret this map $X(f_k)$ as describing the kth n-simplex sitting as a "face on the boundary" of an $n+1$-simplex of X. A similar interpretation applies to the d_k.

A.2.1 Simplicial Sets and Topological Spaces

When working in a homotopical setting, simplicial sets often provide a more tractable approach than topological spaces themselves. In this book, for instance, we describe "spaces of field theories" as simplicial sets. In the text that follows, we sketch how to relate these two kinds of objects.

One can use a simplicial set $X_\bullet$ as the "construction data" for a topological space: each element of X_n labels a distinct n-simplex Δ^n, and the structure maps of $X_\bullet$ indicate how to glue the simplices together. In detail, the *geometric realization* is the quotient topological space

$$|X_\bullet| = \left(\coprod_n X_n \times \Delta^n \right) / \sim$$

under the equivalence relation $\sim$ where $(x, s) \in X_m \times \triangle^m$ is equivalent to $(y, t) \in X_n \times \triangle^n$ if there is a map $f : [m] \rightarrow [n]$ such that $X(f) : X_n \rightarrow X_m$ sends y to x and $f_*(s) = t$.

Lemma A.2.3 *Under the Yoneda embedding, $[n]$ defines a simplicial set*

$$\triangle[n] : [m] \in \triangle^{op} \mapsto \triangle([m], [n]) \in Sets.$$

The geometric realization of $\triangle[n]$ is the n-simplex $\triangle^n$ (more accurately, it is homeomorphic to the n-simplex).

In general, every (geometric) simplicial complex can be obtained by the geometric realization of some simplicial set. Thus, simplicial sets provide an efficient way to study combinatorial topology.

One can go the other way, from topological spaces to simplicial sets: given a topological space X, there is a simplicial set $\text{Sing} \, X$, known as the "singular simplicial set of S." The set of n-simplices $(\text{Sing} \, X)_n$ is simply $\text{Top}(\triangle^n, X)$, the set of continuous maps from the n-simplex $\triangle^n$ into X. The structure maps arise from pulling back along the natural maps of simplices.

Theorem A.2.4 *Geometric realization and the singular functor form an adjunction*

$$| - | : \text{sSet} \rightleftarrows \text{Top} : \text{Sing}$$

between the category of simplicial sets and the category of topological spaces.

Thus, when we construct a simplicial set of Batalin–Vilkovisky (BV) theories $T_\bullet$, we obtain a topological space $|T_\bullet|$.

This theorem suggests as well how to transport notions of homotopy to simplicial sets: the homotopy groups of a simplicial set $X_\bullet$ are the homotopy groups of its realization $|X_\bullet|$, and maps of simplicial sets are homotopic if their realizations are. We would then like to say that these functors $| - |$ and Sing make the categories of simplicial sets and topological spaces equivalent, *in some sense*, particularly when it comes to questions about homotopy type. One sees immediately that some care must be taken, since there are topological spaces very different in nature from simplicial or cell complexes and for which no simplicial set could provide an accurate description. The key is only to think about topological spaces and simplicial sets up to the appropriate notion of equivalence.

Remark: It is more satisfactory to define these homotopy notions directly in terms of simplicial sets and then to verify that these match up with the topological notions. We direct readers to the references for this story, as the details are not relevant to our work in the book. ◊

We say a continuous map $f : X \to Y$ of topological spaces is a *weak equivalence* if it induces a bijection between connected components and an isomorphism $\pi_n(f,x) : \pi_n(X,x) \to \pi_n(Y,f(x))$ for every $n > 0$ and every $x \in X$. Let Ho(Top), the *homotopy category* of Top, denote the category of topological spaces where we localize at the weak equivalences. There is a concrete way to think about this homotopy category. For every topological space, there is some CW complex weakly equivalent to it, under a zig zag of weak equivalences; and by the Whitehead theorem, a weak equivalence between CW complexes is in fact a homotopy equivalence. Thus, Ho(Top) is equivalent to the category of CW complexes with morphisms given by continuous maps modulo homotopy equivalence.

Likewise, let Ho(sSet) denote the homotopy category of simplicial sets, where we localize at the appropriate notion of simplicial homotopy. Then Quillen proved the following wonderful theorem.

Theorem A.2.5 *The adjunction induces an equivalence*

$$| - | : \text{Ho(sSet)} \rightleftarrows \text{Ho(Top)} : \text{Sing}$$

between the homotopy categories. (In particular, they provide a Quillen *equivalence between the standard model category structures on these categories.)*

This theorem justifies the assertion that, from the perspective of homotopy type, we are free to work with simplicial sets in place of topological spaces. In addition, it helpful to know that algebraic topologists typically work with a better behaved categories of topological spaces, such as compactly generated spaces.

Among simplicial sets, those that behave like topological spaces are known as *Kan complexes* or *fibrant simplicial sets*. Their defining property is a simplicial analogue of the homotopy lifting property, which we now describe explicitly. The *horn* for the kth face of the n-simplex, denoted $\Lambda_k[n]$, is the sub-simplicial set of $\Delta[n]$ given by the union of the all the faces $\Delta[n-1] \hookrightarrow \Delta[n]$ except the kth. (As a functor on Δ, the horn takes the $[m]$ to monotonic maps $[m] \to [n]$ that do not have k in the image.) A simplicial set X is a *Kan complex* if for every map of a horn $\Lambda_k[n]$ into X, we can extend the map to the n-simplex $\Delta[n]$. Diagrammatically, we can fill in the dotted arrow to get a commuting diagram. In general, one can always find a "fibrant replacement" of a simplicial set $X_\bullet$ (e.g., by taking Sing $|X_\bullet|$ or via Kan's Ex^∞ functor) that is weakly equivalent and a Kan complex.

A.2.2 Simplicial Sets and Homological Algebra

Our other use for simplicial sets relates to homological algebra. We always work with *co*chain complexes, so our conventions will differ from those who prefer chain complexes. For instance, the chain complex computing the homology of a topological space is concentrated in nonnegative degrees. We work instead with the cochain complex concentrated in nonpositive degrees. (To convert, simply swap the sign on the indices.)

Definition A.2.6 A *simplicial Abelian group* is a simplicial object $A_\bullet$ in the category of Abelian groups, i.e., a functor $A : \Delta^{op} \to \mathrm{Ab}$.

By composing with the forgetful functor $U : \mathrm{Ab} \to \mathrm{Set}$ that sends a group to its underlying set, we see that a simplicial Abelian group has an underlying simplicial set. To define simplicial vector spaces or simplicial R-modules, one simply replaces *Abelian group* by *vector space* or *R-module* everywhere. All the work below will carry over to these settings in a natural way.

There are *two* natural ways to obtain a cochain complex of Abelian groups (respectively, vector spaces) from a simplicial abelian group. The *unnormalized chains* $\mathbf{CA_\bullet}$ is the cochain complex

$$(\mathbf{CA_\bullet})^m = \begin{cases} A_{|m|}, & m \leq 0 \\ 0, & m > 0 \end{cases}$$

with differential

$$\begin{aligned} d : \quad (\mathbf{CA_\bullet})^m &\to \quad (\mathbf{CA_\bullet})^{m+1} \\ a &\mapsto \quad \sum_{k=0}^{|m|}(-1)^k A_\bullet(f_k)(a), \end{aligned}$$

where the f_k run over the coface maps from $[|m| - 1]$ to $[|m|]$. The *normalized chains* $\mathbf{NA_\bullet}$ is the cochain complex

$$(\mathbf{NA_\bullet})^m = \bigcap_{k=0}^{|m|-1} \ker A_\bullet(f_k),$$

where $m \leq 0$ and where the f_k run over the coface maps from $[|m| - 1]$ to $[|m|]$. The differential is $A(f_{|m|})$, the remaining coface map. One can check that the inclusion $\mathbf{NA_\bullet} \hookrightarrow \mathbf{CA_\bullet}$ is a quasi-isomorphism (in fact, a cochain homotopy equivalence).

Example: Given a topological space X, its singular chain complex $C_*(X)$ arises as a composition of three functors in this simplicial world. (Because we prefer cochain complexes, the singular chain complex is, in fact, a cochain complex concentrated in *nonpositive* degrees.) First, we make the simplicial set Sing X, which knows about all the ways of mapping a simplex into X. Then we apply the free Abelian group functor $\mathbb{Z}-$: *Sets* $\to$ *AbGps* levelwise to obtain the simplicial Abelian group

$$\mathbb{Z} \operatorname{Sing} X : [n] \mapsto \mathbb{Z}(\operatorname{Top}(\Delta^n, X)).$$

Then we apply the unnormalized chains functor to obtain the singular chain complex

$$C_*(X) = \mathbf{C}\mathbb{Z} \operatorname{Sing} X.$$

In other words, the simplicial language lets us decompose the usual construction into its atomic components. $\Diamond$

It is clear from the constructions that we only ever obtain cochain complexes concentrated in *nonpositive* degrees from simplicial Abelian groups. In fact, the *Dold–Kan correspondence* tells us that we are free to work with either kind of object – simplicial abelian group or such a cochain complex – as we prefer.

Let $\operatorname{Ch}^{\leq 0}(\mathrm{Ab})$ denote the category of cochain complexes concentrated in nonpositive degrees, and let *sAbGps* denote the category of simplicial Abelian groups.

Theorem A.2.7 **(Dold–Kan correspondence)** *The normalized chains functor*

$$\mathbf{N} : \mathrm{sAb} \to \operatorname{Ch}^{\leq 0}(\mathrm{Ab})$$

is an equivalence of categories. Under this correspondence,

$$\pi_n(A_\bullet) \cong H^{-n}(\mathbf{N}A_\bullet)$$

for all $n \geq 0$, and simplicial homotopies go to chain homotopies.

Throughout the book, we will often work with cochain complexes equipped with algebraic structures (e.g., commutative dg algebras or dg Lie algebras). Thankfully, it is well understood how the Dold–Kan correspondence intertwines with the tensor structures on sAb and $\operatorname{Ch}^{\leq 0}(\mathrm{Ab})$.

Let $A_\bullet$ and $B_\bullet$ be simplicial Abelian groups. Then their *tensor product* $(A \otimes B)_\bullet$ is the simplicial Abelian group with n-simplices

$$(A \otimes B)_n = A_n \otimes_\mathbb{Z} B_n.$$

There is a natural transformation

$$\nabla_{A,B} : \mathbf{C}A \otimes \mathbf{C}B \to \mathbf{C}(A \otimes B),$$

known as the *Eilenberg–Zilber map* or *shuffle map*, which relates the usual tensor product of complexes with the tensor product of simplicial Abelian groups. As $\mathbf{C}A_{\bullet} \otimes \mathbf{C}B_{\bullet}$ is not isomorphic to $\mathbf{C}(A \otimes B)_{\bullet}$, it is not a strong monoidal functor but instead a *lax* monoidal functor. The Eilenberg–Zilber map is, however, always a quasi-isomorphism, and so it preserves weak equivalences.

Theorem A.2.8 *The unnormalized chains functor and the normalized chains functor are both lax monoidal functors via the Eilenberg–Zilber map.*

Thus, with a little care, we can relate algebra in the setting of simplicial modules with algebra in the setting of cochain complexes.

A.2.3 References

Friedman (2012) is a very accessible and concrete introduction to simplicial sets, with lots of intuition and pictures. Weibel (1994) explains clearly how simplicial sets appear in homological algebra, notably for us, the Dold–Kan correspondence and the Eilenberg–Zilber map. As usual, Gelfand and Manin (2003) provides a nice complement to Weibel. The expository article Goerss and Schemmerhorn (2007) provides a lucid and quick discussion of how simplicial methods relate to model categories and related issues. For the standard, modern reference on simplicial sets and homotopy theory, see the thorough and clear Goerss and Jardine (2009).

A.3 Operads and Algebras

An operad is a way of describing the essential structure underlying some class of algebraic objects. For instance, there is an operad *Ass* that captures the essence of associative algebras, and there is an operad *Lie* that captures the essence of Lie algebras. An algebra over an operad is an algebraic object with that kind of structure: a Lie algebra is an algebra over the operad *Lie*. Although their definition can seem abstract and unwieldy at first, operads provide an efficient language for thinking about algebra and proving theorems about large classes of algebraic objects. As a result, they appear throughout mathematics.

A colored operad is a way of describing a collection of objects that interact algebraically. For example, there is a colored operad that describes a pair consisting of an algebra and a module over that algebra. Another name for a colored operad is a symmetric multicategory, which emphasizes a different intuition: it is a generalization of the notion of a category in which we allow maps with multiple inputs and one output.

In the book, we use these notions in several ways:

- We capture the algebraic essence of the Batalin–Vilkovisky notion of a quantization via the *Beilinson–Drinfeld operad*.
- The notion of a prefactorization algebra – perhaps the central notion in the book – is an algebra over a colored operad made from the open sets of a topological space.
- We define colored operads that describe how observables vary under translation and that generalize the notion of a vertex algebra.

The first use has a different flavor from the others, so we begin here by focusing on operads with a linear flavor before we introduce the general formalism of colored operads. We hope that by being concrete in the first part, the abstractions of the second part will not seem arid.

A.3.1 Operads

In the loosest sense – encompassing Lie, associative, commutative, and more – an algebra is a vector space A with some way of combining elements multilinearly. Typically, we learn first about examples determined by a binary operation $\mu : A \otimes A \to A$ such that

- μ has some symmetry under permutation of the inputs (e.g., $\mu(a, b) = -\mu(b, a)$ for a Lie algebra).
- The induced 3-ary operations $\mu \circ (\mu \otimes 1)$ and $\mu \circ (1 \otimes \mu)$ satisfy a linear relation (e.g., associativity or the Jacobi identity).

But we recognize that there should be more elaborate notions involving many different n-ary operations required to satisfy complicated relations. As a basic example, a Poisson algebra has two binary operations.

Before we give the general definition of an operad in (dg) vector spaces, we explain how to visualize such algebraic structures. An n-ary operation $\tau : A^{\otimes n} \to A$ is pictured as a rooted tree with n labeled leaves and one root.

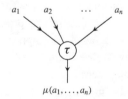

For us, operations move down the page. To compose operations, we need to specify where to insert the output of each operation. We picture this as stacking rooted trees. For example, given a binary operation μ, the composition $\mu \circ (\mu \otimes 1)$ corresponds to the tree

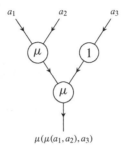

$$\mu(\mu(a_1, a_2), a_3)$$

whereas $\mu \circ (1 \otimes \mu)$ is the tree

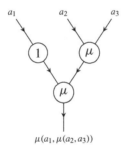

$$\mu(a_1, \mu(a_2, a_3))$$

with the first μ on the other side.

We also allow permutations of the inputs, which rearranges the inputs. Thus the vector space of n-ary operations $\{\tau : A^{\otimes n} \to A\}$ has an action of the permutation group S_n. We want the permutations to interact in the natural way with composition.

We now give the formal definition. We will describe an operad in vector spaces over a field $\mathbb{K}$ of characteristic zero (e.g., $\mathbb{C}$ or $\mathbb{R}$) and use $\otimes$ to denote $\otimes_{\mathbb{K}}$. It is straightforward to give a general definition of an operad in an arbitrary symmetric monoidal category. It should be clear, for instance, how to replace vector spaces with cochain complexes.

Definition A.3.1 An *operad* $\mathcal{O}$ in vector spaces consists of

(i) A sequence of vector spaces $\{\mathcal{O}(n) \mid n \in \mathbb{N}\}$, called the *operations*.

(ii) A collection of multilinear maps

$$\circ_{n;m_1,\ldots,m_n} : \mathcal{O}(n) \otimes (\mathcal{O}(m_1) \otimes \cdots \otimes \mathcal{O}(m_n)) \to \mathcal{O}\left(\sum_{j=1}^{n} m_j\right),$$

called the *composition maps*.

(iii) A *unit element* $\eta : \mathbb{K} \to \mathcal{O}(1)$.

These data are equivariant, associative, and unital in the following way.

(1) The n-ary operations $\mathcal{O}(n)$ have a right action of S_n.
(2) The composition maps are equivariant in the sense that the following diagram commutes,

$$
\begin{array}{ccc}
\mathcal{O}(n) \otimes (\mathcal{O}(m_1) \otimes \cdots \otimes \mathcal{O}(m_n)) & \xrightarrow{\sigma \otimes \sigma^{-1}} & \mathcal{O}(n) \otimes (\mathcal{O}(m_{\sigma(1)}) \otimes \cdots \otimes \mathcal{O}(m_{\sigma(n)})) \\
\downarrow{\circ} & & \downarrow{\circ} \\
\mathcal{O}\left(\sum_{j=1}^{n} m_j\right) & \xrightarrow{\sigma(m_{\sigma(1)},\ldots,m_{\sigma(n)})} & \mathcal{O}\left(\sum_{j=1}^{n} m_j\right)
\end{array}
$$

where $\sigma \in S_n$ acts as a block permutation on the $\sum_{j=1}^{n} m_j$ inputs, and the following diagram also commutes,

$$
\begin{array}{ccc}
& \xrightarrow{\mathrm{id} \otimes (\tau_1 \otimes \cdots \otimes \tau_n)} & \\
\mathcal{O}(n) \otimes (\mathcal{O}(m_1) \otimes \cdots \otimes \mathcal{O}(m_n)) & & \mathcal{O}(n) \otimes (\mathcal{O}(m_{\sigma(1)}) \otimes \cdots \otimes \mathcal{O}(m_{\sigma(n)})) \\
\downarrow{\circ} & & \downarrow{\circ} \\
\mathcal{O}\left(\sum_{j=1}^{n} m_j\right) & \xrightarrow{\tau_1 \oplus \cdots \oplus \tau_n} & \mathcal{O}\left(\sum_{j=1}^{n} m_j\right)
\end{array}
$$

where each τ_j is in S_{m_j} and $\tau_1 \oplus \cdots \oplus \tau_n$ denotes the blockwise permutation in $S_{\sum_{j=1}^{n} m_j}$.

(3) The composition maps are associative in the following sense. Let $n, m_1, \ldots, m_n, \ell_{1,1}, \ldots, \ell_{1,m_1}, \ell_{2,1}, \ldots, \ell_{n,m_n}$ be positive integers, and set $M = \sum_{j=1}^{n} m_j$, $L_j = \sum_{i=1}^{m_j} \ell_{j,i}$, and

$$N = \sum_{i=1}^{n} L_j = \sum_{(j,k)\in M} \ell_{j,k}.$$

Then the diagram

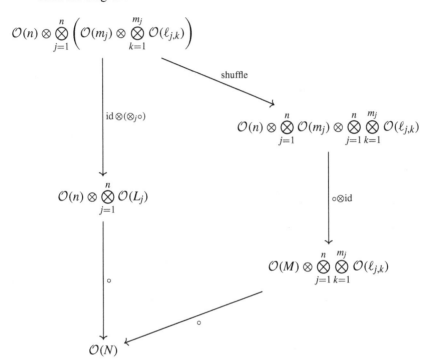

commutes.

(4) The unit diagrams commute:

$$\mathcal{O}(n) \otimes \mathbb{K}^{\otimes n} \xrightarrow[\text{id} \otimes \eta^{\otimes n}]{} \mathcal{O}(n) \otimes \mathcal{O}(1)^{\otimes n} \xrightarrow[\circ]{} \mathcal{O}(n)$$

with the top arrow $\cong$

and

$$\mathbb{K} \otimes \mathcal{O}(n) \xrightarrow[\eta \otimes \text{id}]{} \mathcal{O}(1) \otimes \mathcal{O}(n) \xrightarrow[\circ]{} \mathcal{O}(n) .$$

with the top arrow $\cong$

A *map of operads* $f : \mathcal{O} \to \mathcal{P}$ is a sequence of linear maps

$$\{f(n) : \mathcal{O}(n) \to \mathcal{P}(n)\}$$

interwining in the natural way with all the structure of the operads.

Example: The operad *Com* describing commutative algebras has $Com(n) \cong \mathbb{K}$, with the trivial S_n action, for all n. This is because there is only one way to multiply n elements: even if we permute the inputs, we have the same output.   ◊

Example: The operad *Ass* describing associative algebras has $Ass(n) = \mathbb{K}[S_n]$, the regular representation of S_n, for all n. This is because the product of n elements only depends on their left-to-right ordering, not on a choice of parantheses. We should have exactly one n-ary product for each ordering of n elements.   ◊

Remark: One can describe operads via generators and relations. The two examples above are generated by a single binary operation. In *Ass*, there is a relation between the 3-ary operations generated by that binary operation – the associativity relation – as already discussed. For a careful treatment of this style of description, we direct readers to the references.   ◊

We now explain the notion of an *algebra over an operad*. Our approach is modeled on defining a representation of a group G on a vector space V as a group homomorphism $\rho : G \to GL(V)$. Given a vector space V, there is an operad End_V that contains all imaginable multilinear operations on V and how they compose, just like $GL(V)$ contains all linear automorphisms of V.

Definition A.3.2 The *endomorphism operad* End_V of a vector space V has n-ary operations $End_V(n) = Hom(V^{\otimes n}, V)$ and compositions

$$\circ_{n;m_1,\ldots,m_n} : \mathcal{O}(n) \otimes (\mathcal{O}(m_1) \otimes \cdots \otimes \mathcal{O}(m_n)) \to \mathcal{O}\left(\sum_{j=1}^n m_j\right)$$
$$\mu_n \otimes (\mu_{m_1} \otimes \cdots \otimes \mu_{m_n}) \mapsto \mu_n \circ (\mu_{m_1} \otimes \cdots \otimes \mu_{m_n})$$

are simply composition of multilinear maps.

Definition A.3.3 For $\mathcal{O}$ an operad and V a vector space, an *algebra over* $\mathcal{O}$ is a map of operads $\rho : \mathcal{O} \to End_V$.

For example, an associative algebra A is given by specifying a vector space A and a linear map $\mu : A^{\otimes 2} \to A$ satisfying the associativity relation. These data are equivalent to specifying an operad map from *Ass* to End_A.

Alternatively, an algebra V over $\mathcal{O}$ is a collection of equivariant maps

$$\rho(n) : \mathcal{O}(n) \otimes V^{\otimes n} \to V$$

compatible with the structure maps of $\mathcal{O}$. These maps arise from our definition of an algebra by the Hom-tensor adjunction: a map in $Hom(\mathcal{O}(n), Hom(V^{\otimes n}, V))$ yields a map in $Hom(\mathcal{O}(n) \otimes V^{\otimes n}, V)$.

Dually, we consider the notion of a *coalgebra over an operad*. The *coendomorphism operad coEnd$_V$* of a vector space V has n-ary operations

$$coEnd_V(n) = Hom(V, V^{\otimes n}),$$

and compositions are the obvious multilinear maps. A coalgebra over $\mathcal{O}$ is a map of operads $\rho : \mathcal{O} \to coEnd_V$. Alternatively, it is a collection of equivariant maps $\mathcal{O}(n) \otimes V \to V^{\otimes n}$ with natural compatibilities.

Remark: Dual to these definitions is the notion of a *cooperad*, which intuitively encodes the ways something can decompose into constituent pieces. One can then speak about algebras and coalgebras over a cooperad. $\Diamond$

References
For a very brief introduction to operads, see the "What is" column by Stasheff (2004). Vallette (2014) provides a nice motivation and overview for linear operads and their relation to homotopical algebra. For a systematic treatment with an emphasis on Koszul duality, see Loday and Vallette (2012). In Costello (2004), there is a description of the basics emphasizing a diagrammatic approach: an operad is a functor on a category of rooted trees. Finally, the book Fresse (2016) is wonderful.

A.3.2 The P_0 and BD Operads

Two linear operads play a central role in this book and in the BV formalism. In physics, there is a basic division into classical and quantum, and the BV formalism also breaks down along these lines. The operad controlling the classical BV formalism is the P_0 operad, encoding Poisson-zero algebras, equivalently commutative dg algebras with a compatible Poisson bracket of cohomological degree 1. The operad controlling the quantum BV formalism is the BD operad, encoding *Beilinson–Drinfeld* algebras.

P_0
We will define the shifted Poisson operads P_k in terms of their algebras. These operads live in the category of graded vector spaces over any field of characteristic zero.

Definition A.3.4 A P_k *algebra* is a cochain complex A that possesses a commutative product $\cdot : A \otimes A \to A$ of cohomological degree 0 and a Lie bracket $\{-, -\} : A[k-1] \otimes A[k-1] \to A[k-1]$ on the shifted complex $A[k-1]$ satisfying the Poisson relation: for every $a \in A$, the graded linear map $\{a, -\}$ is a derivation for the commutative product.

A P_1 algebra is precisely an ordinary Poisson dg algebra. The strange-looking indexing that we use is due to the following fact. By taking homology, the E_k operad in spaces provides an operad in graded vector spaces whose n-ary operations are $H_*(E_k(n))$. This operad is P_k when $k \geq 2$, essentially by work of Fred Cohen.

In Volume 2 we develop the classical and quantum BV formalisms in great detail, but we can already give a quick motivation for the appearance of P_0 algebras. As explained in Section 4.1 of Chapter 4, the key motivating example for the BV formalism is the divergence complex of a measure. The classical limit of the divergence complex is the polyvector fields on a manifold equipped with a derivation so as to form a commutative dg algebra. We call this complex the functions on the "derived critical locus" (see Definition 4.1.1). The crucial observation here is that the polyvector fields possess a natural shifted Poisson bracket known as the Schouten bracket. It is simply the Lie bracket of vector fields extended antilinearly and compatibly with the action of vector fields on smooth functions.

BD

Just as we find the P_0 operad by examining the classical limit of divergence complexes and extracting the formal properties, the notion of the *BD* operad arises by extracting formal properties of the divergence complex.

Definition A.3.5 The *BD operad* is an operad in dg modules over $\mathbb{R}[[\hbar]]$, where $\hbar$ has cohomological degree 0, whose underlying graded operad agrees with the P_0 operad (tensored up to $\mathbb{R}[[\hbar]]$). The differential on the binary operations $\cdot$ (in degree 0) and $\{-, -\}$ in degree 1 satisfy the following relation,

$$d(\cdot) = \hbar\{-, -\},$$

and determine the differential on the n-ary operations.

Let's unpack what that condition means for a *BD* algebra. Let (A, d_A) be a cochain complex. To equip it with a *BD* algebra structure is to give a map ρ of dg operads from *BD* to End_A. Forgetting the differentials, we see that the underlying graded vector space $A^\sharp$ is thus a P_0 algebra. Now let's examine what it means that the map $\rho(2) : BD(2) \to Hom(A \otimes A, A)$ is a cochain map. Let $\bullet$ denote the commutative product on A, which is the element $\rho(2)(\cdot)$, and let $\{-, -\}$ denote the Poisson bracket on A, which is the element $\rho(2)(\{-, -\})$. Then, since $\rho(2)$ commutes with the differentials, we must have that

$$d_{Hom}(\rho(2)(\cdot)) = d_A \circ \bullet - \bullet \circ (d_A \otimes 1 + 1 \otimes d_A)$$

equals

$$\rho(2)(\mathrm{d}(\cdot)) = \hbar\{-, -\}.$$

In other words, we need

$$\mathrm{d}_A(x \bullet y) - (\mathrm{d}_A x) \bullet y - x \bullet (\mathrm{d}_A y) = \hbar\{x, y\}$$

for any $x, y \in A$. This equation is precisely the standard relation in the BV formalism.

We should remark that this operad is quite similar to an operad known as the *BV* operad, except that cohomological degrees are different. In the mathematics literature, the *BV* operad typically refers to an operad related to the little two-dimensional discs operad E_2. There is the potential for confusion here, because the BV quantization procedure is related not to the E_2 operad, but to the E_0 operad. (In the original physics literature, where these relations were first introduced, everything was $\mathbb{Z}/2$-graded and so it is impossible to distinguish between these structures.) The role of this *BD* operad was first recognized in Beilinson and Drinfeld (2004), and so we named it after them.

A.3.3 Colored Operads *aka* Multicategories

We now introduce a natural simultaneous generalization of the notions of a category and of an operad. The essential idea is to have a collection of objects that we can "combine" or "multiply." If there is only one object, we recover the notion of an operad. If there are no ways to combine multiple objects, then we recover the notion of a category.

We will give the *symmetric* version of this concept, just as we did with operads.

Let $(\mathcal{C}, \boxtimes)$ be a symmetric monoidal category, such as $(Sets, \times)$ or $(Vect_{\mathbb{K}}, \otimes_{\mathbb{K}})$. We require $\mathcal{C}$ to have all reasonable colimits.

Definition A.3.6 A *multicategory* (or *colored operad*) $\mathcal{M}$ over $\mathcal{C}$ consists of

(i) A collection of *objects* (or *colors*) $Ob\mathcal{M}$.
(ii) For every $n + 1$-tuple of objects $(x_1, \ldots, x_n \mid y)$, an object

$$\mathcal{M}(x_1, \ldots, x_n \mid y)$$

in $\mathcal{C}$ called the *maps* from the x_j to y.
(iii) A *unit element* $\eta_x : 1_{\mathcal{C}} \to \mathcal{M}(x \mid x)$ for every object x.
(iv) A collection of *composition maps* in $\mathcal{C}$

$$\mathcal{M}(x_1, \ldots, x_n \mid y) \boxtimes \left(\mathcal{M}(x_1^1, \ldots, x_1^{m_1} \mid x_1) \boxtimes \cdots \boxtimes \mathcal{M}(x_n^1, \ldots, x_n^{m_n} \mid x_n) \right)$$
$$\to \mathcal{M}(x_1^1, \ldots, x_n^{m_n} \mid y)$$

(v) For every $n + 1$-tuple $(x_1, \ldots, x_n \mid y)$ and every permutation $\sigma \in S_n$ a morphism

$$\sigma^* : \mathcal{M}(x_1, \ldots, x_n \mid y) \to \mathcal{M}(x_{\sigma(1)}, \ldots, x_{\sigma(n)} \mid y).$$

in $\mathcal{C}$.

These data satisfy conditions of associativity, units, and equivariance directly analogous to that of operads. For instance, given $\sigma, \tau \in S_n$, we require

$$\sigma^* \tau^* = (\tau\sigma)^*,$$

so we have an analog of a right S_n action on maps out of n objects. Each unit η_x is a two-sided unit for composition in $\mathcal{M}(x \mid x)$.

Definition A.3.7 A *map of multicategories* (or *functor* between multicategories) $F : \mathcal{M} \to \mathcal{N}$ consists of

(i) An object $F(x)$ in $\mathcal{N}$ for each object x in $\mathcal{M}$, and
(ii) a morphism

$$F(x_1, \ldots, x_n \mid y) : \mathcal{M}(x_1, \ldots, x_n \mid y) \to \mathcal{N}(F(x_1), \ldots, F(x_n) \mid F(y))$$

in the category $(\mathcal{C}, \boxtimes)$ for every tuple $(x_1, \ldots, x_n \mid y)$ of objects in $\mathcal{M}$ such that the structure of a multicategory is preserved (namely, the units, associativity, and equivariance).

There are many familiar examples.

Example: Let $\mathcal{B}$ be an ordinary category, so that each collection of morphisms $\mathcal{B}(x, y)$ is a set. Then $\mathcal{B}$ is a multicategory over the symmetric monoidal category $(Sets, \times)$ where $\mathcal{B}(x \mid y) = \mathcal{B}(x, y)$ and $\mathcal{B}(x_1, \ldots, x_n \mid y) = \emptyset$ for $n > 1$.
$\Diamond$

Similarly, a $\mathbb{K}$-linear category is a multicategory over the symmetric monoidal category $(Vect_{\mathbb{K}}, \otimes_{\mathbb{K}})$ with no compositions between two or more elements.

Example: An operad $\mathcal{O}$, in the sense of the preceding subsection, is a multicategory over the symmetric monoidal category $(Vect_{\mathbb{K}}, \otimes_{\mathbb{K}})$ with a single object $*$ and

$$\mathcal{O}(\underbrace{*, \ldots, *}_{n} \mid *) = \mathcal{O}(n).$$

Replacing $Vect_{\mathbb{K}}$ with another symmetric monoidal category $\mathcal{C}$, we obtain a definition for operad in $\mathcal{C}$.
$\Diamond$

Example: Every symmetric monoidal category $(\mathcal{C}, \boxtimes)$ has an underlying multicategory $\underline{\mathcal{C}}$ with the same objects and with maps

$$\underline{\mathcal{C}}(x_1, \ldots, x_n \mid y) = \mathcal{C}(x_1 \boxtimes \cdots \boxtimes x_n, y).$$

When $\mathcal{C} = Vect_{\mathbb{K}}$, these are precisely all the multilinear maps. $\Diamond$

For every multicategory $\mathcal{M}$, one can construct a *symmetric monoidal envelope* $\mathbf{S}\mathcal{M}$ by forming the left adjoint to the "forgetful" functor from symmetric monoidal categories to multicategories. An object of $\mathbf{S}\mathcal{M}$ is a formal finite sequence $[x_1, \ldots, x_m]$ of colors x_i, and a morphism $f : [x_1, \ldots, x_m] \to [y_1, \ldots, y_n]$ consists of a surjection $\phi : \{1, \ldots, m\} \to \{1, \ldots, n\}$ and a morphism $f_j \in \mathcal{M}(\{x_i\}_{i \in \phi^{-1}(j)} \mid y_j)$ for every $1 \le j \le n$. The symmetric monoidal product in $\mathbf{S}\mathcal{M}$ is simply concatenation of formal sequences.

Finally, an *algebra over a colored operad* $\mathcal{M}$ with values in $\mathcal{N}$ is simply a functor of multicategories $F : \mathcal{M} \to \mathcal{N}$. When we view $\mathcal{O}$ as a multicategory and use the underlying multicategory $\underline{Vect}_{\mathbb{K}}$, then $F : \mathcal{O} \to \underline{Vect}_{\mathbb{K}}$ reduces to an algebra over the operad $\mathcal{O}$ as in the preceding subsection.

References

For a readable discussion of operads, multicategories, and different approaches to them, see Leinster (2004). Beilinson and Drinfeld (2004) develop *pseudotensor categories* – yet another name for this concept – for exactly the same reasons as we do in this book. Lurie (2016) provides a thorough treatment of colored operads compatible with ∞-categories.

A.4 Lie Algebras and Chevalley–Eilenberg Complexes

Lie algebras, and their homotopical generalization L_∞ algebras, appear throughout both volumes in a variety of contexts. It might surprise readers that we never use their representation theory or almost any aspects emphasized in textbooks on Lie theory. Instead, we primarily use dg Lie algebras as a convenient language for describing deformations of some structure, which is a central theme of Volume 2. For the purposes of Volume 1, we need to describe standard homological constructions with dg Lie algebras. For the definition of L_∞ algebras and the relationship with derived geometry, see Volume 2.

A.4.1 A Quick Review of Homological Algebra Lie Algebras

Let $\mathbb{K}$ be a field of characteristic zero. (We always have in mind $\mathbb{K} = \mathbb{R}$ or $\mathbb{C}$.)
A *Lie algebra* over $\mathbb{K}$ is a vector space $\mathfrak{g}$ with a bilinear map $[-, -] : \mathfrak{g} \otimes_{\mathbb{K}} \mathfrak{g} \to \mathfrak{g}$ such that

- $[x, y] = -[y, x]$ (antisymmetry) and
- $[x, [y, z]] = [[x, y], z] + [y, [x, z]]$ (Jacobi rule).

The Jacobi rule is the statement that $[x, -]$ acts as a derivation. A simple example is the space of $n \times n$ matrices $M_n(\mathbb{K})$, usually written as $\mathfrak{gl}_n$ in this context, where the bracket is

$$[A, B] = AB - BA,$$

the commutator using matrix multiplication. Another classic example is given by *Vect(M)*, the vector fields on a smooth manifold M, via the commutator bracket of derivations acting on smooth functions.

A *module* over $\mathfrak{g}$, or *representation* of $\mathfrak{g}$, is a vector space M with a bilinear map $\rho : \mathfrak{g} \otimes_{\mathbb{K}} M \to M$ such that

$$\rho(x \otimes \rho(y \otimes m)) - \rho(y \otimes \rho(x \otimes m)) = \rho([x, y] \otimes m).$$

Usually, we will suppress the notation ρ and simply write $x \cdot m$ or $[x, m]$. Continuing with the examples from above, the matrices $\mathfrak{gl}_n$ acts on $\mathbb{K}^n$ by left multiplication, so $\mathbb{K}^n$ is naturally a $\mathfrak{gl}_n$-module. Analogously, vector fields *Vect(M)* act on smooth functions $C^{\infty}(M)$ as derivations, and so $C^{\infty}(M)$ is a *Vect(M)*-module.

There is a category $\mathfrak{g} - mod$ whose objects are $\mathfrak{g}$-modules and whose morphisms are the natural structure-preserving maps. To be explicit, a map $f \in \mathfrak{g} - mod(M, N)$ of $\mathfrak{g}$-modules is a linear map $f : M \to N$ such that $[x, f(m)] = f([x, m])$ for every $x \in \mathfrak{g}$ and every $m \in M$.

Lie algebra homology and cohomology arise as the derived functors of two natural functors on the category of $\mathfrak{g}$-modules. We define the *invariants* as the functor

$$(-)^{\mathfrak{g}} : \quad \mathfrak{g} - mod \quad \to \quad Vect_{\mathbb{K}}$$
$$M \quad \mapsto \quad M^{\mathfrak{g}}$$

where $M^{\mathfrak{g}} = \{m \mid [x, m] = 0 \ \forall x \in \mathfrak{g}\}$. (A nonlinear analog is taking the fixed points of a group action on a set.) The *coinvariants* is the functor

$$(-)_{\mathfrak{g}} : \quad \mathfrak{g} - mod \quad \to \quad Vect_{\mathbb{K}}$$
$$M \quad \mapsto \quad M_{\mathfrak{g}}$$

where $M_{\mathfrak{g}} = M/\mathfrak{g}M = M/\{[x,m] \mid x \in \mathfrak{g}, \; m \in M\}$. (A nonlinear analog is taking the quotient, or orbit space, of a group action on a set, i.e., the collection of orbits.)

To define the derived functors, we rework our constructions into the setting of modules over *associative* algebras so that we can borrow the Tor and Ext functors. The *universal enveloping algebra* of a Lie algebra $\mathfrak{g}$ is

$$U\mathfrak{g} = \mathrm{Tens}(\mathfrak{g})/(x \otimes y - y \otimes x - [x,y])$$

where $\mathrm{Tens}(\mathfrak{g}) = \oplus_{n \geq 0} \mathfrak{g}^{\otimes n}$ denotes the tensor algebra of $\mathfrak{g}$. Note that the ideal by which we quotient ensures that the commutator in $U\mathfrak{g}$ agrees with the bracket in $\mathfrak{g}$: for all $x, y \in \mathfrak{g}$,

$$x \cdot y - y \cdot x = [x,y],$$

where $\cdot$ denotes multiplication in $U\mathfrak{g}$. It is straightforward to verify that there is an adjunction

$$U : LieAlg_{\mathbb{K}} \leftrightarrows AssAlg_{\mathbb{K}} : Forget,$$

where *Forget*(A) views an associative algebra A over $\mathbb{K}$ as a vector space with bracket given by the commutator of its product. As a consequence, we can view $\mathfrak{g} - mod$ as the category of left $U\mathfrak{g}$-modules, which we'll denote $U\mathfrak{g} - mod$, without harm.

Now observe that $\mathbb{K}$ is a *trivial* $\mathfrak{g}$-module for any Lie algebra $\mathfrak{g}$: $x \cdot k = 0$ for all $x \in \mathfrak{g}$ and $k \in \mathbb{K}$. Moreover, $\mathbb{K}$ is the quotient of $U\mathfrak{g}$ by the ideal $(\mathfrak{g})$ generated by $\mathfrak{g}$ itself, so that $\mathbb{K}$ is a *bi*module over $U\mathfrak{g}$. It is then straightforward to verify that

$$M_{\mathfrak{g}} = \mathbb{K} \otimes_{U\mathfrak{g}} M \quad \text{and} \quad M^{\mathfrak{g}} = \mathrm{Hom}_{U\mathfrak{g}}(\mathbb{K}, M)$$

for every module M.

Definition A.4.1 For M a $\mathfrak{g}$-module, the *Lie algebra homology* of M is

$$H_*(\mathfrak{g}, M) = \mathrm{Tor}_*^{U\mathfrak{g}}(\mathbb{K}, M),$$

and the *Lie algebra cohomology* of M is

$$H^*(\mathfrak{g}, M) = \mathrm{Ext}_{U\mathfrak{g}}^*(\mathbb{K}, M).$$

Notice that H^0 and H_0 recover invariants and coinvariants, respectively. There are concrete interpretations of the lower (co)homology groups, such as that $H^1(\mathfrak{g}, M)$ describes "outer derivations" (i.e., derivations modulo inner derivations) and and $H^2(\mathfrak{g}, M)$ describes extensions of $\mathfrak{g}$ as a Lie algebra by M.

There are standard cochain complexes for computing Lie algebra (co)homology, and their generalizations will play a large role throughout the book. The key is to find an efficient, tractable resolution of $\mathbb{K}$ as a $U\mathfrak{g}$ module. We use again that $\mathbb{K}$ is a quotient of $U\mathfrak{g}$ to produce a resolution:

$$\left(\cdots \to \wedge^n \mathfrak{g} \otimes_{\mathbb{K}} U\mathfrak{g} \to \cdots \to \mathfrak{g} \otimes_{\mathbb{K}} U\mathfrak{g} \to U\mathfrak{g} \right) \xrightarrow{\sim} \mathbb{K},$$

where the final map is given by the quotient and the remaining maps are of the form

$$(y_1 \wedge \cdots \wedge y_n) \otimes (x_1 \cdots x_m) \mapsto \sum_{k=1}^{n} (-1)^{n-k} (y_1 \wedge \cdots \widehat{y_k} \cdots \wedge y_n)$$

$$\otimes (y_k \cdot x_1 \cdots x_m)$$

$$- \sum_{1 \le j < k \le n} (-1)^{j+k-1} ([y_j, y_k] \wedge y_1 \wedge \cdots \widehat{y_j} \cdots \widehat{y_k} \cdots \wedge y_n)$$

$$\otimes (x_1 \cdots x_m),$$

where the hat $\widehat{y_k}$ indicates removal. As this is a free resolution of $\mathbb{K}$, we use it to compute the relevant Tor and Ext groups: for coinvariants we have

$$\mathbb{K} \otimes_{U\mathfrak{g}}^{\mathbb{L}} M \simeq \left(\cdots \to \wedge^n \mathfrak{g} \otimes_{\mathbb{K}} M \to \cdots \to \mathfrak{g} \otimes_{\mathbb{K}} M \to M \right)$$

and for invariants we have

$$\mathbb{R}\,\mathrm{Hom}_{U\mathfrak{g}}(\mathbb{K}, M) \simeq \left(M \to \mathfrak{g}^{\vee} \otimes_{\mathbb{K}} M \to \cdots \to \wedge^n \mathfrak{g}^{\vee} \otimes_{\mathbb{K}} M \to \cdots \right),$$

where $\mathfrak{g}^{\vee} = \mathrm{Hom}_{\mathbb{K}}(\mathfrak{g}, \mathbb{K})$ is the linear dual. These resolutions were introduced by Chevalley and Eilenberg and so their names are attached.

Definition A.4.2 The *Chevalley–Eilenberg complex for Lie algebra homology* of the $\mathfrak{g}$-module M is

$$C_*(\mathfrak{g}, M) = (\mathrm{Sym}_{\mathbb{K}}(\mathfrak{g}[1]) \otimes_{\mathbb{K}} M, d)$$

where the differential d encodes the bracket of $\mathfrak{g}$ on itself and on M. Explicitly, we have

$$d(x_1 \wedge \cdots \wedge x_n \otimes m) = \sum_{1 \le j < k \le n} (-1)^{j+k} [x_j, x_k] \wedge x_1 \wedge \cdots \widehat{x_j} \cdots \widehat{x_k} \cdots \wedge x_n \otimes m$$

$$+ \sum_{j=1}^{n} (-1)^{n-j} x_1 \wedge \cdots \widehat{x_j} \cdots \wedge x_n \otimes [x_j, m].$$

We often call this complex the *Chevalley–Eilenberg chains*.

The *Chevalley–Eilenberg complex for Lie algebra cohomology* of the $\mathfrak{g}$-module M is

$$C^*(\mathfrak{g}, M) = (\mathrm{Sym}_{\mathbb{K}}(\mathfrak{g}^{\vee}[-1]) \otimes_{\mathbb{K}} M, d)$$

where the differential d encodes the linear dual to the bracket of $\mathfrak{g}$ on itself and on M. Fixing a linear basis $\{e_k\}$ for $\mathfrak{g}$ and hence a dual basis $\{e^k\}$ for $\mathfrak{g}^\vee$, we have

$$d(e^k \otimes m) = -\sum_{i<j} e^k([e_i, e_j])e^i \wedge e^j \otimes m + \sum_l e^k \wedge e^l \otimes [e_l, m]$$

and we extend d to the rest of the complex as a derivation of cohomological degree 1 (i.e., use the Leibniz rule repeatedly to reduce to the preceding text). We often call this complex the *Chevalley–Eilenberg cochains*.

When M is the trivial module $\mathbb{K}$, we simply write $C_*(\mathfrak{g})$ and $C^*(\mathfrak{g})$. It is important for us that $C^*(\mathfrak{g})$ is a commutative dg algebra and that $C_*(\mathfrak{g})$ is a cocommutative dg coalgebra. This property suggests a geometric interpretation of the Chevalley–Eilenberg complexes: under the philosophy that every commutative algebra should be interpreted as the functions on some "space," we view $C^*(\mathfrak{g})$ as "functions on a space $B\mathfrak{g}$" and $C_*(\mathfrak{g})$ as "distributions on $B\mathfrak{g}$." Here we interpret the natural pairing between the two complexes as providing the pairing between functions and distributions. This geometric perspective on the Chevalley–Eilenberg complexes motivates the role of Lie algebras in deformation theory, as we explain in the following section.

References

Weibel (1994) contains a chapter on the homological algebra of ordinary Lie algebras, of which we have given a gloss. In Lurie (2009a), Lurie gives an efficient treatment of this homological algebra in the language of model and infinity categories.

A.4.2 dg Lie Algebras

We now quickly extend and generalize homologically the notion of a Lie algebra. Our base ring will now be a commutative algebra R over a characteristic zero field $\mathbb{K}$, and we encourage readers to keep in mind the simplest case: where $R = \mathbb{R}$ or $\mathbb{C}$. Of course, one can generalize the setting considerably, with a little care, by working in a symmetric monoidal category (with a linear flavor); the cleanest approach is to use operads.

Definition A.4.3 A *dg Lie algebra* over R is a $\mathbb{Z}$-graded R-module $\mathfrak{g}$ such that

(1) There is a differential

$$\cdots \xrightarrow{d} \mathfrak{g}^{-1} \xrightarrow{d} \mathfrak{g}^0 \xrightarrow{d} \mathfrak{g}^1 \to \cdots$$

making $(\mathfrak{g}, d)$ into a dg R-module.

(2) There is a bilinear bracket $[-, -] : \mathfrak{g} \otimes_R \mathfrak{g} \to \mathfrak{g}$ such that

- $[x, y] = -(-1)^{|x||y|}[y, x]$ (graded antisymmetry)
- $d[x, y] = [dx, y] + (-1)^{|x|}[x, dy]$ (graded Leibniz rule)
- $[x, [y, z]] = [[x, y], z] + (-1)^{|x||y|}[y, [x, z]]$ (graded Jacobi rule),

where $|x|$ denotes the cohomological degree of $x \in \mathfrak{g}$.

In other words, a dg Lie algebra is an algebra over the operad *Lie* in the category of dg R-modules. In practice – and for the rest of the section – we require the graded pieces $\mathfrak{g}^k$ to be projective R-modules so that we do not need to worry about the tensor product or taking duals.

Here are several examples.

(a) We construct the dg analog of $\mathfrak{gl}_n$. Let (V, d_V) be a cochain complex over $\mathbb{K}$. Let $\mathrm{End}(V) = \oplus_n \mathrm{Hom}^n(V, V)$ denote the graded vector space where Hom^n consists of the linear maps that shift degree by n, equipped with the differential

$$d_{\mathrm{End}\, V} = [d_V, -] : f \mapsto d_V \circ f - (-1)^{|f|} f \circ d_V.$$

The commutator bracket makes $\mathrm{End}(V)$ a dg Lie algebra over $\mathbb{K}$.

(b) For M a smooth manifold and $\mathfrak{g}$ an ordinary Lie algebra (such as $su(2)$), the tensor product $\Omega^*(M) \otimes_{\mathbb{R}} \mathfrak{g}$ is a dg Lie algebra where the differential is simply the exterior derivative and the bracket is

$$[\alpha \otimes x, \beta \otimes y] = \alpha \wedge \beta \otimes [x, y].$$

We can view this dg Lie algebra as living over $\mathbb{K}$ or over the commutative dg algebra $\Omega^*(M)$. This example appears naturally in the context of gauge theory.

(c) For X a simply connected topological space, let $\mathfrak{g}_X^{-n} = \pi_{1+n}(X) \otimes_{\mathbb{Z}} \mathbb{Q}$ and use the Whitehead product to provide the bracket. Then $\mathfrak{g}_X$ is a dg Lie algebra with zero differential. This example appears naturally in rational homotopy theory.

The Chevalley–Eilenberg complexes of Lie algebras admit immediate analogues for dg Lie algebras. The formulas are more complicated, owing to the internal differential of a dg Lie algebra $\mathfrak{g}$ and the signs arising from the brackets, which now depend on the cohomological degree of the inputs. But the constructions are parallel. One starts by providing the natural resolution of $\mathbb{K}$ as a $U\mathfrak{g}$-module: view the resolution given for an ordinary Lie algebra as a chain complex of chain complexes (i.e., as a double complex) and then take the totalization. Then one uses this resolution to compute the derived hom and tensor products over $U\mathfrak{g}$. For a thorough reference on dg Lie algebras and homological algebra with them, see Félix et al. (2001).

A.5 Sheaves, Cosheaves, and Their Homotopical Generalizations

Sheaves appear throughout geometry and topology because they capture the idea of gluing together local data to obtain something global. *Cosheaves* are an equally natural construction that are nonetheless used less frequently. We will give a very brief discussion of these ideas. As we always work with sheaves and cosheaves of a linear nature, we give definitions in that setting.

A.5.1 Sheaves

Definition A.5.1 A *presheaf* of vector spaces on a space X is a functor $\mathcal{F}$: $\mathrm{Opens}_X^{op} \to Vect_{\mathbb{K}}$ where Opens_X is the category encoding the partially ordered set of open sets in X (i.e., the objects are open sets in X and there is a map from U to V exactly when $U \subset V$).

In other words, a presheaf $\mathcal{F}$ assigns a vector space $\mathcal{F}(U)$ to each open U and a *restriction map* $res_{V \supset U} : \mathcal{F}(V) \to \mathcal{F}(U)$ whenever $U \subset V$. Bear in mind the following two standard examples. The *constant presheaf* $\mathcal{F} = \mathbb{K}$ has $\mathbb{K}(U) = \mathbb{K}$ (hence it assigns the same vector space to every open) and its restriction map is always the identity. The *presheaf of continuous functions* C_X^0 assigns the vector space $C_X^0(U)$ of continuous functions from U to $\mathbb{R}$ (or $\mathbb{C}$, as one prefers) and the restriction map $res_{V \supset U}$ consists precisely of restricting a continuous function from V to a smaller open U.

A sheaf is a presheaf whose value on an open is determined by its behavior on smaller opens.

Definition A.5.2 A *sheaf* of vector spaces on a space X is a presheaf $\mathcal{F}$ such that for every open U and every cover $\mathfrak{U} = \{V_i\}_{i \in I}$ of U, we have

$$\mathcal{F}(U) \xrightarrow{\cong} \lim \left(\prod_{i \in I} \mathcal{F}(V_i) \rightrightarrows \prod_{i,j \in I} \mathcal{F}(V_i \cap V_j) \right),$$

where the map out of $\mathcal{F}(U)$ is the product of the restriction maps for the inclusion of each V_i into U and where, in the limit diagram, the top arrow is restriction for the inclusion of $V_i \cap V_j$ into V_i and the bottom arrow is restriction for the inclusion of $V_i \cap V_j$ into V_j.

This *gluing condition* says that an element $s \in \mathcal{F}(U)$, called a *section of* $\mathcal{F}$ *on* U, is given by sections on the cover, $(s_i \in \mathcal{F}(V_i))_{i \in I}$, that agree on the overlapping opens $V_i \cap V_j$. It captures in a precise way how to reconstruct $\mathcal{F}$ on an open in terms of data on a cover.

It is a good exercise to verify that C_X^0 is always a sheaf and that $\underline{\mathbb{K}}$ is *not* a sheaf on a disconnected space.

In this book, our spaces are nearly always smooth manifolds, and most of our sheaves arise in the following way. Let $E \to X$ be a vector bundle on a smooth manifold. Let $\mathcal{E}$ denote the presheaf where $\mathcal{E}(U)$ is the vector space of smooth sections $E|_U \to U$ on the open set U. It is quick to show that $\mathcal{E}$ is a sheaf.

A.5.2 Cosheaves

We now discuss the dual notion of a *cosheaf*.

Definition A.5.3 A *precosheaf* of vector spaces on a space X is a functor $\mathcal{G} :$ Opens$_X \to$ *Vect*. A *cosheaf* is a precosheaf such that for every open U and every cover $\mathfrak{U} = \{V_i\}_{i \in I}$ of U, we have

$$\text{colim}\left(\coprod_{i,j \in I} \mathcal{G}(V_i \cap V_j) \rightrightarrows \coprod_{i \in I} \mathcal{G}(V_i) \right) \xrightarrow{\cong} \mathcal{G}(U),$$

where the map into $\mathcal{G}(U)$ is the coproduct of the *extension* maps from the V_i to U and where, in the colimit diagram, the top arrow is extension from $V_i \cap V_j$ to V_i and the bottom arrow is extension from $V_i \cap V_j$ to V_j.

The crucial example of a cosheaf (for us) is the functor $\mathcal{E}_c$ that assigns to the open U, the vector space $\mathcal{E}_c(U)$ of *compactly supported* smooth sections of E on U. If $U \subset V$, we can *extend* a section $s \in \mathcal{E}_c(U)$ to a section $ext_{U \subset V}(s) \in \mathcal{E}_c(V)$ on V by setting it equal to zero on $V \setminus U$.

A closely related example is the cosheaf of compactly supported distributions dual to the sheaf $\mathcal{E}$ of smooth sections of a bundle. When $\mathcal{E}$ denotes the sheaf of fields for a field theory, this cosheaf describes the linear observables on the fields and, importantly for us, organizes them by their support.

A.5.3 Homotopical Versions

Often, we want our sheaves or cosheaves to take values in categories of a homotopical nature. For instance, we might assign a cochain complex to each open set, rather than a mere vector space. In this homotopical setting, one typically works with a modified version of the gluing axioms above. The modifications are twofold:

(i) We work with all finite intersections of the opens in the cover (i.e., not just overlaps $V_i \cap V_j$ but also $V_{i_1} \cap \cdots \cap V_{i_n}$).
(ii) We use the *homotopy* limit (or colimit).

The first modification is straightforward – and familiar to anyone who has seen a Čech complex – but the second is more subtle. (We highly recommend Dugger (2008) for an introduction and development of these notions.)

In our main examples, we give explicit complexes that encode the relevant information.

For completeness' sake, we develop the homotopical version explicitly. A *precosheaf* with values in dg vector spaces is a functor $\mathcal{G} : \text{Opens}_X \to dgVect$. Given a cover cover $\mathfrak{U} = \{V_i\}_{i \in I}$ of an open U, the *Čech complex* $\check{C}(\mathfrak{U}, \mathcal{G})$ is the totalization of the following double cochain complex: take the cochain complex, with values in dg vector spaces, whose $-n$th term is

$$\bigoplus_{\vec{i} \in I^{n+1}} \mathcal{G}(\cap_{j=0}^{n} V_{i_j}),$$

which has an internal differential inherited from $\mathcal{G}$, and whose "external" differential is the alternating sum of the structure maps for the inclusion of an $n+1$-fold intersection into an n-fold intersection. We write

$$\check{C}(\mathfrak{U}, \mathcal{G}) = \text{Tot}^{\oplus} \left(\bigoplus_{n=0}^{\infty} \bigoplus_{\vec{i} \in I^{n+1}} \mathcal{G}(\cap_{j=0}^{n} V_{i_j})[n] \right),$$

as a compact description. Observe that there is a canonical map from $\check{C}(\mathfrak{U}, \mathcal{G})$ to $\mathcal{G}(U)$ given by the sum of the structure maps from $\mathcal{G}(U_i)$ to $\mathcal{G}(U)$. (In other words, one can extend the double complex by inserting $\mathcal{G}(U)$ as the first term.)

Definition A.5.4 A *homotopy cosheaf* with values in dg vector spaces is a precosheaf $\mathcal{G} : \text{Opens}_X \to dgVect$ such that for every open U and every cover $\mathfrak{U} = \{V_i\}_{i \in I}$ of U, the natural map

$$\check{C}(\mathfrak{U}, \mathcal{G}) \xrightarrow{\simeq} \mathcal{G}(U),$$

is a quasi-isomorphism.

We say that a homotopy cosheaf $\mathcal{G}$ satisfies Čech (co)descent.

Remark: In our preceding definition, we use implicitly the fact that the Čech complex actually provides a representative of the homotopy colimit. For dg vector spaces (or, more generally, cochain complexes of modules over a ring), one can deduce this assertion from standard results in homotopical algebra. We provide an argument in the more general setting of cochain complexes in a Grothendieck Abelian category in Section C.5 in Appendix C.   ◊

A.5.4 Partitions of Unity and the Čech Complex

We review here why compactly supported sections of a vector bundle form both a cosheaf and a homotopy cosheaf. Our arguments are simple modifications of the standard arguments for why smooth sections of a vector bundle form both a sheaf and a homotopy sheaf. (See, for instance, Bott and Tu (1982).)

Definition A.5.5 A *partition of unity* on a smooth manifold M subordinate to an open cover $\{U_i\}_{i \in I}$ is a collection $\{\rho_i\}_{i \in I}$ of nonnegative smooth functions such that

(1) Every point $x \in M$ has a neighborhood in which all but finitely many ρ_i vanish.
(2) The sum $\sum_i \rho_i = 1$.
(3) supp $\rho_i \subset U_i$ for all $i \in I$.

For every open cover, there is a partition of unity subordinate to it.

Let $\pi : V \to M$ be a smooth vector bundle. Let $\mathcal{V}_c$ denote the precosheaf that assigns to the open $U \subset M$ the vector space $\mathcal{V}_c(U)$ of compactly supported smooth sections of V on U. Note that $\mathcal{V}_c(U)$ is a module for $C^\infty(U)$. Partitions of unity then make it easy to verify the cosheaf gluing condition.

Lemma A.5.6 *$\mathcal{V}_c$ is a cosheaf.*

Proof Let $\{U_i\}_{i \in I}$ be an open cover of an open $U \subset M$. Let $ext_i : \mathcal{V}_c(U_i) \to \mathcal{V}_c(U)$ denote the extension-by-zero maps defining the precosheaf structure of $\mathcal{V}_c$. Then the map

$$\sum_i ext_i : \bigoplus_i \mathcal{V}_c(U_i) \to \mathcal{V}_c(U)$$

induces a map

$$ext_I : \mathrm{colim}\left(\bigoplus_{i,j \in I} \mathcal{V}_c(U_i \cap U_j) \rightrightarrows \bigoplus_{i \in I} \mathcal{V}_c(U_i) \right) \to \mathcal{V}_c(U) \qquad (\dagger)$$

since the extension from $U_i \cap U_j$ to U factors injectively through both U_i and U_j.

To show that $\mathcal{V}_c$ is a cosheaf, we need to show that the map ext_I is an isomorphism. A choice of partition of unity allows us to split the map ext_I, and thus it becomes straightforward to verify the isomorphism.

Fix a partition of unity $\{\rho_i\}_{i \in I}$ subordinate to the cover. Then we obtain a map

$$\tilde{\rho}_i : \mathcal{V}_c(U) \to \mathcal{V}_c(U_i)$$

sending v to $\rho_i v$. (Note that here we use the fact that if the support of a section on U is contained in U_i, it is in the image of ext_i.) The map

$$\sum_i \tilde{\rho}_i : \mathcal{V}_c(U) \to \bigoplus_i \mathcal{V}_c(U_i)$$

has the property that

$$\left(\sum_i ext_i \right) \circ \left(\sum_i \tilde{\rho}_i \right) = \mathrm{id}_{\mathcal{V}_c(U)}$$

by the definition of a partition of unity.

We now want to show that this map $\sum_i \tilde{\rho}_i$ induces an inverse to ext_I. Note that the colimit appearing in (†) is a quotient of $\oplus_i \mathcal{V}_c(U_i)$. If we denote this quotient space by Q, then we have an exact sequence

$$\bigoplus_{i,j} \mathcal{V}_c(U_i \cap U_j) \xrightarrow{f} \bigoplus_i \mathcal{V}_c(U_i) \to Q \to 0,$$

where f denotes the difference of the two maps in the colimit appearing in (†). Consider the composition of $\sum_i \tilde{\rho}_i$ followed by the projection to A. To show it is a left inverse to ext_I, we will show that every element v in $\oplus_i \mathcal{V}_c(U_i)$ is in the same equivalence class as $\sum_i \tilde{\rho}_i \circ \sum_i ext_i(v)$, i.e., they map to the same element in Q.

It suffices to verify this for a particularly simple class of elements that span $\oplus_i \mathcal{V}_c(U_i)$. Fix an index $k \in I$ and let $v = (v_i) \in \oplus_i \mathcal{V}_c(U_i)$ have the property that $v_i = 0$ if $i \neq k$. Let $\tilde{v} = (\tilde{v}_i)$ be $\sum_i \tilde{\rho}_i \circ \sum_i ext_i(v)$. Then $\tilde{v}_i = \rho_i v_k$. We want to show that v and $\tilde{v}$ map to the same element in Q, so we need their difference to live in image of f. Consider the element $w = (w_{ij})$ in $\oplus_{i,j} \mathcal{V}_c(U_i \cap U_j)$ where $w_{ik} = \rho_i v_k$ for all i and $w_{ij} = 0$ for $j \neq k$. Then $f(w) = (f(w)_i)$ is given by $f(w)_i = \sum_j (w_{ij} - w_{ji})$ and so $f(w)_i = \rho_i v_k$ for $i \neq k$ and

$$f(w)_k = \rho_k v_k - \sum_j \rho_j v_k = \rho_k v_k - v_k.$$

Hence $f(w) = \tilde{v} - v$. As this holds for any k, we know it holds for all vectors v. $\qquad\square$

The preceding argument is interesting because it shows not only that $\mathcal{V}_c$ is a cosheaf but exhibits an explicit decomposition of a section into sections on the cover, by using a partition of unity. In fact, the argument can be carried further.

Lemma A.5.7 *Let* $\mathfrak{U} = \{U_i\}_{i \in I}$ *be an open cover of an open* $U \subset M$, *and let* $\{\rho_i\}_{i \in I}$ *be a partition of unity subordinate to the cover. There is a cochain homotopy equivalence between* $\check{C}(\mathfrak{U}, \mathcal{V}_c)$ *and* $\mathcal{V}_c(U)$ *where the cochain map*

$$\sigma = \sum_i ext_{U_i \subset U} : \check{C}(\mathfrak{U}, \mathcal{V}_c) \to \mathcal{V}_c(U)$$

is determined by the extension maps. Explicitly, σ *vanishes on an element of* $\mathcal{V}_c(U_{i_0} \cap \cdots \cap U_{i_n})$ *for* $n > 0$ *and it sends an element of* $\mathcal{V}_c(U_i)$ *to its extension by zero.*

This lemma immediately implies the following.

Corollary A.5.8 $\mathcal{V}_c$ *is a homotopy cosheaf.*

Proof of lemma Recall that a cochain map induces a cochain homotopy equivalence if and only if the mapping cone admits a contracting cochain homotopy. We will thus show instead that the "augmented" cochain complex given by the cone of σ admits a contracting homotopy K. (Those who prefer simplicial constructions will recognize that we are giving a simplicial cochain complex with extra degeneracies.)

A degree $-n$ element v of $\check{C}(\mathfrak{U}, \mathcal{V}_c)$ can be expressed in terms of its components $(v_{\vec{i}})_{\vec{i} \in I^{n+1}}$, where each $v_{\vec{i}} \in \mathcal{V}_c(U_{i_0} \cap \cdots \cap U_{i_n})$. The cone is obtained by putting $\mathcal{V}_c(U)$ in degree 1 and making σ the differential from degree 0 to degree 1.

The contraction K sends a section $v \in \mathcal{V}_c(U)$ (i.e., a degree 1 element) to $K(v) = (K(v)_i)_{i \in I}$ where $K(v)_i = \rho_i v$. (Note that this is precisely the map $\tilde{\rho}$ from the proof that $\mathcal{V}_c$ is a cosheaf.) To a degree $-n$ element $v = (v_{\vec{i}})_{\vec{i} \in I^{n+1}}$, the contraction K assigns

$$K(v)_{\vec{j}} = \rho_{j_0} v_{\vec{j}_{>0}},$$

where $\vec{j} \in I^{n+2}$ and $\vec{j}_{>0} = (j_1, \ldots, j_{n+1})$. (Note that this is a natural extension of the construction of the element w.) Recall that for $\vec{i} \in I^{n+1}$, we have

$$d(v)_{\vec{i}} = \sum_{j=0}^{n+1} (-1)^j \sum_{k \in I} v_{c_j(k,\vec{i})}$$

where $c_j(k, \vec{i}) = (i_0, \ldots, i_{j-1}, k, i_j, \ldots, i_n)$. (There is no internal differential here to worry about, just the Čech differential.) Then we find, with a little manipulation, that

$$((dK + Kd)v)_{\vec{i}} = \rho_{i_0} \sum_{k \in I} \left(\sum_{j=0}^{n} (-1)^j v_{c_j(k,\vec{i}_{>0})} + \sum_{j=1}^{n} (-1)^j v_{c_j(k,\vec{i})_{>0}} \right)$$
$$+ \sum_{k \in I} \rho_{c_0(k,\vec{i})} v_{c_0(k,\vec{i})}.$$

On the right-hand side, the term multiplied by ρ_{i_0} vanishes, and the remaining term simplifies as

$$\sum_{k \in I} \rho_{c_0(k,\vec{i})} v_{c_0(k,\vec{i})} = \left(\sum_{k \in I} \rho_k \right) v_{\vec{i}} = v_{\vec{i}}.$$

Hence $[d, K] = \text{id}$. $\qquad\qquad\qquad\qquad\qquad\qquad\qquad\qquad\qquad\qquad\qquad\square$

These arguments are minor modifications of those in Bott and Tu (1982). In their section 8, just after Proposition 8.5, they construct the homotopy equivalence we use in the case of the de Rham complex. It is quick to see that the construction applies to any smooth graded vector bundle. In the case of the de Rham complex, they explicitly provide the cochain homotopy as the "Collating Formula," Proposition 9.5.

References

Nearly any modern book on differential or algebraic geometry contains an introduction to sheaves. See, for instance, Griffiths and Harris (1994) or Ramanan (2005). For a clear exposition (and more) of cosheaves, see Curry (2014). The classic text Bott and Tu (1982) explores in depth the sheaf of smooth sections of a vector bundle, but typically via the example of the de Rham complex. For systematic development of sheaves on sites and sheaves of modules over a sheaf of rings, see Kashiwara and Schapira (2006) or Stacks Project Authors (2016).

A.6 Elliptic Complexes

Classical field theory involves the study of systems of partial differential equations (or generalizations), which is an enormously rich and sophisticated subject. We focus in this book on a tractable and well-understood class of partial differential equations that appear throughout differential geometry: elliptic complexes. Here we will spell out the basic definitions and some examples that suffice for our work in this book.

A.6.1 Ellipticity

We start with a local description of differential operators before homing in on elliptic operators. Let U be an open set in $\mathbb{R}^n$. A linear *differential operator* is an $\mathbb{R}$-linear map $L : C^\infty(U) \to C^\infty(U)$ of the form

$$L(f) = \sum_{\alpha \in \mathbb{N}^n} a_\alpha(x) \partial^\alpha f(x),$$

where we use the multi-index $\alpha = (\alpha_1, \ldots, \alpha_n)$ to efficiently denote

$$\partial^\alpha = \frac{\partial^{\alpha_1}}{\partial x_1^{\alpha_1}} \cdots \frac{\partial^{\alpha_n}}{\partial x_n^{\alpha_n}},$$

where the coefficients a_α are smooth functions, and where only finitely many of the a_α are nonzero functions. We call $|\alpha| = \sum_j \alpha_j$ the *order* of the index, and thus the *order of L* is the maximum order $|\alpha|$ among the nonzero coefficients a_α. (For example, the order of the Laplacian $\sum_j \partial^2/\partial x_j^2$ is two.) The *principal symbol* (or *leading symbol*) of a kth order differential operator L is the "fiberwise polynomial"

$$\sigma_L(\xi) = \sum_{|\alpha|=k} a_\alpha(x) i^k \xi^\alpha$$

obtained by summing over the indices of order k and replacing the partial derivative $\partial/\partial x_j$ by variables $i\xi_j$, where i is the usual square root of 1. (It is the standard convention to include the factor of i, due to the role of Fourier transforms in motivating many of these constructions and definitions.) It is natural to view the principal symbol as a function on the cotangent bundle T^*U that is a homogeneous polynomial of degree k along the cotangent fibers, where ξ_j is the linear functional dual to dx_j. The principal symbol controls the qualitative behavior of L. It also behaves nicely under changes of coordinates, transforming as a section of the bundle $\mathrm{Sym}^k(T_U) \to U$.

Remark: Elsewhere in the text, we talk about differential operators in a more abstract, homological setting. For instance, we say the BV Laplacian is a second-order differential operator. Our use of the differential operators in this homological setting is inspired by their use in analysis, but the definitions are modified, of course. For instance, we view the odd directions as geometric in the BV formalism, so the BV Laplacian is second order. $\Diamond$

It is straightforward to extend these notions to a smooth manifold and to smooth sections of vector bundles on that manifold. Let $E \to X$ be a rank m vector bundle and let $F \to X$ be a rank n vector bundle. Let $\mathscr{E}$ denote the smooth global sections of E and let $\mathscr{F}$ denote the smooth global sections of F.

A *differential operator* from E to F is an $\mathbb{R}$-linear map $L : \mathscr{E} \to \mathscr{F}$ such that for a local choice of coordinates on X and trivializations of E and F,

$$(Ls)_i = \sum_{\alpha \in \mathbb{N}^n} a_\alpha^{ij}(x) \partial^\alpha s_j(x),$$

where $s = (s_1, \ldots, s_m)$ is a section of $\mathscr{E}(U) \cong C^\infty(U)^m$ on a small neighborhood U in X, $Ls = ((Ls)_1, \ldots, (Ls)_n)$ is a section of $\mathscr{F}(U) \cong C^\infty(U)^n$ on that small neighborhood, and where the $a_\alpha^{ij}(x)$ are smooth functions. In other words, locally we have a matrix-valued differential operator. The *principal symbol* σ_L of a kth order differential operator L defines a section of the bundle $\mathrm{Sym}^k(T_X) \otimes E^\vee \otimes F \to X$. Thus, by pulling back along the canonical projection $\pi : T^*X \to X$, we can view the principal symbol as a map of *vector bundles* $\sigma_L : \pi^*E \to \pi^*F$ over the cotangent bundle T^*X.

We now introduce one of our key definitions.

Definition A.6.1 An *elliptic operator* $L : \mathscr{E} \to \mathscr{F}$ is a differential operator whose principal symbol $\sigma_L : \pi^*E \to \pi^*F$ is an isomorphism of vector bundles on $T^*X \setminus X$, the cotangent bundle with the zero section removed.

This definition says that the principal symbol is an *invertible* linear operator after evaluating at any nonzero covector. As an example, consider the Laplacian on $\mathbb{R}^n$: its principal symbol is $\sum \xi_j^2$, which vanishes only when all the ξ_j are zero.

A.6.2 Functional Analytic Consequences

Ellipticity is purely local and is an easy property to check in practice. Globally, it has powerful consequences, of which the following is the most famous.

Theorem A.6.2 *For X a closed manifold (i.e., compact and boundaryless), an elliptic operator $Q : \mathscr{E} \to \mathscr{F}$ is Fredholm, so that its kernel and cokernel are finite-dimensional vector spaces.*

Thus, an elliptic operator on a closed manifold is invertible up to a finite-dimensional "error." Moreover, there is a rich body of techniques for constructing these partial inverses, especially for classical operators such as the Laplacian.

The preceding notions extend naturally to cochain complexes. A *differential complex* on a manifold X is a $\mathbb{Z}$-graded vector bundle $\oplus_n E^n \to X$ (with finite total rank) and a differential operator $Q^n : \mathscr{E}^n \to \mathscr{E}^{n+1}$ for each integer n such that $Q^{n+1} \circ Q^n = 0$. We typically denote this by $(\mathscr{E}, Q)$. There is an associated *principal symbol complex* (π^*E, σ_Q) on T^*X by taking the principal symbol of each operator Q^n.

Definition A.6.3 An *elliptic complex* is a differential complex whose principal symbol complex is exact on $T^*X \setminus X$ (i.e., the cohomology of the symbol complex vanishes away from the zero section of the cotangent bundle).

Every elliptic operator $Q : \mathscr{E} \to \mathscr{F}$ defines a two-term elliptic complex

$$\cdots \to 0 \to \mathscr{E} \xrightarrow{Q} \mathscr{F} \to 0 \to \cdots.$$

The other standard examples of elliptic complexes are the de Rham complex $(\Omega^*(X), d)$ of a smooth manifold and the Dolbeault complex $(\Omega^{0,*}(X), \bar{\partial})$ of a complex manifold.

The analog of the Fredholm result above is the following, sometimes known as the formal Hodge theorem (by analogy to the Hodge theorem, which is usually focused on the de Rham complex).

Theorem A.6.4 *Let $(\mathscr{E}, Q)$ be an elliptic complex on a closed manifold X. Then the cohomology groups $H^k(\mathscr{E}, Q)$ are finite-dimensional vector spaces.*

In fact, the proof does something better: it constructs a continuous cochain homotopy equivalence between the elliptic complex and its cohomology. Below, we will explain some ingredients of the proof, emphasizing aspects that play a role in our constructions of observables.

A.6.3 Parametrices

In proving these theorems, one constructs a "partial inverse" to Q with nice, geometric properties. We will not give a full definition here, because we do not want to delve into pseudodifferential operators, but will state the important properties. We will invoke the analysis as a black box.

Recall that every continuous linear operator $F : \mathscr{E} \to \mathscr{F}$ between smooth sections of a vector bundles possesses a Schwartz kernel, K_F, a section of the bundle $F \boxtimes (E^\vee \otimes \mathrm{Dens})$ on $X \times X$ that is smooth along the first copy of X (i.e., in the F direction) and distributional along the second copy of X (i.e., in the E direction). Then

$$F(s)(x) = \int_{y \in X} K_F(x,y)s(y).$$

The notation is a suggestive way of writing the composition

$$\mathscr{E} \xrightarrow{F \otimes \mathrm{id}} \mathscr{F} \widehat{\otimes}_\pi \mathscr{E}^* \widehat{\otimes}_\pi \mathscr{E} \xrightarrow{\mathrm{id} \otimes \mathrm{ev}} \mathscr{F}$$
$$s \mapsto \qquad F \otimes s \qquad \mapsto \quad F(s)$$

where $\widehat{\otimes}_\pi$ denotes the completed projective tensor product, which is the natural tensor to use in this context. This notation emphasizes the formal similarity to matrix multiplication: the position y provides an index for a kind of "basis," $s(y)$ denotes the coefficient for the vector s of that basis element, $K_F(x, y)$ describes the matrix coefficient, and we sum (i.e., integrate) over a common index y to compute the multiplication by the linear operator F.

If the kernel K_F is smooth over all of $X \times X$, then F is called a *smoothing* operator. For X a closed manifold, a smoothing operator F is compact (when viewed as an operator between Sobolev space completions of the smooth sections).

For $Q : \mathscr{E} \to \mathscr{F}$ an elliptic operator, a *parametrix* is an operator $P : \mathscr{F} \to \mathscr{E}$ satisfying

(i) $\mathrm{id}_{\mathscr{E}} - PQ = S$ for some smoothing operator S,
(ii) $\mathrm{id}_{\mathscr{F}} - QP = T$ for some smoothing operator T, and
(iii) the Schwartz kernel of P is smooth away from the diagonal $X \subset X \times X$.

Thus, a parametrix P for Q is a partial, or approximate, inverse to Q with a particularly nice Schwartz kernel. The theory of pseudodifferential operators provides a construction of such a parametrix for any elliptic operator.

Armed with a parametrix P, we deduce that Q is Fredholm, as follows. First, note that the Fredholm operators are stable under compact perturbations: if T is Fredholm and C is compact, then $T + C$ is also Fredholm. In particular, $\mathrm{id} + C$ is always Fredholm. Second, because $\mathrm{id}_{\mathscr{E}} - S$ is Fredholm, we see that $\mathrm{Ker}(PQ)$ is finite dimensional and hence its subspace $\mathrm{Ker}(Q)$ is finite dimensional. Similarly, we deduce that $\mathrm{Coker}(Q)$ is finite dimensional.

A.6.4 Formal Hodge Theory

We want to explain how to generalize the parametrix approach to elliptic *complexes*. The essential idea to make a retraction of the elliptic complex $\mathscr{E}$ onto its cohomology $H^* \mathscr{E}$, viewed as a complex with zero differential:

$$\eta \, \overset{\curvearrowright}{\bigcirc} \, \mathscr{E} \underset{\iota}{\overset{\pi}{\rightleftarrows}} H^* \mathscr{E} \, .$$

Recall that a retraction means that ι includes the cohomology as a subcomplex, π projects away everything but cohomology, and η is a degree -1 map on $\mathscr{E}$, such that these maps satisfy

$$\mathrm{id}_{H^* \mathscr{E}} = \pi \circ \iota \quad \text{and} \quad \mathrm{id}_{\mathscr{E}} - \iota \circ \pi = [Q, \eta] = Q\eta + \eta Q.$$

Thus, a retraction is a homological generalization of a partial inverse.

We want the retractions for elliptic complexes to have two important properties. First, we require all the maps above to be continuous with respect to the usual Fréchet topologies. Second, the operator $C = \iota \circ \pi$ should be a cochain map that is smoothing, which ensures that $\mathrm{id}_{\mathscr{E}} - C$ is Fredholm.

The existence of such a retraction implies Theorem A.6.4, for the following reason. The operator $C : \mathscr{E} \to \mathscr{E}$ sends any cocycle s to a distinguished representative $C(s)$ of its cohomology class. In fact, $C^2 = C$, so it is a projection operator whose image is isomorphic to the cohomology of $\mathscr{E}$. Moreover, C annihilates exact cocycles (since π does) and elements that are not cocycles. As $\mathrm{id}_{\mathscr{E}} - C$ is an idempotent Fredholm operator, we know the kernel of this operator is finite dimensional, and so the image of C is finite-dimensional.

Finding the necessary retraction η exploits the existence of parametrices for elliptic operators. In the situations relevant to this book, the general idea is simple. First, one finds a differential operator Q^* of degree -1 such that the commutator $D = [Q, Q^*]$ is a generalized Laplacian, meaning its principal symbol looks like that of a Laplacian. (One way to find Q^* is to pick inner products on the bundles E^j (hermitian, if complex bundles) and a Riemannian metric on X. This provides an adjoint to Q, i.e., a differential operator Q^* of degree -1.) In our setting, we require the existence of such a Q^*. This operator D is Fredholm by Theorem A.6.2, so we know it has finite-dimensional kernel.

Second, let G denote the parametrix for D (where G is for "Green's function"). Set $\eta = Q^* G$. We need to show it a contracting homotopy with the desired properties. Let

$$\mathrm{id} - DG = S \quad \text{and} \quad \mathrm{id} - GD = T,$$

where S and T are smoothing endomorphisms of cohomological degree 0. Note that

$$QD = QQ^* Q = DQ,$$

so

$$G(QD)G = G(DQ)G \implies (GQ)(\mathrm{id} - S) = (\mathrm{id} - T)(QG).$$

Hence we find $GQ = QG - U$, where $U = QS - TQG$ is smoothing because Q is a differential operator. In consequence,

$$
\begin{aligned}
[Q, \eta] &= QQ^* G + Q^* GQ \\
&= QQ^* G + Q^* QG - Q^* U \\
&= DG - Q^* U \\
&= \mathrm{id} - (S + Q^* U).
\end{aligned}
$$

The final term in parantheses is smoothing. We have thus verified that η has the desired properties.

A.6.5 References

As usual, Atiyah and Bott (1967) explain beautifully the essential ideas of elliptic complexes and pseudodifferential techniques and show how to use them efficiently. For an accessible development of the analytic methods in the geometric setting, we recommend Wells (2008). The full story and much more is available in the classic works of Hörmander (2003).

Appendix B

Functional Analysis

B.1 Introduction

The goal of this appendix is to introduce several types of vector spaces and explain how they are related. In the end, most of the vector spaces we work with – which are built out of smooth or distributional sections of vector bundles – behave nicely in whichever framework one chooses to use, but it is important to have a setting where abstract constructions behave well. In particular, we will do homological algebra with infinite-dimensional vector spaces, and that requires care. Below, we introduce the underlying "functional analysis" that we need (i.e., we describe here just the vector spaces and discuss the homological issues in a separate appendix). For a briefer overview, see Section 3.5 in Chapter 3.

There are four main categories of vector spaces that we care about:

- LCTVS, the category of locally convex Hausdorff topological vector spaces
- BVS, the category of bornological vector spaces
- CVS, the category of convenient vector spaces
- DVS, the category of differentiable vector spaces

The first three categories are vector spaces equipped with some extra structure, like a topology or bornology, satisfying some list of properties. The category DVS consists of sheaves of vector spaces on the site of smooth manifolds, equipped with some extra structure that allows us to differentiate sections (hence the name).

The main idea is that DVS provides a natural place to compare and relate vector spaces that arise in differential geometry and physics. As a summary of the relationships between these categories, we have the following diagram of functors:

$$\begin{array}{ccc}
\text{BVS} & \xrightarrow{\ inc_\beta\ } & \text{LCTVS} \ . \\
\Big\downarrow c^\infty & \searrow^{dif_\beta} & \Big\downarrow dif_t \\
\text{CVS} & \xrightarrow[\ dif_c\ ]{} & \text{DVS}
\end{array}$$

All the functors into DVS preserve limits. The functors inc_β and c^∞ out of BVS are left adjoints. The functors into DVS all factor as a right adjoint followed by the functor dif_β. For example, dif_t : LCTVS $\to$ DVS is the composition $dif_\beta \circ born$, where the bornologification functor $born$: LCTVS $\to$ BVS is the right adjoint to the inclusion inc_β : BVS $\to$ LCTVS.

For our work, it is also important to understand multilinear maps between vector spaces (in the categories mentioned earlier). Of course, it is pleasant to have a tensor product that represents bilinear maps. In the infinite-dimensional setting, however, the tensor product is far more complicated than in the finite-dimensional setting, with many different versions of the tensor product, each possessing various virtues and defects. We will discuss certain natural tensor products that appear on the aforementioned categories and how the functors intertwine with them.

B.2 Differentiable Vector Spaces

We begin with a sheaf-theoretic approach, introducing the category of *differentiable vector spaces*. This category will provide a setting in which we can compare other approaches.

Definition B.2.1 The *smooth site* Mfld will denote the category where an object is a smooth manifold, a morphism is a smooth map, and a cover f : $X \to Y$ is a surjective local diffeomorphism.

Note that "manifold" means Hausdorff, second-countable, and without boundary.

Our basic objects of study will be sheaves on this site. Thus, from the beginning, everything is built to vary nicely in families over smooth manifolds. For instance, a *smooth vector space* is a sheaf of vector spaces on Mfld.

Remark: As we develop the relevant facts about differentiable vector spaces in this section, it is important to bear in mind some useful facts about (pre)sheaves on a site. For a development of these assertions and much more, see Chapters 17 and 18 of Kashiwara and Schapira (2006) or Chapter 7 of Stacks Project Authors (2016).

Suppose C is a category containing limits and colimits, and let X be a site. Then the inclusion of sheaves on X with values in C into presheaves is a right adjoint whose left adjoint is the sheafification functor. When C is the category of sets Set, then sheafification also preserves finite limits, in addition to all colimits.

Suppose that there is a "forgetful functor $F : C \to$ Set that is faithful, reflects isomorphisms, and commutes with limits and filtered colimits. (For example, consider the usual forgetful functor from vector spaces to sets.) Then a presheaf V with values in C is a sheaf with values in C if and only if $F(V)$ is a sheaf with values in Set, and sheafification commutes with F. These results make it possible to work easily with "sheaves with algebraic structure" as simply sheaves of sets whose openswise algebraic structure intertwines with the structure maps. $\Diamond$

Our focus here will be on smooth vector spaces with some extra structure that makes our main constructions possible.

Definition B.2.2 Let C^∞ denote the smooth vector space that assigns $C^\infty(X)$ to each smooth manifold X. A C^∞-*module* is a smooth vector space V equipped with the structure of a module over C^∞. That is, it is equipped with a map of sheaves $\cdot : C^\infty \times V \to V$, called the scalar multiplication map, satisfying the usual compatibilities with the addition map $+ : V \times V \to V$.

A *map of C^∞-modules* is map of sheaves that respects the structure as a C^∞-module. We denote the category of C^∞-modules by Mod_{C^∞}.

As suggestive notation, we will typically denote the vector space $V(X)$ of sections on X of V by $C^\infty(X, V)$, to emphasize that we think of these as the smooth functions from X to the smooth vector space V. A map $\phi : V \to W$ then consists of a map $\phi_X : C^\infty(X, V) \to C^\infty(X, W)$ for every manifold X such that for every smooth map $f : X \to Y$, we have $f_W^* \circ \phi_Y = \phi_X \circ f_V^*$, where we use $f_V^* : C^\infty(Y, V) \to C^\infty(X, V)$ to denote the natural pullback map.

Remark: Given a presheaf F on Mfld that is a module over C^∞ (in the category of presheaves of vector spaces), the sheafification is also naturally a C^∞-module, and this sheafification can be computed as simply a presheaf of sets or vector spaces. Moreover, the forgetful functor from C^∞-modules to smooth vector spaces is right adjoint to the tensor product $C^\infty \otimes -$, i.e., base-change behaves as expected from ordinary algebra in this sheaf-theoretic context. See Kashiwara and Schapira (2006) or Stacks Project Authors (2016) for more. $\Diamond$

All the natural vector bundles in differential geometry provide examples of C^∞-modules. For instance, consider differential forms. Let Ω^k denote the sheaf that assigns $\Omega^k(X)$ to each smooth manifold X. Given a C^∞-module V, we define the k-forms with values in V as

$$\Omega^k(-, V) := \Omega^k \otimes_{C^\infty} V,$$

so that on a manifold X, we have

$$\Omega^k(X, V) = \Omega^k(X) \otimes_{C^\infty(X)} C^\infty(X, V).$$

With this definition in hand, we can define our main object of interest.

Definition B.2.3 A *differentiable vector space* is a C^∞-module V equipped with a flat connection

$$\nabla_{X,V} : C^\infty(X, V) \to \Omega^1(X, V)$$

for every smooth manifold X such that pullback commutes with the connections,

$$f^* \circ \nabla_{Y,V} = \nabla_{X,V} \circ f^*,$$

for every smooth map $f : X \to Y$.

To say that $\nabla_{X,V}$ is a connection means that it satisfies the Leibniz rule,

$$\nabla_{X,V}(f \cdot v) = (\mathrm{d}f)v + f\nabla_{X,V}v,$$

where $f \in C^\infty(X)$ and $v \in C^\infty(X, V)$. To say that it is flat means that the curvature

$$F(\nabla_{X,V}) = (\nabla_{X,V})^2 : C^\infty(X, V) \to \Omega^2(X, V)$$

vanishes.

The flat connection $\nabla_{X,V}$ thus allows us to differentiate sections of V on X. If $\mathcal{X}$ is a vector field on X, we define

$$\mathcal{X}(v) := \langle \mathcal{X}, \nabla_{X,V}v \rangle \in C^\infty(X, V),$$

where $\langle -, - \rangle$ denotes the $C^\infty(X)$-pairing between vector fields and 1-forms

$$\langle -, - \rangle : \mathcal{T}(X) \times \Omega^1(X, V) \to C^\infty(X, V).$$

Moreover, because the curvature vanishes, the Lie bracket of vector fields goes to the commutator:

$$[\mathcal{X}, \mathcal{Y}](v) = \mathcal{X}(\mathcal{Y}(v)) - \mathcal{Y}(\mathcal{X}(v)).$$

Thus we can take as many derivatives as we would like.

This ability to do calculus with differentiable vector spaces is crucial to many of our constructions. See, for instance, the sections on translation-invariant factorization algebras, holomorphically translation-invariant factorization algebras, and equivariant factorization algebras.

Definition B.2.4 A *map of differentiable vector spaces* $\phi : V \to W$ is a map of C^∞-modules such that

$$\nabla_{X,W} \circ \phi_X = (\mathrm{id}_{\Omega^1(X)} \otimes \phi_X) \circ \nabla_{X,V}$$

for every smooth manifold X.

We denote the category of differentiable vector spaces by DVS. We denote the vector space of morphisms from V to W by $\mathrm{DVS}(V, W)$.

In the remainder of this section, we will introduce a large class of examples of differentiable vector spaces, which we use throughout the book. We will also discuss categorical properties of DVS, such as the existence of limits and colimits. Finally, we describe the natural multicategory structure arising from multilinear maps.

B.2.1 Differentiable Vector Spaces from Sections of a Vector Bundle

Let M be a smooth manifold and $p : E \to M$ a smooth vector bundle. Let $\Gamma(M, E)$ denote the vector space of smooth sections of E. Let $\Gamma_c(M, E)$ denote the vector space of compactly supported smooth sections of E. There is a natural way to construct differentiable vector spaces whose value on a point recovers these familiar vector spaces.

Definition B.2.5 For a smooth manifold X, let $\pi_M : X \times M \to M$ denote the projection map. Let $C^\infty(X, \mathscr{E}(M))$ denote the vector space of smooth sections of $\pi_M^* E$ over $X \times M$. This smooth vector space $\mathscr{E}(M)$ is naturally a C^∞-module, as $C^\infty(X, \mathscr{E}(M))$ is a module over $C^\infty(X)$.

We say $s \in C^\infty(X, \mathscr{E}(M))$ has *proper support over* X if the composition

$$\mathrm{Supp}(s) \hookrightarrow X \times M \xrightarrow{\pi_X} X$$

is a proper map. Let $C^\infty(X, \mathscr{E}_c(M))$ denote the vector space of smooth sections with proper support over X. Then $\mathscr{E}_c(M)$ is a C^∞-module.

We still need to equip these sheaves with flat connections to make them *differentiable* vector spaces. There is a natural choice, because the pullback bundle $\pi_M^* E$ is trivial in the X-direction.

Definition B.2.6 Let X be a smooth manifold. Equip the pullback bundle $\pi_M^* E$ on $X \times M$ with the natural flat connection along the fibers of the projection map π_M (i.e., we differentiate only in X-directions). We thus obtain a map

$$\nabla_{X,\mathscr{E}} : \Gamma(X \times M, \pi_M^* E) \rightarrow \Gamma(X \times M, T_X^* \boxtimes \pi_M^* E),$$

or equivalently a map

$$\nabla_{X,\mathscr{E}} : C^\infty(X, \mathscr{E}(M)) \rightarrow \Omega^1(X, \mathscr{E}(M)).$$

This map defines a flat connection on $\mathscr{E}(M)$ and hence gives it the structure of a differentiable vector space.

As this flat connection does not increase support, it preserves the subspace $\mathscr{E}_c(M)$ of sections with proper support over X, and it gives $\mathscr{E}_c(M)$ the structure of a differentiable vector space.

Remark: Typically, one works with $\Gamma(M, E)$ as a complete locally convex topological vector space, via the Fréchet topology. In Section B.3, we will explain a result of Kriegl and Michor that shows that every locally convex topological vector space V naturally produces a differentiable vector space. Moreover, $\Gamma(M, E)$ goes to $\mathscr{E}$ under this functor. Hence, the differentiable vector space $\mathscr{E}(M)$ arises from the standard topology on $\Gamma(M, E)$. (An identical comment applies to $\Gamma_c(M, E)$.) $\diamond$

We are also interested in distributional sections of a vector bundle $p : E \rightarrow M$. We will show that these also form a differentiable vector space, after setting up the preliminaries about distributions.

Let $\mathcal{D}(M)$ denote the vector space of distributions on the smooth manifold M. That is, $\mathcal{D}(M)$ is the continuous linear dual to the vector space $C_c^\infty(M)$. Let $\mathcal{D}_c(M)$ denote the vector space of compactly supported distributions on M, i.e., the continuous linear dual of $C^\infty(M)$. This space is also a $C^\infty(M)$-module.

Definition B.2.7 Let $\mathcal{D}(M)$ denote the C^∞-module whose smooth sections $C^\infty(X, \mathcal{D}(M))$ on the manifold X are the continuous linear maps from $C_c^\infty(M)$ to $C^\infty(X)$.

Similarly, let $\mathcal{D}_c(M)$ denote the C^∞-module whose smooth sections on the manifold X, denoted $C^\infty(X, \mathcal{D}_c(M))$, are the continuous linear maps from $C^\infty(M)$ to $C^\infty(X)$.

Note that when X is a point, we recover the usual notion of a distribution. The definition above arises by asking that a smooth map ϕ from X to $\mathcal{D}(M)$ correspond to a smooth family of distributions $\{\phi_x\}_{x \in X}$ on M. If one evaluates on $f \in C_c^\infty(M)$, then $\{\phi_x(f)\}_{x \in X}$ should be a smooth function X.

We now equip these C^∞-modules with a natural flat connection.

Definition B.2.8 The vector fields $\mathcal{T}_X$ on X act in a natural way on the vector space $C^\infty(X, \mathcal{D}(M))$ by

$$\mathcal{X} \cdot \phi : f \mapsto \mathcal{X}(\phi(f)),$$

where $f \in C_c^\infty(M)$, $\phi \in C^\infty(X, \mathcal{D}(M))$, and $\mathcal{X} \in \mathcal{T}_X$. This action is compatible with the Lie bracket of vector fields, and hence it makes $C^\infty(X, \mathcal{D}(M))$ into a representation of $\mathcal{T}_X$. This data is equivalent to equipping $C^\infty(X, \mathcal{D}(M))$ with a flat connection and hence the structure of a differentiable vector space.

This action clearly preserves the subspace $C^\infty(X, \mathcal{D}_c(M))$, and hence gives it the structure of a differentiable vector space.

Let $p : E \to M$ be a smooth vector bundle. We wish to construct a differentiable vector space of distributional sections of E.

Note that $\mathcal{D}(M)$ is a module over $C^\infty(M)$. The vector space of distributional sections of E is

$$\Gamma(M, E) \otimes_{C^\infty(M)} \mathcal{D}(M),$$

using the algebraic tensor product. We are simply allowing sections of E with distributional coefficients. We extend this object to a C^∞-module by using the same logic as we used for $\mathcal{D}(M)$.

Recall that $\mathscr{E}$ denotes the differentiable vector space associated to the smooth sections $\Gamma(M, E)$.

Definition B.2.9 Let $\overline{\mathscr{E}}(M)$ denote the C^∞-module where

$$C^\infty(X, \overline{\mathscr{E}}(M)) := C^\infty(X, \mathscr{E}(M)) \otimes_{C^\infty(X \times M)} C^\infty(X, \mathcal{D}(M)).$$

We call this sheaf the *distributional sections of the vector bundle E*. Similarly, the *compactly supported distributional sections of E* is the C^∞-module $\overline{\mathscr{E}}_c(M)$ where

$$C^\infty(X, \overline{\mathscr{E}}_c(M)) := C^\infty(X, \mathscr{E}_c) \otimes_{C^\infty(X, C_c^\infty(M))} C^\infty(X, \mathcal{D}_c(M)).$$

Note that setting X to be a point, we recover the original notion of the distributional sections.

There is a natural flat connection on $\overline{\mathscr{E}}(M)$ and $\overline{\mathscr{E}}_c(M)$ arising from a natural action of the vector fields $\mathcal{T}_X$ on these spaces, just as in the case of $\mathcal{D}(M)$ and $\mathcal{D}_c(M)$. Hence, these are differentiable vector spaces.

Holomorphic Vector Bundles and Sections

Let M be a complex manifold and $p : E \to M$ a holomorphic vector bundle. The holomorphic sections also form a differentiable vector space; we will show this in the simplest case, as the general case is completely parallel.

Definition B.2.10 For M a complex manifold and X a smooth manifold, let $f \in C^\infty(X, \mathcal{O}(M))$ denote a smooth function on the product manifold $X \times M$ such that $f_x(m) := f(x, m)$ is holomorphic on M for every $x \in X$.

Note that $C^\infty(-, \mathcal{O}(M))$ is a subsheaf of $C^\infty(-, C^\infty(M))$, using the definition of the differentiable vector space $C^\infty(M)$ from above. It is straightforward to see that the flat connection on $C^\infty(X, C^\infty(M))$ preserves the subspace $C^\infty(X, \mathcal{O}(M))$ for every X. Hence $C^\infty(-, \mathcal{O}(M))$ is a differentiable vector space that we will denote $\mathcal{O}(M)$.

In Proposition B.3.6, we explain how to obtain a differentiable vector space from each locally convex topological vector space. The construction there applied to $\mathcal{O}(M)$ coincides with this construction.

B.2.2 Stalks and Local Properties

Let V be a differentiable vector space. Restricting to $\mathbb{R}^n$, we obtain a sheaf $V|_{\mathbb{R}^n}$ on $\mathbb{R}^n$. We can thus define the *n-dimensional stalk of V*:

$$\mathrm{Stalk}_n(V) = \mathrm{colim}_{0 \in U \subset \mathbb{R}^n} V(U).$$

The colimit above is taken over open subsets of $\mathbb{R}^n$ containing the origin and is computed in the category of vector spaces.

Note that the stalk of V at a point in any manifold can be defined in the same way, but the stalk at a point in a n-dimensional manifold is the same as the stalk at the origin in $\mathbb{R}^n$. Because of this "uniformity" of differentiable vector spaces, a local property (i.e., a property verified stalkwise) needs to be checked only in countably many cases. As an example, consider how convenient the standard criterion for the exactness of a sequence becomes in our setting.

Lemma B.2.11 *A sequence of differentiable vector spaces*

$$0 \to A \to B \to C \to 0$$

is exact if and only if, for all n, the sequence

$$0 \to \mathrm{Stalk}_n(A) \to \mathrm{Stalk}_n(B) \to \mathrm{Stalk}_n(C) \to 0$$

of vector spaces is exact.

More generally, it is a standard fact that the functor of taking stalks preserves colimits and finite limits. The following corollary will be useful for us.

Corollary B.2.12 *A map $f : A \to B$ of differentiable vector spaces is an iso-morphism if and only if, for all n, the map $\mathrm{Stalk}_n(f) : \mathrm{Stalk}_n(A) \to \mathrm{Stalk}_n(B)$ is an isomorphism.*

B.2.3 Categorical Properties

The category DVS is well behaved in the sense that all the usual constructions make sense. For example, we have the following.

Lemma B.2.13 DVS *contains all finite products.*

Proof Given a finite collection $\{V_\alpha\}_{\alpha \in A}$ of differentiable vector spaces, the product $\prod_{\alpha \in A} V_\alpha$ is the functor

$$X \mapsto C^\infty\left(X, \prod_{\alpha \in A} V_\alpha\right) = \prod_{\alpha \in A} C^\infty(X, V_\alpha),$$

where on the right-hand side we use the product as vector spaces.

Note that finite products coincide with finite coproducts in C^∞-modules, and the tensor product commutes with coproducts so that

$$\Omega^1 \otimes_{C^\infty} \left(\prod_{\alpha \in A} V_\alpha\right) \cong \prod_{\alpha \in A} \left(\Omega^1 \otimes_{C^\infty} V_\alpha\right)$$

for A a finite set. Hence the product of the connections for each V_α provides the desired connection

$$\nabla_{X, \prod V_\alpha} : C^\infty\left(X, \prod_{\alpha \in A} V_\alpha\right) \to \Omega^1\left(X, \prod_{\alpha \in A} V_\alpha\right).$$

It is straightforward to check that this construction satisfies the universal property of a product. $\square$

Note in the proof that we used crucially the tensoring with Ω^1 commutes with finite products of C^∞-modules. There is no reason to expect tensoring commutes with *infinite* products. Below we give an indirect proof that DVS contains infinite products as well.

We now provide kernels.

Lemma B.2.14 *For any map $\phi : V \to W$ of differentiable vector spaces, the kernel, given by*

$$\ker \phi : X \mapsto \ker\left(\phi(X) : C^\infty(X, V) \to C^\infty(X, W)\right).$$

is a differentiable vector space by restricting the connection $\nabla_{X,V}$ to $\ker \phi(X)$.

Proof The definition of $\ker \phi$ clearly provides a C^∞-module, since $\phi(X)$ is $C^\infty(X)$-linear for every smooth manifold X. It is also manifestly the kernel in the category of C^∞-modules.

As ϕ commutes with the flat connection, we see that if $v \in \ker \phi(X) \subset C^\infty(X, V)$, then

$$\phi(\nabla_{X,V} v) = \nabla_{X,W}(\phi(v)) = 0,$$

so $\langle \mathcal{X}, \nabla_{X,V} v \rangle = 0$ for any vector field $\mathcal{X}$ on X. $\qquad \square$

Corollary B.2.15 DVS *contains all finite limits.*

Proof Every finite limit can be constructed out of finite products and equalizers. Since we are in a linear setting, we can construct an equalizer as a kernel. $\qquad \square$

Note that one reason this situation for limits is simple is that limits of sheaves are simply computed objectwise (i.e., separately on each manifold $X \in$ Mfld), and we know how to compute limits of vector spaces.

Colimits are a bit more complicated.

Lemma B.2.16 DVS *contains all coproducts.*

Proof Given a collection $\{V_\alpha\}_{\alpha \in A}$ of differentiable vector spaces, the coproduct $\bigoplus_{\alpha \in A} V_\alpha$ is the sheafification of the functor

$$X \mapsto \bigoplus_{\alpha \in A} C^\infty(X, V_\alpha),$$

where on the right-hand side we use the coproduct as vector spaces. (We use the notation $\oplus$ for the coproduct, as usual for vector spaces.) This sheaf is the coproduct in C^∞-modules, since the presheaf defined earlier is the coproduct in presheaves of C^∞-modules and sheafification is a left adjoint.

The connection

$$\nabla_{X, \oplus V_\alpha} : C^\infty \left(X, \bigoplus_{\alpha \in A} V_\alpha \right) \to \Omega^1 \left(X, \bigoplus_{\alpha \in A} V_\alpha \right)$$

is the unique connection that restricts to ∇_{X, V_α} for each V_α. It is manifest that any cocone in DVS for this collection $\{V_\alpha\}$ factors through the differentiable vector space just constructed. $\qquad \square$

Remark: For finite coproducts, the presheaf formula is already a sheaf and provides the coproduct in C^∞-modules. The natural connection makes it the coproduct of differentiable vector spaces. $\qquad \diamond$

Lemma B.2.17 *For any map* $\phi : V \to W$ *of differentiable vector spaces, the cokernel* coker ϕ *is given by sheafifying the presheaf*

$$\widetilde{\text{coker}}\,\phi : X \mapsto \text{coker}\left(\phi(X) : C^\infty(X, V) \to C^\infty(X, W)\right).$$

This C^∞ module is a differentiable vector space by a natural connection induced from the connection $\nabla_{X,W}$.

Proof Let C_ϕ denote the sheafification of the presheaf $\widetilde{\text{coker}}\,\phi$. (Recall Remark B.2: the sheafification can be as a presheaf of sets or vector spaces or C^∞-modules, as they all provide the same object, after suitable forgetting of module structure.) This sheaf C_ϕ provides the cokernel in C^∞-modules.

We need to produce the flat connection on C_ϕ. As Ω^1 is a projective C^∞-module, it is a flat C^∞-module so that

$$\Omega^1 \otimes_{C^\infty} V \xrightarrow{1\otimes\phi} \Omega^1 \otimes_{C^\infty} W \xrightarrow{1\otimes q_\phi} \Omega^1 \otimes_{C^\infty} C_\phi \longrightarrow 0$$

is exact. As flat connections are $\mathbb{C}$-linear but not C^∞-linear (i.e., they are maps of smooth vector spaces but not of C^∞-modules), the commutative diagram

$$
\begin{array}{ccccccc}
\Omega^1 \otimes_{C^\infty} V & \xrightarrow{1\otimes\phi} & \Omega^1 \otimes_{C^\infty} W & \xrightarrow{1\otimes q_\phi} & \Omega^1 \otimes_{C^\infty} C_\phi & \longrightarrow & 0 \\
\uparrow{\scriptstyle\nabla_V} & & \uparrow{\scriptstyle\nabla_W} & & & & \\
V & \xrightarrow{\phi} & W & \xrightarrow{q_\phi} & C_\phi & \longrightarrow & 0
\end{array}
$$

lives in the category of smooth vector spaces. Since smooth vector spaces form an abelian category, there is naturally a $\mathbb{C}$-linear map f_{C_ϕ} from C_ϕ to $\Omega^1 \otimes_{C^\infty} C_\phi$ inherited from the connection on W. We need to show that f_{C_ϕ} is a flat connection, but this assertion follows from the fact that ϕ intertwines the flat connections on V and W. $\qquad\square$

Corollary B.2.18 DVS *contains all colimits. Hence it is a cocomplete category.*

Proof Every colimit can be constructed out of coproducts and coequalizers. Since we are in a linear setting, we can construct an coequalizer as a cokernel. $\qquad\square$

In fact, DVS contains all limits and is hence complete, but the argument is indirect. We first prove that DVS is locally presentable, and by Corollary 1.28 of Adámek and Rosický (1994), we know that every locally presentable category is complete.

Proposition B.2.19 DVS *is locally presentable.*

Corollary B.2.20 DVS *contains all limits, and so it is a complete category.*

The proof of this proposition depends on some technical results in category theory, rather orthogonal from the rest of this book, so we will not give expository background. Our main reference are Adámek and Rosický (1994) and Kashiwara and Schapira (2006).

Proof of Proposition B.2.19 As we know that DVS is cocomplete, it suffices to prove that DVS is accessible (see Corollary 2.47 of Adámek and Rosický (1994)). Our strategy is to present DVS as constructed out of known accessible categories by operations that produce accessible categories. To be more precise, there is a 2-category *ACC* of accessible categories, and it possesses certain "limits" understood in the correct 2-categorical sense. In particular, *ACC* is closed under "inserters" and "equifiers." In the text that follows, we will define and use precisely these constructions.

First, we must show that our initial inputs are accessible. Recall that the module sheaves for a sheaf of rings on a site is presentable and hence accessible. (See, for instance, Theorem 18.1.6 of Kashiwara and Schapira (2006).) Thus the category of smooth vector spaces Vect_{sm} – i.e., sheaves of vector spaces on the site Mfld – is accessible. Likewise, the category Mod_{C^∞} of C^∞-modules is accessible.

Consider the following two functors:

$$F : \text{Mod}_{C^\infty} \to \text{Vect}_{sm},$$

the forgetful functor, and

$$G := G \circ \Omega^1 \otimes_{C^\infty} : \text{Mod}_{C^\infty} \to \text{Vect}_{sm},$$

which sends a C^∞-module V to the 1-forms valued in V, $\Omega^1 \otimes_{C^\infty}$, and then views it as a smooth vector space. The forgetful functor F preserves filtered colimits by Theorem 18.1.6 of Kashiwara and Schapira (2006); hence F is accessible. The functor $\Omega^1 \otimes_{C^\infty} -$ is a left adjoint from Mod_{C^∞} to itself and hence preserves all colimits, so the composite G also preserves filtered colimits and is hence accessible.

The *inserter* of F and G is the following category, denoted $Ins(F, G)$. An object of $Ins(F, G)$ is a morphism $f : F(V) \to G(V)$ with V an object in Mod_{C^∞}. A morphism in $Ins(F, G)$ is a commuting square

$$\begin{array}{ccc}
F(V) & \xrightarrow{f} & G(V) \\
{\scriptstyle F(\phi)}\downarrow & & \downarrow{\scriptstyle G(\phi)} \\
F(V') & \xrightarrow{f'} & G(V')
\end{array}$$

with $\phi : V \to V'$ a morphism in Mod_{C^∞}. For our situation, this inserter amounts to a choice of "pre-connection" $\nabla : V \to \Omega^1 \otimes_{C^\infty} V$ on a C^∞-module V. It is accessible because the inserter of two accessible functors is accessible (see Theorem 2.72 of Adámek and Rosický (1994)).

A pre-connection ∇ is a connection if it satisfies the Leibniz rule:

$$\nabla(f \cdot s) = \mathrm{d}f \otimes s + f \cdot \nabla s$$

for f a smooth function and s a section. This equation can be used to produce the category of C^∞-modules with connection, as follows.

Consider the following two functors:

$$F' : Ins(F, G) \to \text{Vect}_{sm},$$

the functor which sends a pre-connection $\nabla : V \to \Omega^1 \otimes_{C^\infty} V$ to the smooth vector space $C^\infty \otimes F(V)$, where this tensor product is in Vect_{sm}; and

$$G' : Ins(F, G) \to \text{Vect}_{sm},$$

which sends ∇ to $F(\Omega^1 \otimes_{C^\infty} V)$, the underlying smooth vector space of the 1-forms valued in V. Note that the forgetful functor $P : Ins(F, G) \to \text{Mod}_{C^\infty}$ is itself accessible (see Remark 2.73 of Adámek and Rosický (1994)). These functors F', G' are accessible because they are compositions of P with the accessible functors F, G, and tensoring with C^∞.

Consider the following two natural transformations $\alpha, \beta : F' \Rightarrow G'$. The transformation α sends a pre-connection (V, ∇) to the map of smooth vector spaces

$$\alpha(\nabla): \quad \begin{array}{ccc} C^\infty \otimes F(V) & \to & F(\Omega^1 \otimes_{C^\infty} V) \\ f \otimes s & \mapsto & \nabla(f \cdot s) - (\mathrm{d}f \otimes s + f \cdot \nabla s) \end{array} .$$

The transformation β sends a pre-connection (V, ∇) to the map of smooth vector spaces

$$\beta(\nabla): \quad \begin{array}{ccc} C^\infty \otimes F(V) & \to & F(\Omega^1 \otimes_{C^\infty} V) \\ f \otimes s & \mapsto & 0 \end{array} .$$

Note that ∇ is a connection precisely if $\alpha(\nabla) = \beta(\nabla)$, which is the Leibniz rule.

The *equifier* of F', G', α, and β, denoted $Eq(F', G', \alpha, \beta)$, is the full sub-category of $Ins(F, G)$ consisting of objects ∇ on which $\alpha(\nabla) = \beta(\nabla)$. In other words, $Eq(F', G', \alpha, \beta)$ is the category of connections. By Lemma 2.76 of Adámek and Rosický (1994), we see that the equifier of accessible functors is accessible.

The category of C^∞-modules with *flat* connections can likewise we constructed as an equifier, since a connection is flat exactly when its curvature is zero. Here the functors are $F'', G'' : Eq(F', G', \alpha, \beta) \rightarrow \text{Vect}_{sm}$, where $F''(\nabla : V \rightarrow \Omega^1 \otimes_{C^\infty} V) = F(V)$ the underlying smooth vector space of V and $G''(\nabla) = F(\Omega^2 \otimes_{C^\infty} V)$. These are accessible functors, for reasons identical to the preceding constructions. The natural transformations $\alpha', \beta' : F'' \Rightarrow G''$ are $\alpha'(\nabla) = \Omega_\nabla$, the curvature of the connection, and $\beta'(\nabla) = 0$. The equifier then amounts to those connections with zero curvature. $\square$

B.2.4 Tensoring over C^∞

Standard constructions on vector spaces, such as tensor product and formation of internal hom spaces, carry over to vector bundles, and also to vector bundles with connection. We now spell out how those constructions work in DVS.

Let V, W be C^∞-modules. Then we define a C^∞-module $V \otimes_{C^\infty} W$ by sheafifying the presheaf $X \mapsto C^\infty(X, V) \otimes_{C^\infty(X)} C^\infty(X, W)$. When V, W are differentiable, we equip this tensor product with the flat connection

$$\nabla_{X, V \otimes W} := \nabla_{X, V} \otimes \text{Id}_W + \text{Id}_V \otimes \nabla_{X, W},$$

which is the usual formula for connections.

Recall now the following standard construction. Let V, W be C^∞-modules. Let $\mathcal{H}om_{C^\infty}(V, W)$ denote the "sheaf Hom": to each manifold X, we assign

$$C^\infty(X, \mathcal{H}om_{C^\infty}(V, W)) := \text{Hom}_{C^\infty(X)}(C^\infty(X, V), C^\infty(X, W)),$$

the vector space of $C^\infty(X)$-linear maps between $C^\infty(X, V)$ and $C^\infty(X, W)$. This presheaf is, in fact, a sheaf, and it is naturally a C^∞-module. If V, W are differentiable, then $\mathcal{H}om_{C^\infty}(V, W)$ is naturally a differentiable vector space, with

$$(\mathcal{X} \cdot \phi)(v) = \mathcal{X}(\phi(v)) - \phi(\mathcal{X}(v)),$$

where $\mathcal{X}$ is a vector field on X, ϕ is a section of $C^\infty(X, \mathcal{H}om_{C^\infty}(V, W))$, and v is a section of $C^\infty(X, V)$. By definition, this connection is flat. To emphasize that it has a differentiable structure, we use $\mathcal{H}om_{DVS}$ to denote this $\mathcal{H}om_{C^\infty}$.

This symmetric monoidal structure, however, does not have several properties that we desire. For example, "global sections" of $\mathcal{H}om_{DVS}(V, W)$, namely its value on a point $C^\infty(*, \mathcal{H}om_{DVS}(V, W))$, is not $DVS(V, W)$. Instead, it is $\text{Mod}_{C^\infty}(V, W)$, merely the smooth maps and not the connection-respecting maps. This is the "wrong answer" for certain purposes. We explain a resolution of this issue in Section B.6.

As another example of a "defect" of this monoidal structure, we would like $C^\infty(M) \otimes_C^\infty C^\infty(N) \cong C^\infty(M \times N)$ in DVS, but it is not true. Consider evaluating both sheaves on a point $*$: the sections of the tensor product is

$$C^\infty(*, C^\infty(M) \otimes_C^\infty C^\infty(N)) = C^\infty(M) \otimes_\mathbb{R} C^\infty(N)$$

and that is not equal to $C^\infty(M \times N)$. The usual remedy to this issue is to take the appropriate completion of this algebraic tensor product, using natural topologies on these spaces. Alternatively, relying on ideas from the next sections of this appendix, one can note that these vector spaces come from bornological vector spaces, and that their tensor product in DVS agrees with the tensor product in BVS. But we know $C^\infty(X) \widehat{\otimes}_\beta C^\infty(Y) \cong C^\infty(X \times Y)$, which is the completion of the bornological tensor product to a convenient vector space. The bornological tensor product needs to be completed to obtain what we want. Hence a strategy coherent with our emphasis on DVS is to "complete" the tensor product $\otimes_{C^\infty}$ and develop "convenient differentiable vector spaces," but we will not pursue this construction here.

B.2.5 Multilinear Maps and the Multicategory Structure

There is a natural way to form "many-to-one" maps between ordinary vector spaces: consider the vector space of multilinear maps $\phi : V_1 \times \cdots \times V_n \to W$. Multilinear maps compose naturally, by feeding the output of one multilinear map into one of the inputs of the next. Altogether, this rich structure is formalized in the notion of a multicategory (see Section A.3.3 in Appendix A). We have a similar story for differentiable vector spaces.

Proposition B.2.21 DVS *forms a multicategory in which*

$$\mathrm{DVS}(V, \ldots, V_n \mid W)$$

denotes the collection of multilinear maps of differentiable vector spaces. A multilinear map

$$\phi : V_1 \times \cdots \times V_n \to W$$

is a multilinear map of C^∞-modules compatible with the flat connections: for each manifold X,

(1) *Given sections $f_i \in C^\infty(X, V_i)$, $\phi(f_1, \ldots, f_n) \in C^\infty(X, W)$, and*
(2) *If $\mathcal{X}$ is a vector field on X, then*

$$\mathcal{X}(\phi(f_1, \ldots, f_n)) = \sum_{i=1}^n \phi(f_1, \ldots, \mathcal{X}(f_i), \ldots, f_n).$$

B.3 Locally Convex Topological Vector Spaces

This section is brief, and it aims simply to explain a result of Kriegl and Michor (1997) that shows how locally convex Hausdorff topological vector spaces provide differentiable vector spaces. In other words, we explain how a very large and important class of topological vector spaces fits inside our preferred approach.

Definition B.3.1 Let LCTVS denote the category where an object is a locally convex Hausdorff topological vector space and a morphism is a continuous linear map.

Remark: In the remainder of this section, we will use *topological vector space* to mean *locally convex Hausdorff topological vector space*, for simplicity. ◊

The key idea of Kriegl and Michor is simple and compelling: the notion of a smooth curve is well behaved, so build up everything from that notion.

In particular, it is possible to define smooth curves $\gamma : \mathbb{R} \to V$ in a topological vector space in a very simple way, in the style of elementary calculus. One then defines a smooth map $f : V \to W$ between topological vector spaces to be a map that sends smooth curves in V to smooth curves in W. This notion extends naturally to define smooth maps from a finite-dimensional smooth manifold X into a topological vector space V. Hence, we obtain a smooth vector space (i.e., a sheaf on Mfld) from V: to the manifold X, we assign the set of smooth maps $C^\infty(X, V)$.

Remark: In fact, a map $f : M \to V$ is smooth in this sense if and only if it is smooth in the usual sense. See Lemma B.3.7. ◊

Definition B.3.2 Let $V \in$ LCTVS. A curve $\gamma : \mathbb{R} \to V$ is *differentiable* if its *derivative*

$$\gamma'(t) := \lim_{s \to 0} \frac{\gamma(t + s) - \gamma(t)}{s}$$

exists at $t \in \mathbb{R}$ for all t.

A curve γ is *smooth* if all iterated derivatives exist.

With this definition in hand, we introduce the notion of a smooth map between topological vector spaces, following Kriegl and Michor.

Definition B.3.3 Let $V, W \in$ LCTVS. A (not necessarily linear) function $f : V \to W$ is *smooth* if for each smooth curve $\gamma : \mathbb{R} \to V$, the composition $f \circ \gamma : \mathbb{R} \to W$ is smooth.

As Kriegl and Michor explain, this notion arises by taking seriously the perspective of variational calculus.

Remark: We note that not every smooth function between topological vector spaces, in this sense, is necessarily continuous. As Kriegl and Michor point out, this *a priori* unappealing mismatch is not so strange: smoothness is about nonlinear phenomena, and the usual topology on a topological vector space is focused on linear phenomena. (For many nice topological vector spaces, however, a smooth function is continuous.)   ◊

In particular, for an open $U \subset \mathbb{R}^n$ and V a topological vector space, we say that $f : U \to V$ is smooth if any smooth curve $\gamma : \mathbb{R} \to U$ goes to a smooth curve $f \circ \gamma : \mathbb{R} \to V$. For X a smooth manifold, a function $f : X \to V$ is smooth if its restriction to every chart is smooth. Hence the notion is local on X, so that we obtain the following definition.

Definition B.3.4 Let $V \in$ LCTVS. We define a C^∞-module by

$$X \in \mathrm{Mfld}^{op} \mapsto \{f : X \to V \mid f \text{ smooth}\}.$$

We denote this functor by $C^\infty(-, V)$.

Let $mod_t :$ LCTVS $\to$ Mod$_{C^\infty}$ denote the functor that sends V to $C^\infty(-, V)$. It has the following useful property.

Lemma B.3.5 *The functor mod_t preserves limits.*

Proof The forgetful functor from sheaves of C^∞-modules to sheaves of sets preserves limits, because it is a right adjoint. Likewise, the forgetful functor from sheaves of sets to presheaves of sets preserves limits, since it is a right adjoint. But a limit of presheaves is computed objectwise, since it is simply a limit in a functor category. We thus merely need to show that

$$C^\infty(X, mod_t(\varprojlim_I V)) \cong C^\infty(X, \varprojlim_I mod_t(V))$$

for every smooth manifold X.

Now observe that a map from a manifold X into a product of locally convex topological vector spaces is smooth if and only if it it is smooth into each component separately (i.e., if its projection to each component is smooth). Hence, products are preserved. Similarly, a map from X into a kernel ker$(\phi : V \to W)$ in LCTVS is smooth if and only if it is smooth into V and the composition into W is the constant zero map. As limits are generated by kernels and products, we are done.   □

But this functor $C^\infty(-, V)$ is even better. By construction, these maps $C^\infty(X, V)$ are differentiable, so we find that $C^\infty(-, V)$ is naturally a differentiable vector space. To be explicit, for each smooth map $f : X \to V$, there is a derivative $Tf : TX \to V \oplus V$, by the definition of "smooth." Hence the flat connection is

$$\nabla_X : \quad C^\infty(X, V) \quad \to \quad \Omega^1(X, V)$$
$$f \quad \mapsto \quad df = \pi_2 \circ Tf \quad,$$

where $\pi_2 : V \oplus V \to V$ denotes projection onto the second summand and $\pi_1 \circ Tf = f$. Note that this construction is precisely the usual construction of the natural flat connection on a trivialized vector bundle.

Proposition B.3.6 *There is a faithful functor dif_t : LCTVS $\to$ DVS sending V to $C^\infty(-, V)$, equipped with its natural flat connection. It preserves limits.*

Proof We already described the construction that produces the functor, so it remains to check that dif_t preserves limits. This result is not formal because the forgetful functor U from DVS to Mod_{C^∞} does *not* preserve all limits. (One problem is that the same C^∞-module might admit multiple flat connections, and hence distinct objects in DVS might have the same image.) But $mod_t = U \circ dif_t$, so the situation simplifies.

Given a diagram $V : I \to$ LCTVS, we know that $dif_t (\lim_I V)$ is simply

$$C^\infty \left(-, \lim_I V \right) = mod_t \left(\lim_I V \right) \cong \lim_I mod_t \circ V,$$

equipped with its canonical flat connection. Since every map of differentiable vector spaces is also C^∞-linear, the map from $\lim_I dif_t \circ V$ to the diagram $dif_t \circ V$ factors through $dif_t (\lim_I V)$. Hence $\lim_I dif_t \circ V \cong dif_t (\lim_I V)$. $\qquad\square$

This functor dif_t factors through the category of bornological vector spaces (and in consequence is not full, as we will see). Thus, in Section B.4, we will describe the bornological vector space associated to every complete locally convex topological vector space and how every bornological vector space provides a differentiable vector space.

Finally, we elaborate on Remark B.3.

Lemma B.3.7 (Kriegl and Michor (1997), Lemma 3.14) *For $U \subset \mathbb{R}^n$ an open set and $V \in$ LCTVS, a function $f : U \to V$ is smooth in the sense of Definition B.3.3 if and only if all iterated partial derivatives $\partial^\mu f / \partial x^\mu$ exist and are smooth.*

This lemma is due to Boman.

B.3.1 Categorical Properties

Lemma B.3.8 LCTVS *admits all limits.*

Proof We show the existence of products and of kernels. Together, these guarantee the existence of limits.

Let $\{V_\alpha \mid \alpha \in A\}$ be a collection in LCTVS. Consider the product of the V_α *as topological spaces*, i.e., equipped with the usual product topology. As any product of convex sets is convex, the topology on this topological product space is also generated by convex sets. One can check that the induced scalar multiplication and vector addition are continuous with respect to this topology.

Kernels are similar. Being the kernel of a continuous linear map is a closed condition, and a closed linear subspace of a locally convex vector space is a locally convex vector space.   □

Lemma B.3.9 LCTVS *admits all colimits.*

Proof We show the existence of cokernels and coproducts.

Cokernels are straightforward. For a continuous linear map $\phi : V \to W$, the cokernel is $W/\overline{\phi(V)}$, the quotient of W by the closure of the image of ϕ. We are forced to use the closure as we work with Hausdorff spaces.

Coproducts are a little trickier. The underlying vector space is simply the direct sum, but we equip it with the "diamond" topology. (We learned this term from the helpful online lecture notes of Paul Garrett.) We construct a collection of convex neighborhoods of the origin out of convex neighborhoods of the constituents as follows.

Let $i_\alpha : V_\alpha \hookrightarrow \bigoplus_{\alpha \in A} V_\alpha$ denote the canonical inclusion. Given a choice of convex neighborhood U_α of V_α for every $\alpha \in A$, take the convex hull of the union $\bigcup_{\alpha \in A} i_\alpha(U_\alpha)$. Declare such a convex hull to be an open. We take the vector space topology induced by all such opens.   □

B.4 Bornological Vector Spaces

There is another natural approach to taming infinite-dimensional vector spaces, via the notion of a bornology. Instead of specifying a topology on V, one specifies the *bounded* subsets of V.

Definition B.4.1 A *bornology* on a vector space V (over a field $K = \mathbb{R}$ or $\mathbb{C}$) is a collection of subsets $\mathcal{B}$ such that

(1) $\mathcal{B}$ covers V, i.e., $V = \bigcup_{B \in \mathcal{B}} B$.
(2) $\mathcal{B}$ is closed under inclusions, i.e., if $B \in \mathcal{B}$ and $B' \subset B$, then $B' \in \mathcal{B}$.

(3) $\mathcal{B}$ is closed under finite unions, i.e., if $B_1, \ldots, B_n \in \mathcal{B}$, then $B_1 \cup \cdots \cup B_n \in \mathcal{B}$.

Moreover, these bounded sets are compatible with the vector space structure:

(4) $\mathcal{B}$ is closed under translation, i.e., if $v \in V$ and $B \in \mathcal{B}$, then $v + B \in \mathcal{B}$.
(5) $\mathcal{B}$ is closed under dilation, i.e., if $\lambda \in K$ and $B \in \mathcal{B}$, then $\lambda \cdot B \in \mathcal{B}$.
(6) $\mathcal{B}$ is closed under the formation of balanced hulls, i.e., if $B, B' \in \mathcal{B}$, then $B + B' \in \mathcal{B}$.

A topological vector space V has an associated bornology, as follows. We say $B \subset V$ is *bounded* if for every open set $U \subset V$ containing the origin, there is a real number $\lambda > 0$ such that $B \subset \lambda U$. When V is locally convex, a set B is bounded if and only if every continuous semi-norm of V is bounded on B. Note that distinct topologies may have the identical associated bornologies.

We are interested only in such bornologies, so we work with the following category.

Definition B.4.2 A *bornological vector space* $(V, \mathcal{B})$ is a vector space whose bornology is obtained from some locally convex Hausdorff topology on V.

A linear map $f : V \to W$ is *bounded* if the image of every bounded set in V is a bounded set in W.

We denote by BVS the category where an object is a bornological vector space and a morphism is a bounded linear map.

Note that a *continuous* linear map is always bounded, but the converse is false in general. (Consider an infinite-dimensional Banach space V, and let V_w denote the underlying vector space equipped with the weak topology. The identity map id : $V_w \to V$ is bounded but not continuous.)

Consider the functor inc_β : BVS $\to$ LCTVS that assigns to a bornological vector space the finest locally convex topology with the same underlying bounded sets. By Lemma 4.2 of Kriegl and Michor (1997), this functor embeds BVS as a full subcategory of LCTVS. In particular, a linear map $f : V \to W$ between bornological vector spaces is bounded if and only if $f : inc_\beta V \to inc_\beta W$ is continuous.

Corollary B.4.3 *The functor* inc_β : BVS $\hookrightarrow$ LCTVS *preserves all colimits.*

Proof LCTVS contains all colimits. Given a diagram $\delta : D \to$ BVS, let colim $inc_\beta \circ \delta$ denote its colimit in LCTVS. If there exists a colimit colim δ in BVS, then we know there is a canonical map

$$\text{colim } inc_\beta \circ \delta \to inc_\beta \text{ colim } \delta.$$

Hence, it suffices to show that colim $inc_\beta \circ \delta$ is in the image of inc_β to obtain the claim.

For each object $d \in D$, we have a continuous linear map

$$inc_\beta \delta(d) \to \text{colim} \, inc_\beta \circ \delta,$$

which is thus a bounded linear map. Hence, we know that this map factors through the underlying vector space of colim $inc_\beta \circ \delta$, but now equipped with the finest locally convex topology with the same bounded sets. Hence, the colimit colim $inc_\beta \circ \delta$ is in the image of inc_β. □

By the very definition of inc_β, we see that there is a natural right adjoint.

Corollary B.4.4 *Consider the functor*

$$born : \text{LCTVS} \to \text{BVS}$$

sending a locally convex Hausdorff topological vector space to its underlying bornological vector space. This functor born is right adjoint to inc_β.

It is important to know that it is the bornology of a topological vector space that matters for smoothness, not the topology itself. The following two results from Kriegl and Michor (1997) make this assertion precise.

Lemma B.4.5 (Kriegl and Michor (1997), Corollary 1.8) *For $V \in$ LCTVS, a curve $\gamma : \mathbb{R} \to V$ is smooth if and only if $\gamma : \mathbb{R} \to inc_\beta(born(V))$ is smooth.*

Lemma B.4.6 (Kriegl and Michor (1997), Corollary 2.11) *A linear map $L : V \to W$ of locally convex vector spaces is bounded if and only if it maps smooth curves in V to smooth curves in W.*

In consequence, we have the following result.

Proposition B.4.7 *The functor*

$$dif_\beta : \quad \text{BVS} \quad \to \quad \text{DVS}$$
$$V \quad \mapsto \quad dif_t(inc_\beta(V))$$

embeds BVS as a full subcategory of DVS. It preserves all limits.

Note that BVS is therefore equivalent to the essential image of LCTVS in DVS.

Proof Only the assertion about limits remains to be proved. Note that *born* preserves limits, so that since BVS is a full subcategory of LCTVS, we can compute a limit in LCTVS and then apply *born*. As dif_t preserves limits and only cares about the underlying bornology, we are finished. □

B.4.1 Categorical Properties

We have seen that inc_β realizes BVS as a reflective subcategory of LCTVS. Thus, we know the following.

Lemma B.4.8 BVS *admits all limits and colimits.*

Proof LCTVS contains all limits, and *born* : LCTVS $\to$ BVS is a right adjoint and hence preserves limits. A colimit of BVS is computed by applying *born* to the colimit computed in LCTVS. ☐

B.4.2 Multilinear Maps and the Bornological Tensor Product

As in any category of linear objects, we can discuss multilinear maps. We can also ask if multilinear maps out of a tuple of vector spaces is co-represented by a vector space, called the tensor product. For bornological vector spaces, there is such a natural tensor product with all the properties we love about the tensor product of finite-dimensional vector spaces.

Definition B.4.9 Let $BVS(V_1, \ldots, V_n \mid W)$ denote the set of all bounded multilinear maps from the set-theoretic product of bornological vector spaces $V_1 \times \cdots \times V_n$ to the bornological vector space W. Equip it with the bornology of uniform convergence on bounded sets. This bornology makes $BVS(V_1, \ldots, V_n \mid W)$ into a bornological vector space.

The following result is the first step to showing that BVS forms a closed symmetric monoidal category.

Proposition B.4.10 (Kriegl and Michor (1997), Proposition 5.2) *There are natural bornological isomorphisms*

$$BVS(V_1, \ldots, V_{n+m} \mid W) \cong BVS(V_1, \ldots, V_n \mid BVS(V_{n+1}, \ldots, V_{n+m} \mid W)).$$

Hence, BVS *forms a multicategory where multimorphisms are themselves bornological vector spaces and composition is bounded.*

In BVS, there is a natural bornology to put on the algebraic tensor product $V \otimes_{\text{alg}} W$ that co-represents bilinear maps. Simply put, we equip $V \otimes_{\text{alg}} W$ with the finest locally convex topology such that the canonical map $V \times W \to V \otimes_{\text{alg}} W$ is bounded. We denote this bornological vector space by $V \otimes_\beta W$.

Theorem B.4.11 (Kriegl and Michor (1997), Theorem 5.7) *The bornological tensor product* $V \otimes_\beta - : BVS \to BVS$ *is left adjoint to the* Hom-*functor* $BVS(V, -) : BVS \to BVS$.

In particular, we have the following bornological isomorphisms:

$$BVS(V \otimes_\beta W, U) \cong BVS(V, W \mid U) \cong BVS(V, BVS(W, U)),$$
$$V \otimes_\beta \mathbb{R} \cong V,$$
$$V \otimes_\beta W \cong W \otimes_\beta V,$$
$$(V \otimes_\beta W) \otimes_\beta U \cong V \otimes_\beta (W \otimes_\beta U).$$

Moreover, $\otimes_\beta$ preserves all colimits separately in each variable.

To summarize, we have the following.

Proposition B.4.12 BVS *is a closed symmetric monoidal category via $\otimes_\beta$. This symmetric monoidal structure agrees with the multicategory structure on* BVS *induced from the embedding inc_β :* BVS $\to$ DVS.

Proof What remains is to show that the multicategory structure on BVS agrees with that on DVS. Lemma 5.5 of Kriegl and Michor (1997) show that a multilinear map of bornological vector spaces is bounded if and only if it is smooth, and so this multicategory structure agrees with the induced multicategory structure from DVS. □

B.5 Convenient Vector Spaces

The power of calculus often involves the interplay of differentiation with integration, and we will find it convenient to focus on a class of bornological vector spaces where integration along curves is well behaved. As integration involves, speaking casually, infinite sums, we can view the existence of integrals as a kind of completeness property. We use the notion developed and applied to great effect by Kriegl and Michor (1997).

Definition B.5.1 A bornological vector space $V \in$ BVS is c^∞-complete if one of the following equivalent conditions holds:

(a) For any smooth curve $\gamma : \mathbb{R} \to V$, there is a smooth curve $\Gamma : \mathbb{R} \to V$ such that $\Gamma' = \gamma$. We call Γ an *antiderivative*.
(b) A curve $\gamma : \mathbb{R} \to V$ is smooth if and only if for every bounded linear functional $\lambda : V \to \mathbb{R}$, the composition $\lambda \circ \gamma$ is smooth.

A c^∞-complete bornological vector space will be called a *convenient vector space*. We denote by CVS the full subcategory of BVS whose objects are convenient vector spaces.

Several other equivalent definitions of c^∞-complete are given in Theorem 2.14 of Kriegl and Michor (1997).

The following result ensures that the inclusion functor $inc_c : \text{CVS} \to \text{BVS}$ is a right adjoint.

Theorem B.5.2 (Kriegl and Michor (1997), Theorem 2.15) *The full subcategory* $\text{CVS} \subset \text{BVS}$ *is closed under limits.*

We denote the left adjoint by $c^\infty : \text{BVS} \to \text{CVS}$. It sends a bornological vector space to its c^∞ completion.

As dif_β preserves limits, we see that the composition

$$
\begin{array}{rccc}
dif_c : & \text{CVS} & \to & \text{DVS} \\
& V & \mapsto & dif_\beta \circ inc_c(V)
\end{array}
$$

also preserves limits. Moreover, because CVS is a full subcategory of BVS and BVS is a full subcategory of DVS, we have the following.

Lemma B.5.3 *The functor* $dif_c : \text{CVS} \to \text{DVS}$ *embeds* CVS *as a full subcategory that is closed under limits.*

We note that while dif_c does not preserve all colimits, it does preserve some.

Proposition B.5.4 *The functor* $dif_c : \text{CVS} \to \text{DVS}$ *preserves countable coproducts and sequential colimits of closed embeddings.*

Proof A countable coproduct is a special case of a sequential colimit of closed embeddings, so we will prove only the latter property.

Given a sequence

$$V_1 \to V_2 \to \cdots$$

of closed embeddings, let V denote $\text{colim } V_i$. We need to show that

$$\eta : \text{colim } dif_t(V_i) \to dif_t(V)$$

is an equivalence of differentiable vector spaces. By Corollary B.2.12, it is enough to show that this natural transformation is an isomorphism on stalks, as these are sheaves. In other words, the problem is local and we can restrict our attention to the neighborhood of the origin in each Euclidean space $\mathbb{R}^n$.

First, by Lemma 3.8 of Kriegl and Michor (1997), because V_i is a closed subspace of V, a curve in V_i is smooth if and only if the composition into V is smooth. Hence, $V_i \hookrightarrow V$ is a smooth map. As $C^\infty(X, V_i) \hookrightarrow C^\infty(X, V)$ for each i, we see that the map η_X above is an inclusion for any manifold X.

Now we want to show that each smooth map $f : \mathbb{R}^n \to V$ factors, in some sufficiently small neighborhood of the origin, through some V_i. Pick an open set $0 \in U \subset \mathbb{R}^n$ whose closure $\overline{U}$ is compact. As $f : \mathbb{R}^n \to V$ is continuous by Lemma B.3.7, $f(\overline{U})$ is also compact and hence bounded. By Result 52.8 of Kriegl and Michor (1997), a subset of V is bounded if and only if it is a bounded subset of some V_i. Hence f restricted to U maps into this V_i. □

B.5.1 Examples from Differential Geometry

Let us explain some further examples. Let E be a vector bundle on a manifold M, and, as before, let $\mathscr{E}, \mathscr{E}_c, \overline{\mathscr{E}}, \overline{\mathscr{E}}_c$ refer to sections of E that are smooth, compactly supported, distributional, or compactly supported and distributional. All of these spaces have natural topologies and so can be viewed as bornological vector spaces, and they are all c^∞-complete.

We explained in Section B.2.1 how to view these vector spaces as differentiable. It is easy to check that the smooth structure discussed there is the same as the one that arises from the topology. Indeed, the Serre–Swan theorem tells us that any vector bundle is a direct summand of a trivial vector bundle, so we can reduce to the case that E is trivial. Then results of Grothendieck, summarized in Grothendieck (1952), allow one to describe smooth maps to these various vector spaces using the theory of nuclear vector spaces; in this way we arrive at the description given earlier.

Lemma B.5.5 *For any vector bundle E on a manifold M, the standard nuclear topologies on $\overline{\mathscr{E}}_c$, $\mathscr{E}$ and $\mathscr{E}_c$ are bornological. It follows that these spaces are convenient (because completeness in the locally convex sense is stronger than c^∞-completeness).*

Proof Because every vector bundle is a summand of a trivial one, it suffices to prove the statement for the trivial vector bundle. According to Kriegl and Michor (1997), 52.29, the strong dual of a Fréchet Montel space is bornological. The space $C^\infty(M)$ of smooth functions on a manifold is Fréchet Montel, because every nuclear space is Schwartz (see pages 579–581 of Kriegl and Michor (1997)) and every Fréchet Schwarz space is Montel. Thus the strong dual of $C^\infty(M)$ is bornological, as desired.

Next, we will see that any Fréchet space is bornological. This property follows immediately from Proposition 14.8 of Trèves (1967) (see also the corollary on the following page). It follows that $C^\infty(M)$ is bornological, and that the same holds for $C_K^\infty(M)$ for any compact subset $K \subset M$.

Since bornological spaces are closed under formation of colimits, the same holds for $C_c^\infty(M)$. □

B.5.2 Multilinear Maps and the Completed Tensor Product

We can take the bornological tensor product of any two convenient vector spaces V and W, but $V \otimes_\beta W$ is rarely convenient itself. The completion of a tensor product $c^\infty(V \otimes_\beta W)$ is convenient, however, and equips CVS with a natural symmetric monoidal structure. We denote this completed tensor product by $\widehat{\otimes}_\beta$. That is,

$$V \widehat{\otimes}_\beta W := c^\infty(V \otimes_\beta W).$$

(Strictly speaking, we should write $c^\infty(inc_c(V) \otimes_\beta inc_c(W))$ to make clear where everything lives.)

Lemma B.5.6 *The multicategory structure on* CVS *induced by the functor* $dif_c :$ CVS $\to$ DVS *is represented by* $\widehat{\otimes}_\beta$.

Proof We have

$$\begin{aligned}
\mathrm{CVS}(V\widehat{\otimes}_\beta W, U) &= \mathrm{CVS}(c^\infty(inc_c(V) \otimes_\beta inc_c(W)), U) \\
&\cong \mathrm{BVS}(inc_c(V) \otimes_\beta inc_c(W), inc_c(U)) \\
&\cong \mathrm{DVS}(dif_\beta(inc_c(V)), dif_\beta(inc_c(W)) \mid dif_\beta(inc_c(U))) \\
&\cong \mathrm{DVS}(dif_c(V), dif_c(W) \mid dif_c(U)).
\end{aligned}$$

Hence we see that $\widehat{\otimes}_\beta$ encodes multilinear maps of the underlying differentiable vector spaces. $\qquad\square$

But the situation is even better. Via $\widehat{\otimes}_\beta$, CVS forms a closed symmetric monoidal category, as we now explain.

Lemma B.5.7 *For any two convenient vector spaces* V, W, *the bornological vector space* BVS(V, W) *is a convenient vector space.*

Note that $\mathrm{BVS}(V, W) = \mathrm{CVS}(V, W)$ as CVS is a full subcategory, by definition.

Proof Recall that $C^\infty(V, W)$ denotes the set of functions that sends smooth curves in V to smooth curves in W. We equip it with the initial topology such that for any smooth $\gamma : \mathbb{R} \to V$, the pullback $\gamma^* : C^\infty(V, W) \to C^\infty(\mathbb{R}, W)$ is continuous. This topology makes $C^\infty(V, W)$ into a locally convex topological vector space.

Then $\mathrm{BVS}(V, W) \subset C^\infty(V, W)$ is a *closed* set because $f \in C^\infty(V, W)$ is linear if and only if

$$0 = (\mathrm{ev}_x + \lambda \mathrm{ev}_y - \mathrm{ev}_{x+\lambda y})f = f(x) + \lambda f(y) - f(x + \lambda y)$$

for every $x, y \in V$ and $\lambda \in \mathbb{R}$. As a closed linear subspace of a convenient vector space is convenient (because CVS is closed under limits), it suffices to show that $C^\infty(V, W)$ is convenient.

By Lemma 3.7 of Kriegl and Michor (1997), the set of smooth curves $C^\infty(\mathbb{R}, W)$ is a convenient vector space, as W is convenient. By Lemma 3.11 of Kriegl and Michor (1997), the space $C^\infty(V, W)$ is the limit of $C^\infty(\mathbb{R}, W)$ over all the smooth curves $\gamma : \mathbb{R} \to V$, where the morphisms in the diagram category are given by pullback along the reparametrizations $\gamma_1 = c^* \gamma_0$, with $c \in C^\infty(\mathbb{R}, \mathbb{R})$. (This result should be plausible because a function $f : V \to W$ is smooth precisely when it sends smooth curves into V to smooth curves in W.) Putting these facts together with Theorem B.5.2, we see that $C^\infty(V, W)$ is convenient. $\qquad\square$

Thanks to Lemma B.5.7, we can prove the following.

Lemma B.5.8 *The functor* $c^\infty : \mathrm{BVS} \to \mathrm{CVS}$ *is symmetric monoidal.*

Proof Let $V, W \in \mathrm{BVS}$ and $U \in \mathrm{CVS}$. Then

$$
\begin{aligned}
\mathrm{CVS}(c^\infty(V \otimes_\beta W), U) &\cong \mathrm{BVS}(V \otimes_\beta W, inc_c U) \\
&\cong \mathrm{BVS}(V, \mathrm{BVS}(W, inc_c U)) \\
&\cong \mathrm{BVS}(V, inc_c \mathrm{CVS}(c^\infty W, U)) \\
&\cong \mathrm{CVS}(c^\infty V, \mathrm{CVS}(c^\infty W, U)) \\
&\cong \mathrm{BVS}(c^\infty V, \mathrm{BVS}(c^\infty W, inc_c U)) \\
&\cong \mathrm{BVS}(c^\infty V \otimes_\beta c^\infty W, inc_c U) \\
&\cong \mathrm{CVS}(c^\infty(c^\infty V \otimes_\beta c^\infty W), U) \\
&\cong \mathrm{CVS}(c^\infty V \widehat{\otimes}_\beta c^\infty W, U).
\end{aligned}
$$

In moving from the fourth to the fifth line, we use the fact that $\mathrm{CVS}(X, Y) = \mathrm{BVS}(X, Y)$ for all $X, Y \in \mathrm{CVS}$, since it is a full subcategory of BVS. $\qquad\square$

One nice consequence of these observations is the following.

Corollary B.5.9 *The category* CVS *admits all colimits, and* $\widehat{\otimes}_\beta$ *commutes with all colimits separately in each variable.*

Proof As the category BVS admits all colimits, we obtain the colimit of a diagram $D \to \mathrm{CVS}$ by computing the colimit first in BVS and then applying the completion functor c^∞, which preserves colimits because it is a left adjoint. Completion is also symmetric monoidal. As $\otimes_\beta$ preserves all colimits, we are finished. $\qquad\square$

Here is a summary of the main result.

Proposition B.5.10 CVS *is a closed symmetric monoidal category via* $\widehat{\otimes}_\beta$.
Moreover, this symmetric monoidal structure agrees with the multicategory structure induced by the functor dif_c : CVS $\to$ DVS.

In the next section, we show that the inner hom in CVS is compatible with the correct self-enrichment of DVS. In other words, there is a notion of the differentiable vector space of differentiable maps between convenient vector spaces, developed below, and it agrees with the differentiable vector space of convenient maps between convenient vector spaces.

B.6 The Relevant Enrichment of DVS over Itself

Our goal in this section is to enhance DVS to a category in which there is a differentiable vector space of maps between any two differentiable spaces V and W. Here is the first important result of this section, whose proof appears in the first two subsections.

Theorem B.6.1 *There exists a category* $\mathbb{D}$VS, *enriched over* DVS, *whose objects are differentiable vector spaces and whose morphism-sheaves* $\mathbb{H}om_{DVS}(V, W)$ *satisfy*

$$C^\infty(X, \mathbb{H}om_{DVS}(V, W)) = DVS(V, W(X \times -))$$

for any manifold X. In particular, the value on a point $C^\infty(*, \mathbb{H}om_{DVS}(V, W))$
equals DVS(V, W).

We use this enriched category in our definition of equivariant prefactorization algebras (see Section 3.7 in Chapter 3) and of holomorphically translation-invariant prefactorization algebras (see Section 5.2 in Chapter 5).

Another compelling feature of this category is that it matches nicely with the internal hom in CVS, the category of convenient vector spaces, which is the source of most of the vector spaces that we actually work with.

Lemma B.6.2 *For any manifold M and convenient vector spaces* V, W, *there is an isomorphism*

$$C^\infty(M, CVS(V, W)) = CVS(V, C^\infty(M, W)),$$

and it is natural in M. This identification is compatible with the flat connection possessed by the $C^\infty(M)$ modules on each side. Hence

$$\mathrm{CVS}(V, W) \cong \mathbb{H}\mathrm{om}_{\mathrm{DVS}}(V, W)$$

as differentiable vector spaces.

The proof appears in the third subsection.

B.6.1 Constructions with Families

To start, we want to construct, for every manifold X, a sheaf on Mfld that returns, intuitively speaking, "families over X of horizontal sections of the differentiable vector space V." To do this, we need to specify what we mean by such families.

Definition B.6.3 For a manifold X and a differentiable vector space V, let $C^\infty(X, V)$ denote the presheaf sending a manifold Σ to $C^\infty(\Sigma \times X, V)$.

In other words, $C^\infty(X, V)$ is the composite of the product functor $- \times X$ followed by evaluating the sheaf V.

Lemma B.6.4 *For any manifold X and any $V \in \mathrm{DVS}$, the presheaf $C^\infty(X, V)$ is canonically a differentiable vector space.*

Proof It is straightforward to see that $C^\infty(X, V)$ is a sheaf. Any cover of Σ pulls back to a cover over $\Sigma \times X$ along the projection $\pi_\Sigma : \Sigma \times X \to \Sigma$, and since V is a sheaf, we can then reconstruct the value of V on $\Sigma \times X$ using this cover.

Likewise, as $C^\infty(\Sigma, C^\infty(X, V)) = C^\infty(\Sigma \times X, V)$ is a $C^\infty(\Sigma \times V)$-module, it is a $C^\infty(\Sigma)$-module via the pullback map $\pi_\Sigma^* : C^\infty(\Sigma) \to C^\infty(\Sigma \times X)$.

It remains to produce the natural flat connection. By definition, we have a flat connection

$$\nabla_{\Sigma \times X, V} : C^\infty(\Sigma \times X, V) \to \Omega^1(\Sigma \times X, V),$$

but we need to produce a flat connection

$$\nabla_{\Sigma, C^\infty(X, V)} : C^\infty(\Sigma, C^\infty(X, V)) \to \Omega^1(\Sigma, C^\infty(X, V)).$$

Observe that on $\Sigma \times X$, the cotangent bundle $T_{\Sigma \times X}^*$ naturally splits as a direct sum $\pi_X^* T_X^* \oplus \pi_\Sigma^* T_\Sigma^*$, where $\pi_X : \Sigma \times X \to X$ and $\pi_\Sigma : \Sigma \times X \to \Sigma$ are the natural projection maps. In consequence,

$$\Omega^1(\Sigma \times X) \cong \Omega^1(X) \otimes_{C^\infty(X)} C^\infty(\Sigma \times X) \oplus \Omega^1(\Sigma) \otimes_{C^\infty(\Sigma)} C^\infty(\Sigma \times X),$$

where, for example, $C^\infty(\Sigma \times X)$ is a module over $C^\infty(X)$ via the projection $\pi_X : \Sigma \times X \to X$. Hence we have a decomposition

$$\Omega^1(\Sigma \times X, V) \cong \Omega^1(\Sigma \times X) \otimes_{C^\infty(\Sigma \times X)} C^\infty(\Sigma \times X, V)$$

$$\cong \Omega^1(X) \otimes_{C^\infty(X)} C^\infty(\Sigma \times X, V) \oplus \Omega^1(\Sigma)$$

$$\otimes_{C^\infty(\Sigma)} C^\infty(\Sigma \times X, V)$$

$$\cong \Omega^1(X, C^\infty(\Sigma, V)) \oplus \Omega^1(\Sigma, C^\infty(X, V)).$$

Define $\nabla_{\Sigma, C^\infty(X,V)}$ as the composition of $\nabla_{\Sigma \times X, V}$ followed by projection to $\Omega^1(\Sigma, C^\infty(X, V))$. On $\Sigma \times X$, we know that $\nabla_{\Sigma \times X, V}$ is a connection on $\Sigma \times X$ and hence satisfies the Leibniz rule, so the projection onto $\Omega^1(\Sigma, C^\infty(X, V))$ also satisfies the Leibniz rule. Likewise, the curvature of $\nabla_{\Sigma, C^\infty(X,V)}$ over Σ is the projection of the curvature of $\nabla_{\Sigma \times X, V}$ to $\Omega^2(\Sigma, C^\infty(X, V))$, and hence zero. $\qquad \square$

We also need a slight generalization of the mapping space $C^\infty(X, V)$. Note that $C^\infty(X)$ provides an algebra object in DVS and that $C^\infty(X, V)$ is a module sheaf over $C^\infty(X)$.

Definition B.6.5 For $E \to X$ a vector bundle on a smooth manifold and V a differentiable vector space, let $C^\infty(X, E \otimes V)$ denote the differentiable vector space

$$\mathscr{E}(X) \otimes_{C^\infty(X)} C^\infty(X, V).$$

The value of this sheaf on a manifold Σ is

$$C^\infty(\Sigma, C^\infty(X, E \otimes V)) = C^\infty(\Sigma, \mathscr{E}(X)) \otimes_{C^\infty(\Sigma, C^\infty(X))} C^\infty(\Sigma, C^\infty(X, V))$$

$$= \Gamma(\Sigma \times X, \pi_X^* E) \otimes_{C^\infty(\Sigma \times X)} C^\infty(\Sigma \times X, V).$$

We use $\mathbf{\Omega}^1(X, V)$ to denote $C^\infty(X, T_X^* \otimes V)$.

Tensoring over $C^\infty(X)$ is well behaved here as $\mathscr{E}(X)$ is finite rank and projective over $C^\infty(X)$.

We now prove a key result for constructing the Enriched Hom $\mathbb{H}\text{om}_{\text{DVS}}$.

Lemma B.6.6 *For any differentiable vector spaces V and for any manifold X, there is a natural map of sheaves*

$$\nabla_{C^\infty(X,V)} : C^\infty(X, V) \to \mathbf{\Omega}^1(X, V)$$

making $C^\infty(X, V)$ a $C^\infty(X)$-module sheaf with a flat connection.

Proof The flat connection on $C^\infty(X, V)$ arises from the composition

$$C^\infty(\Sigma, C^\infty(X, V)) = C^\infty(\Sigma \times X, V) \xrightarrow{\nabla_{X \times \Sigma, V}} \Omega^1(\Sigma \times X, V)$$
$$\to \Omega^1(\Sigma, C^\infty(X, V)),$$

where the final projection map arises from the splitting $T^*_{\Sigma \times X} \cong \pi_X^* T_X^* \oplus \pi_\Sigma^* T_\Sigma^*$. This flat connection $\nabla_{\Sigma, C^\infty(X,V))}$ is $C^\infty(X)$-linear, since it only takes derivatives in the Σ-directions. This construction is functorial in Σ.

We now show that the flat connection on V induces a natural map of sheaves

$$\nabla_{C^\infty(X,V)} : C^\infty(X, V) \to \Omega^1(X, V).$$

This map is a flat connection for $C^\infty(X, V)$ viewed as a $C^\infty(X)$-module in Mod_C^∞. It is *not* a map in DVS because it anticommutes with the flat connections on $C^\infty(X, V)$ and $\Omega^1(X, V)$. Nonetheless, as we will see, this property is enough to obtain the lemma.

Let Σ be an arbitrary manifold. The splitting $T^*_{\Sigma \times X} \cong \pi_X^* T_X^* \oplus \pi_\Sigma^* T_\Sigma^*$ produces a connection in the X-direction by projection. More explicitly, the composition of the connection over $\Sigma \times X$ followed by projection onto the first summand,

$$C^\infty(\Sigma \times X, V) \xrightarrow{\nabla_{X \times \Sigma, V}} \Omega^1(\Sigma \times X, V) \to \Gamma(\Sigma \times X, \pi_X^* T_X^*)$$
$$\otimes_{C^\infty(\Sigma \times X)} C^\infty(\Sigma \times X, V),$$

gives a map

$$\nabla_{C^\infty(X,V)} : C^\infty(\Sigma, C^\infty(X, V)) \to C^\infty(\Sigma, \Omega^1(X, V))$$
$$= C^\infty(\Sigma, C^\infty(X, T_X^* \otimes V)).$$

Note that this map is $C^\infty(\Sigma)$-linear since it only takes derivatives in the X-direction. Moreover, this construction is clearly natural in Σ and thus defines a map as C^∞-modules.

This map $\nabla_{C^\infty(X,V)}$ is a flat connection with respect to $C^\infty(X)$, viewed as a C^∞-module. On $\Sigma \times X$, we know that $\nabla_{\Sigma \times X, V}$ is a connection on $\Sigma \times X$ and hence satisfies the Leibniz rule, so the projection onto $C^\infty(\Sigma, C^\infty(X, T_X^* \otimes V))$ also satisfies the Leibniz rule. Likewise, the curvature of $\nabla_{C^\infty(X,V)}$ over Σ is the projection of the curvature of $\nabla_{\Sigma \times X, V}$ to $C^\infty(\Sigma, \Omega^2(X, V))$, and hence zero. In short, we know that

$$\nabla_{\Sigma \times X, V} = \nabla_{\Sigma, C^\infty(X,V)} + \nabla_{C^\infty(X,V)},$$

because they both arise as $\nabla_{\Sigma \times X, V}$ followed by projections onto the two summands of $\Omega^1(\Sigma \times X)$. As $\nabla_{\Sigma \times X, V}$ is flat, we obtain the for each summand separately.

Note that this decomposition also implies that

$$0 = \nabla_{\Sigma, \Omega^1(X,V)} \circ \nabla_{C^\infty(X,V)} + \nabla_{C^\infty(X,V)} \circ \nabla_{\Sigma, C^\infty(X,V)},$$

justifying our claim that $\nabla_{C^\infty(X,V)}$ anticommutes with the flat connections and hence is not a map of differentiable vector spaces. $\qquad\square$

B.6.2 The Enriched Hom

We now introduce the appropriate notion of "families of maps from V to W."

Definition B.6.7 For V and W differentiable vector spaces, let $\mathbb{H}om_{\mathrm{DVS}}(V, W)$ denote the following presheaf. The sections of the presheaf $\mathbb{H}om_{\mathrm{DVS}}(V, W)$ on a manifold X are

$$\mathbb{H}om_{\mathrm{DVS}}(V, W)(X) = \mathrm{DVS}(V, C^\infty(X, W)).$$

Note that, by definition, a section F assigns to each manifold Σ a map

$$F_\Sigma : C^\infty(\Sigma, V) \to C^\infty(\Sigma, C^\infty(X, W)) = C^\infty(\Sigma \times X, W),$$

and these maps F_Σ intertwine with pullbacks along maps $\phi : \Sigma \to \Sigma'$. To a smooth map $f : Y \to X$, the presheaf $\mathbb{H}om_{\mathrm{DVS}}(V, W)$ assigns the natural pullback

$$f^* : \mathrm{DVS}(V, C^\infty(X, W)) \to \mathrm{DVS}(V, C^\infty(Y, W)),$$

where for each manifold Σ,

$$(f^*F)_\Sigma = (\mathrm{id}_\Sigma \times f)^*_W \circ F_\Sigma$$

with $F \in \mathrm{DVS}(V, C^\infty(X, W))$ and $(\mathrm{id}_\Sigma \times f)^*_W : C^\infty(\Sigma \times X, W) \to C^\infty(\Sigma \times Y, W)$ the pullback map for the sheaf W.

Note that by definition, $\mathbb{H}om_{\mathrm{DVS}}(V, W)(*) = \mathrm{DVS}(V, W)$, so our definition has one property necessary to provide an enrichment of DVS. We now see that it naturally lifts to an element DVS.

Lemma B.6.8 *For any* $V, W \in \mathrm{DVS}$, *the presheaf* $\mathbb{H}om_{\mathrm{DVS}}(V, W)$ *is a differentiable vector space.*

Proof First, we explain why $\mathbb{H}om_{\mathrm{DVS}}(V, W)$ is a sheaf. Let $\mathcal{U} = \{U_i\}$ be a cover of the manifold X in the site Mfld. Then, for any manifold Σ, there is a natural product cover $\{\Sigma \times U_i\}$ of $\Sigma \times X$, and

$$C^\infty(\Sigma \times X, W) \cong \lim\left(\prod_i C^\infty(\Sigma \times U_i, W) \rightrightarrows \prod_{i,j} C^\infty(\Sigma \times (U_i \cap U_j), W)\right),$$

as it is a sheaf. Hence, to give a map $F_\Sigma : C^\infty(\Sigma, V) \to C^\infty(\Sigma \times X, W)$ is equivalent to giving maps $F_\Sigma^{(i)} : C^\infty(\Sigma, V) \to C^\infty(\Sigma \times U_i, W)$ for every U_i such that for every i and j, the postcompositions with pullback to the intersections $\Sigma \times (U_i \cap U_j)$ agree. As this equivalence is natural in Σ, we see that

$$\mathrm{DVS}(V, C^\infty(X, W)) \cong \lim$$

$$\times \left(\prod_i \mathrm{DVS}(V, C^\infty(U_i, W)) \rightrightarrows \prod_{i,j} \mathrm{DVS}(V, C^\infty(U_i \cap U_j, W)) \right),$$

which is the sheaf condition.

Now let's see why $\mathbb{H}\mathrm{om}_{\mathrm{DVS}}(V, W)$ is a C^∞-module. Observe that for any manifolds Σ and X, the projection map $\Sigma \times X \to X$ produces an algebra map $C^\infty(X) \to C^\infty(\Sigma \times X)$, and so for any differentiable vector space W, there is a canonical $C^\infty(X)$-module structure on $C^\infty(\Sigma \times X, W)$. Given $f \in C^\infty(X)$, let m_f denote the multiplication-by-f operator on $C^\infty(\Sigma \times X, W)$. Consider now an element $F \in \mathrm{DVS}(V, C^\infty(X, W))$. For each Σ, the map $F_\Sigma : C^\infty(\Sigma, V) \to C^\infty(\Sigma \times X, W)$ is $C^\infty(\Sigma)$-linear and commutes with differentiation by vector fields on Σ. (We are interested here only in the flat connection along Σ of $C^\infty(\Sigma, C^\infty(X, W))$.) The composite $m_f \circ F_\Sigma$ is also $C^\infty(\Sigma)$-linear and commutes with differentiation by vector fields on Σ, since m_f is constant in the Σ-direction. As m_f is also natural in Σ, we can define $f \cdot F$ by $(f \cdot F)_\Sigma = m_f \circ F_\Sigma$. This construction makes $\mathrm{DVS}(V, C^\infty(X, W))$ a $C^\infty(X)$-module, and as the construction is natural in X, we see that $\mathbb{H}\mathrm{om}_{\mathrm{DVS}}(V, W)$ is a C^∞-module.

Finally, recall from Lemma B.6.6 that there is a natural flat connection

$$\nabla_{C^\infty(X,W)} : C^\infty(X, W) \to \mathbf{\Omega}^1(X, W),$$

which induces a map

$$\nabla_{C^\infty(X,V)*} : \mathrm{DVS}(V, C^\infty(X, W)) \to \mathrm{DVS}(V, \mathbf{\Omega}^1(X, W))$$

by post-composition: $F \mapsto \nabla_{C^\infty(X,W)} \circ F$. We want, however, a flat connection

$$\nabla_X : \mathrm{DVS}(V, C^\infty(X, W)) \to \Omega^1(X) \otimes_{C^\infty(X)} \mathrm{DVS}(V, C^\infty(X, W)),$$

so we begin by showing that there is an isomorphism

$$\mu_X : \Omega^1(X) \otimes_{C^\infty(X)} \mathrm{DVS}(V, C^\infty(X, W)) \xrightarrow{\cong} \mathrm{DVS}(V, \mathbf{\Omega}^1(X, W)),$$

where $\mu_X(\alpha \otimes F)_\Sigma$ sends $v \in C^\infty(\Sigma, V)$ to $\alpha \otimes F_\Sigma(v)$.

As we are working with sheaves, we can work locally in X, so it suffices to check only for $X = \mathbb{R}^n$. In that situation, we fix a frame $\{dx_1, \ldots, dx_n\}$ for

$\Omega^1(X)$, and hence obtain an isomorphism

$$\Omega^1(X) \cong C^\infty(X)^{\oplus n},$$

which induces an isomorphism

$$\Omega^1(X, W) \cong C^\infty(X, W)^{\oplus n}.$$

We then see that

$$\Omega^1(X) \otimes_{C^\infty(X)} \mathrm{DVS}(V, C^\infty(X, W)) \cong \mathrm{DVS}(V, C^\infty(X, W))^{\oplus n}$$
$$\cong \mathrm{DVS}(V, C^\infty(X, W)^{\oplus n})$$
$$\cong \mathrm{DVS}(V, \Omega^1(X, W)),$$

and direct inspection identifies this composite as μ_X. An analogous argument shows that

$$\Omega^2(X) \otimes_{C^\infty(X)} \mathrm{DVS}(V, C^\infty(X, W)) \cong \mathrm{DVS}(V, \Omega^2(X, W)).$$

The composite $\mu_X^{-1} \circ \nabla_{C^\infty(X,V)*}$ thus provides a natural flat connection on $\mathrm{DVS}(V, C^\infty(X, W))$. $\qquad\square$

We want to show that $\mathbb{H}\mathrm{om}_{\mathrm{DVS}}$ provides differentiable vector spaces with the structure of a category. Thus, we need to explain how to compose such "families of maps"; the definition is natural but somewhat elaborate upon first encounter.

Let U, V, W be differentiable vector spaces, and let X be a manifold. There is a natural composition

$$\bullet_X : \mathrm{DVS}(V, C^\infty(X, W)) \times \mathrm{DVS}(U, C^\infty(X, V)) \to \mathrm{DVS}(U, C^\infty(X, W))$$

constructed as follows. Let $G \in \mathrm{DVS}(V, C^\infty(X, W))$ and $F \in \mathrm{DVS}(U, C^\infty(X, V))$, and recall that $F_\Sigma : C^\infty(\Sigma, U) \to C^\infty(\Sigma \times X, V)$ denote how F acts on sections for the manifold Σ. For each input manifold Σ, set

$$(G \bullet_X F)_\Sigma = (\mathrm{id}_\Sigma \times \Delta_X)^* \circ G_{\Sigma \times X} \circ F_\Sigma : C^\infty(\Sigma, V) \to C^\infty(\Sigma \times X, W),$$

where $\Delta_X : X \to X \times X$ is the diagonal map. Observe that for any map $\phi : \Sigma \to \Sigma'$, we have

$$(G \bullet_X F)_\Sigma \circ \phi_U^* = \phi_W^* \circ (G \bullet_X F)_{\Sigma'},$$

as each constituent of $(G \bullet_X F)_\Sigma$ commutes with pullbacks. Hence, $\bullet_X$ is well defined. Moreover, by construction, $\bullet_X$ is $C^\infty(X)$-bilinear, and $\bullet_X$ respects the flat connections, since its constituents do.

Definition B.6.9 Let U, V, W be differentiable vector spaces. The *composition in families*

$$\bullet : \mathbb{H}\text{om}_{\text{DVS}}(V, W) \times \mathbb{H}\text{om}_{\text{DVS}}(U, V) \to \mathbb{H}\text{om}_{\text{DVS}}(U, W)$$

is a bilinear map of differentiable vector spaces where the map on sections over a manifold X is

$$\bullet_X : \text{DVS}(V, \boldsymbol{C}^\infty(X, W)) \times \text{DVS}(U, \boldsymbol{C}^\infty(X, V)) \to \text{DVS}(U, \boldsymbol{C}^\infty(X, W)).$$

To see that this operation $\bullet$ is well defined, observe that for any map $f : Y \to X$ of manifolds, we find

$$(f^*G) \bullet_Y (f^*F) = f^* \circ (G \bullet_X F)$$

as each constituent plays naturally with pullbacks such as $f^* : \text{DVS}(U, \boldsymbol{C}^\infty(X, W)) \to \text{DVS}(U, \boldsymbol{C}^\infty(Y, W))$. Hence $\bullet$ defines a map of presheaves, in fact, a bilinear map of differentiable vector spaces.

Lemma B.6.10 *This composition operation is associative:*

$$- \bullet (- \bullet -) = (- \bullet -) \bullet -.$$

Proof We need to show that for a manifold X and morphisms $F \in \text{DVS}(T, \boldsymbol{C}^\infty(X, U))$, $G \in \text{DVS}(U, \boldsymbol{C}^\infty(X, V))$, and $H \in \text{DVS}(V, \boldsymbol{C}^\infty(X, W))$,

$$H \bullet_X (G \bullet_X F) = (H \bullet_X G) \bullet_X F.$$

Fix an input manifold Σ. Observe that $(H \bullet_X (G \bullet_X F))_\Sigma$ is given by

$$(\text{id}_\Sigma \times \Delta_X)^* \circ H_{\Sigma \times X} \circ ((\text{id}_\Sigma \times \Delta_X)^* \circ G_{\Sigma \times X} \circ F_\Sigma)$$

and that $((H \bullet_X G) \bullet_X F)_\Sigma$ is given by

$$(\text{id}_\Sigma \times \Delta_X)^* \circ ((\text{id}_{\Sigma \times X} \times \Delta_X)^* \circ H_{(\Sigma \times X) \times X} \circ G_{\Sigma \times X}) \circ F_\Sigma.$$

Using associativity of (ordinary) composition, we see that it suffices to show

$$H_{\Sigma \times X} \circ (\text{id}_\Sigma \times \Delta_X)^* = (\text{id}_{\Sigma \times X} \times \Delta_X)^* \circ H_{(\Sigma \times X) \times X},$$

but this equality holds because H is a map of sheaves on manifolds, and hence intertwines the pullback of sections. $\square$

In consequence, we obtain the desired result.

Theorem B.6.11 *There is a category $\mathbb{D}\text{VS}$ whose objects are differentiable vector spaces and in which the differentiable vector space of morphisms from*

V to W is $\mathbb{H}om_{DVS}(V, W)$. *As*

$$C^\infty(*, \mathbb{H}om_{DVS}(V, W)) = DVS(V, W),$$

we recover the category DVS *via the global sections functor.*

It is also possible to enrich the multicategory structure of DVS. We define $\mathbb{H}om_{DVS}(V_1, \ldots, V_n|W)$ as the differentiable vector space whose sections on X are

$$C^\infty(X, \mathbb{H}om_{DVS}(V_1, \ldots, V_n|W)) = DVS(V_1, \ldots, V_n|C^\infty(X, W)),$$

which is analogous to our definition of the enriched hom.

B.6.3 Compatibility with CVS

Most of the differentiable vector spaces we work with in this book appear naturally as convenient vector spaces. Thus, we are often interested in describing a convenient vector space as a differentiable vector space, i.e., in describing smooth maps to a convenient vector space. Since CVS is self-enriched due to its closed symmetric monoidal structure, we can ask how $CVS(V, W)$, viewed as a differentiable vector space, compares to $\mathbb{H}om_{DVS}(V, W)$, where V, W are convenient. The answer is the best we can hope for.

Lemma B.6.12 *For any manifold M and convenient vector spaces V, W there is an isomorphism*

$$C^\infty(M, CVS(V, W)) = CVS(V, C^\infty(M, W)),$$

and it is natural in M. This identification is compatible with the flat connections along M on the $C^\infty(M)$-modules for each side. Hence

$$CVS(V, W) \cong \mathbb{H}om_{DVS}(V, W)$$

as differentiable vector spaces.

Proof Let us begin by verifying the final assertion – the enriched homs in CVS and DVS agree – assuming the first isomorphism. Recall the exponential law: there is an isomorphism of convenient vector spaces

$$C^\infty(N, C^\infty(M, W)) \cong C^\infty(N \times M, W)$$

for any manifolds N, M and convenient vector space W. (See Theorem 3.12 of Kriegl and Michor (1997).) Hence we have an isomorphism of differentiable vector spaces

$$dif_c C^\infty(M, W) \cong C^\infty(M, dif_c W),$$

where dif_c is the fully faithful functor embedding CVS into DVS. (We have suppressed that notation so far.) Thus for every manifold M and every $V, W \in$ CVS, we have

$$C^\infty(M, \mathrm{CVS}(V, W)) \cong \mathrm{CVS}(V, C^\infty(M, W))$$
$$\cong \mathrm{DVS}(dif_c V, dif_c C^\infty(M, W))$$
$$\cong \mathrm{DVS}(dif_c V, C^\infty(M, dif_c W))$$
$$\cong C^\infty(M, \mathbb{Hom}_{\mathrm{DVS}}(dif_c V, dif_c W)),$$

and so $dif_c \mathrm{CVS}(V, W) \cong \mathbb{Hom}_{\mathrm{DVS}}(dif_c V, dif_c W)$ as differentiable vector spaces.

We now turn to proving the main isomorphism.

Lemma 3.7 of Kriegl and Michor (1997) states that the set of smooth curves $C^\infty(\mathbb{R}, W)$ is a convenient vector space when W is convenient. By Lemma 3.11 of Kriegl and Michor (1997), the space $C^\infty(V, W)$ is the limit of $C^\infty(\mathbb{R}, W)$ over all the smooth curves $\gamma : \mathbb{R} \to V$, where the morphisms in the diagram category are given by pullback along the reparametrizations $\gamma_1 = c^* \gamma_0$, with $c \in C^\infty(\mathbb{R}, \mathbb{R})$. (This result should be plausible because a function $f : V \to W$ is smooth precisely when it sends smooth curves into V to smooth curves in W.) Putting these facts together with the fact that CVS is closed under limits (see Theorem B.5.2,) we see that $C^\infty(V, W)$ is convenient.

Let us now analyze the smooth maps from a manifold M to $C^\infty(V, W)$. We will repeatedly use the fact that the functor sending a convenient vector space E to $C^\infty(M, E)$ preserves limits. (This claim follows from combining Lemma 3.8 with Lemma 3.11 of Kriegl and Michor (1997) and using the fact that limits commute.) We compute

$$C^\infty(M, C^\infty(V, W)) = C^\infty(M, \lim_{\gamma \in C^\infty(\mathbb{R}, V)} C^\infty(\mathbb{R}, W))$$
$$= \lim_{\gamma \in C^\infty(\mathbb{R}, V)} C^\infty(M, C^\infty(\mathbb{R}, W))$$
$$= \lim_{\gamma \in C^\infty(\mathbb{R}, V)} C^\infty(M \times \mathbb{R}, W)$$
$$= \lim_{\gamma \in C^\infty(\mathbb{R}, V)} C^\infty(\mathbb{R}, C^\infty(M, W))$$
$$= C^\infty(V, C^\infty(M, W)).$$

(Alternatively, one can simply cite the exponential law, Theorem 3.12 of Kriegl and Michor (1997).) Now the linear maps are a closed subset of the smooth maps

$$\mathrm{CVS}(V, W) \subset C^\infty(V, W)$$

cut out by linear equations that enforce linearity of the smooth maps. (See the proof of Lemma B.5.7.) It follows that a smooth map from M to $\mathrm{CVS}(V, W)$ is a smooth map from M to $C^\infty(V, W)$ that is pointwise in $\mathrm{CVS}(V, W)$. Similarly an element of $C^\infty(V, C^\infty(M, W))$ is linear if for each $m \in M$ the corresponding map from V to W is linear. It follows that

$$C^\infty(M, \mathrm{CVS}(V, W)) = \mathrm{CVS}(V, C^\infty(M, W))$$

as desired. □

B.7 Vector Spaces Arising from Differential Geometry

We describe here some aspects of the vector spaces that feature in our work, namely, sections of vector bundles on manifolds.

B.7.1 Comparison with Tensor Product of Nuclear Spaces

We have shown that the tensor products on BVS and CVS encode precisely the multilinear maps of the associated differentiable vector spaces. We have avoided, however, discussing how these tensor products compare to the various flavors of tensor products for *topological* vector spaces. Unfortunately, there are no simple general statements. Thankfully, for the kinds of vector spaces that we will work with, things are much better behaved.

Recall that the vector space of smooth sections of a vector bundle is typically equipped with a Fréchet topology. (That is, it is a complete locally convex topological vector space whose topology is metrizable.)

Proposition B.7.1 (Kriegl and Michor (1997), Proposition 5.8) *If* $V, W \in$ LCTVS *are metrizable, then*

$$V \otimes_\pi W = V \otimes_\beta W,$$

where $\otimes_\pi$ *denotes the projective tensor product.*

To be more pedantic, we might write that

$$inc_\beta(born(V) \otimes_\beta born(W)) = V \otimes_\pi W,$$

although we will leave these functors implicit from hereon.

As a corollary, we have the following crucial result.

Corollary B.7.2 *For* V, W *Fréchet spaces,* $V \widehat{\otimes}_\pi W = V \widehat{\otimes}_\beta W$.

Here it is important to note that a Fréchet space is convenient, by Theorem 4.11 of Kriegl and Michor (1997).

Proof By Theorem 2.14 of Kriegl and Michor (1997), a locally convex vector space is convenient if and only if it is Mackey complete.

By Result 52.23 of Kriegl and Michor (1997), we know that in a metrizable locally convex vector space, the convergent sequences coincide with the Mackey convergent sequences. Hence, the usual metric space completion agrees with the Mackey completion.

Taking the metric space completion of $V \otimes_\pi W$ is thus equivalent to the completion to a convenient vector space. Because $V \otimes_\pi W = V \otimes_\beta W$, we obtain the result. □

We are now in a position to prove a result that is crucial for our constructions.

For $p : E \to M$ a smooth vector bundle, let $\mathscr{E}(M)$ denote the space of smooth sections, which we will momentarily view as a Fréchet space in LCTVS or as a convenient vector space, compatibly via our functors. Let $E^!$ denote the vector bundle $E^\vee \otimes \mathrm{Dens}_M$, obtained by fiberwise tensor product of the fiberwise linear dual $E^\vee$ with the density line bundle Dens_M. The fiberwise evaluation pairing between E and $E^\vee$ induces a pairing

$$\langle -, - \rangle : \quad \mathscr{E}^!_c(M) \times \mathscr{E}(M) \quad \to \quad \mathbb{R}$$
$$(\lambda, s) \quad \mapsto \quad \int_M \mathrm{ev}(\lambda, s),$$

since fiberwise evaluation produces a compactly supported smooth density on M. Extending this pairing to the compactly supported *distributional* sections of $E^!$, we see that the compactly distributional sections $\overline{\mathscr{E}}^!_c(M)$ are canonically isomorphic to the strong dual $\mathscr{E}(M)^*$ of $\mathscr{E}(M)$.

Proposition B.7.3 *Let* $\mathscr{E}_1(M_1)$, ..., $\mathscr{E}_n(M_n)$, *and* $\mathscr{F}(N)$ *denote smooth sections of vector bundles (possibly on different manifolds). For concision, we now suppress the manifolds from the notation.*

Then we have an isomorphism of differentiable vector spaces:

$$dif_c(\mathrm{CVS}(\mathscr{E}_1 \widehat{\otimes}_\beta \cdots \widehat{\otimes}_\beta \mathscr{E}_n, \mathscr{F})) \cong dif_t(\overline{\mathscr{E}}^!_{1,c} \widehat{\otimes}_\pi \cdots \widehat{\otimes}_\pi \overline{\mathscr{E}}^!_{n,c} \widehat{\otimes}_\pi \mathscr{F}).$$

Proof We use the preceding corollary. For Fréchet spaces, we know the completed projective tensor product agrees with the completed bornological tensor product, so we have another isomorphism

$$\mathrm{CVS}(\mathscr{E}_1 \widehat{\otimes}_\beta \cdots \widehat{\otimes}_\beta \mathscr{E}_n, \mathscr{F}) \cong \mathrm{CVS}(\mathscr{E}_1 \widehat{\otimes}_\pi \cdots \widehat{\otimes}_\pi \mathscr{E}_n, \mathscr{F}).$$

For complete nuclear spaces V, W, we know

$$\mathrm{LCTVS}(V, W) \cong V^* \widehat{\otimes}_\pi W,$$

where V^* denotes the strong dual. Hence, we have

$$\mathrm{CVS}(\mathscr{E}_1 \hat{\otimes}_\pi \ldots \hat{\otimes}_\pi \mathscr{E}_n, \mathscr{F}) \cong (\mathscr{E}_1 \hat{\otimes}_\pi \cdots \hat{\otimes}_\pi \mathscr{E}_n)^* \hat{\otimes}_\pi \mathscr{F}$$
$$\cong \mathscr{E}_1^* \hat{\otimes}_\pi \cdots \hat{\otimes}_\pi \mathscr{E}_n^* \hat{\otimes}_\pi \mathscr{F}.$$

For smooth sections of a vector bundle $\mathscr{E}$, we know $\mathscr{E}^* = \overline{\mathscr{E}}_c^!$. $\qquad\qquad\square$

Let us now explain an application of this result that we use repeatedly in this book. Viewing $\mathscr{E}$ as a vector space, it is natural to try to define the algebra of functions on $\mathscr{E}$ as something like the symmetric algebra $\mathrm{Sym}\,\mathscr{E}^*$ on the dual to $\mathscr{E}$. We need to be careful about what we mean by taking the symmetric algebra, however, so we give our preferred definition.

Definition B.7.4 For $\mathscr{E}$ the convenient vector space of smooth sections of a vector bundle E, the uncompleted *algebra of functions on $\mathscr{E}$* is the convenient vector space

$$\bigoplus_{n=0}^{\infty} \mathrm{CVS}(\mathscr{E}^{\hat{\otimes}_\beta n}, \mathbb{R})_{S_n}.$$

The subscript indicates that one takes the coinvariants with respect to the natural action of the symmetric group S_n on the n-fold tensor product.

This vector space is convenient. "Uncompleted" here refers to using the direct sum rather than product; it is the uncompleted symmetric algebra – namely, polynomials – rather than the completed symmetric algebra – namely, formal power series.

We will often work with the *completed* algebra of functions

$$\mathscr{O}(\mathscr{E}) := \prod_{n=0}^{\infty} \mathrm{CVS}(\mathscr{E}^{\hat{\otimes}_\beta n}, \mathbb{R})_{S_n},$$

particularly in the setting of interacting field theories.

Thanks to our work above, we know that we could also work with the completed projective tensor product instead of the completed bornological tensor product.

Corollary B.7.5 *We have*

$$\mathscr{O}(\mathscr{E}) \cong \prod_{n=0}^{\infty} \left((\overline{\mathscr{E}}_c^!)^{\hat{\otimes}_\pi n} \right)_{S_n}$$

as convenient vector spaces.

In Section 3.5.7, we provide an alternative description of these algebras using the enriched hom $\mathbb{H}\mathrm{om}_{\mathrm{DVS}}$. See, in particular, Lemma 3.5.15.

B.7.2 Sections of a Vector Bundle as (Co)sheaves of Differentiable Vector Spaces

Let M be a manifold and E a vector bundle on M. Recall the following notations:

- $\mathscr{E}(M)$ denotes the vector space of smooth sections of E over M.
- $\mathscr{E}_c(M)$ denotes the vector space of compactly supported smooth sections of E over M.
- $\overline{\mathscr{E}}(M)$ denotes the vector space of distributional sections of E over M.
- $\overline{\mathscr{E}}_c(M)$ denotes the vector space of compactly supported distributional sections of E over M.

We can view these spaces as living in LCTVS, BVS, CVS, or DVS.

We now discuss what happens as one varies over the opens of M.

Lemma B.7.6 *The functor* $\mathscr{E}$: $\mathrm{Opens}_M^{op} \to \mathrm{DVS}$ *sending U to $\mathscr{E}(U)$ is both a sheaf and a homotopy sheaf. Likewise the functor* $\overline{\mathscr{E}}$: $\mathrm{Opens}_M^{op} \to \mathrm{DVS}$ *sending U to $\overline{\mathscr{E}}(U)$ is both a sheaf and a homotopy sheaf.*

The functor $\mathscr{E}_c$: $\mathrm{Opens}_M \to \mathrm{DVS}$ *sending U to $\mathscr{E}_c(U)$ is both a cosheaf and a homotopy cosheaf. Likewise the functor* $\overline{\mathscr{E}}_c$: $\mathrm{Opens}_M \to \mathrm{DVS}$ *sending U to $\overline{\mathscr{E}}_c(U)$ is both a cosheaf and a homotopy cosheaf.*

As explained in the proof, one can also view these as (co)sheaves with values in CVS.

Proof We explain the precosheaf cases; the sheaf cases are parallel.

In Section A.5.4 of Appendix A, we verified explicitly the gluing axiom. In particular, we used a partition of unity for a cover of an open U to produce a map from the compactly supported sections on an open U to the colimit diagram for the cover (or Čech complex, in the homotopy cosheaf case). The maps we used are maps of convenient vector spaces, since they consist either of addition or of multiplication by a smooth function. As CVS is a full subcategory of DVS, these yield maps in DVS as well. Hence, we obtain cochain homotopy equivalences in DVS as well.

We want to emphasize now why it was so useful that we provided cochain homotopy equivalences in Section A.5.4 in Appendix A and not just a quasi-isomorphism. The functor dif_c preserves cochain homotopy equivalences. By contrast, it need not preserve quasi-isomorphism, since taking cohomology involves taking a cokernel, which need not to be preserved by dif_c. $\qquad\square$

Appendix C

Homological Algebra in Differentiable Vector Spaces

C.1 Introduction

In the study of field theories, one works with vector spaces of an analytical nature, like the space of smooth functions or distributions on a manifold. To use the Batalin–Vilkovisky formalism, we need to perform homological algebra in this setting. The standard approach to working with objects of this nature is to treat them as topological vector spaces, but it is not obvious how to set up a well-behaved version of homological algebra with topological vector spaces.

Our approach here breaks the problem into two steps. First, our cochain complexes are constructed out of very nice topological vector spaces that are already convenient vector spaces. Hence, we view them as cochain complexes of convenient vector spaces, since CVS is a better-behaved category (for our purposes) than LCTVS. Second, we apply the functor dif_c to view them as cochain complexes of differentiable vector spaces; as CVS is a full subcategory of DVS, nothing drastic has happened. The benefit, however, is that DVS is a Grothendieck Abelian category, so that standard homological algebra applies immediately.

C.1.1 Motivation

There are a few important observations to make about this approach.

Recall that dif_c does not preserve cokernels, so that dif_c need not preserve cohomology. Given a complex C^* in CVS, the cohomology group $H^k C^*$ is a cokernel computed in CVS. Hence $H^k(dif_c C^*)$ could be different from $dif_c(H^k C^*)$. In consequence, dif_c need not preserve quasi-isomorphisms. We will view quasi-isomorphisms as *differentiable* cochain complexes as the correct notion and avoid discussing quasi-isomorphisms as *convenient* cochain complexes.

The functor dif_c does preserve cochain homotopy equivalences, however. Thus, certain classical results – such as the Atiyah–Bott lemma (see Appendix D) or the use of partitions of unity (see Section A.5.4 in Appendix A) – play a crucial role for us. They establish explicit cochain homotopy equivalences for convenient cochain complexes, which go to cochain homotopy equivalences of differentiable cochain complexes. Later constructions, such as the observables of BV theories, involve deforming the differentials on these differentiable cochain complexes. In almost every situation, these deformed cochain complexes are filtered in such a way that the first page of the spectral sequence is the original, undeformed differential. Thus we can leverage the classical result in the new deformed situation.

In short, many constructions in geometry and analysis are already done in CVS (or cochain complexes thereof), but the homological algebra is better done in DVS. We should explain why it's better to do the algebra in DVS.

As discussed in Appendix B, convenient vector spaces can be understood via their associated sheaves on the site of smooth manifolds. (In particular, morphisms between convenient vector spaces are precisely the maps as such sheaves of vector spaces.) In other words, doing convenient linear algebra is equivalent to thinking about smooth families of points in the convenient spaces.

For instance, in thinking about a differential operator $D : E \to F$, one can think about smoothly varying the inputs and studying the associated family of outputs. It is natural then to understand the kernel by thinking about smooth families of solutions to $D\phi = 0$. As dif_c preserves kernels, this way of thinking agrees with simply picking out the vector subspace annihilated by D.

These ways of thinking differ when it comes to the cokernel. In DVS, one studies the cokernel of D by thinking about when two smooth families in F differ by a family of outputs from D. The cokernel in DVS just amounts to studying such equivalence relations locally, i.e., in very small smooth families. In other words, we patch together equivalence classes; we need to sheafify the cokernel as precosheaves. This approach reflects our initial impulse to study something by varying inputs and outputs. From this perspective, the cokernel in CVS is constructed by a more opaque procedure. The cokernel in DVS maps to the cokernel in CVS, viewed as a differentiable vector space, so the two approaches communicate.

Remark: For another approach to these issues, see Wallbridge (2015), where a monoidal combinatorial model category structure is put on the category Ch(CVS) of unbounded chain complexes of convenient vector spaces.   ◊

C.1.2 Outline of the Appendix

After establishing foundational properties of DVS, we develop some homological techniques that we will use repeatedly throughout the text. Most are simple reworkings of standard results of homological algebra with sheaves. First, we will discuss spectral sequences, which we often use to verify some map of differentiable cochain complexes is a quasi-isomorphism. Second, we discuss the category of differentiable pro-cochain complexes, which is a somewhat technical definition but which plays a crucial role for observables of interacting theories. Finally, we will discuss homotopy colimits and explicit methods for constructing them, which we use in our definition of factorization algebras.

C.2 Linear Algebra and Homological Algebra in DVS

The category DVS behaves formally like a category of modules, as we establish in the first part. Thus, it is no surprise that cochain complexes in DVS is also straightforward, which we explain in the second part.

C.2.1 Linear Algebra

In Section B.2.3 in Appendix B, we showed that DVS is complete and cocomplete. Going further, we showed DVS is a locally presentable category. With a little more work, we obtain the following.

Proposition C.2.1 *The category* DVS *is Abelian.*

Proof The category of C^∞-modules is Abelian, as it is the category of modules for a sheaf of rings on a site. (See, for instance, Theorem 18.1.6 of Kashiwara and Schapira (2006).) As map of differentiable vector spaces $\phi : V \to W$ is C^∞-linear, it possesses a kernel $i_\phi : K_\phi \to V$ and a cokernel $q_\phi : W \to C_\phi$ in the category of C^∞-modules. We have already shown that these sheaves naturally inherit flat connections and hence provide the desired kernel and cokernel in differentiable vector spaces.

It remains to verify that there is a natural isomorphism from the coimage of ϕ (i.e., the cokernel of "$_\phi$") to the image of ϕ (i.e., the kernel of q_ϕ). In the Abelian category of C^∞-modules, there is such an isomorphism, and because ϕ intertwines the connections, this map also intertwines the connections, and hence provides a map of differentiable vector spaces.   □

To do homological algebra, however, we typically want an Abelian category with much stronger properties, first enumerated by Grothendieck (1957). As we already know that DVS possesses limits and colimits, it remains to check the exactness of filtered colimits. (This property already ensures that standard formulas from homological algebra satisfy ∞-categorical universal properties to define our desired homotopy colimits. See Proposition C.5.3.)

Lemma C.2.2 *Let I denote a filtered category. Given an exact sequence of I-diagrams*

$$0 \to A \xrightarrow{f} B \xrightarrow{g} C \to 0,$$

with diagrams $A, B, C : I \to$ DVS, the colimits also form an exact sequence

$$0 \to \operatorname*{colim}_I A \to \operatorname*{colim}_I B \to \operatorname*{colim}_I C \to 0.$$

In other words, DVS *is an AB5 Abelian category.*

Proof As in the previous proof, it is convenient first to "forget" down to the category of C^∞-modules, which is an *AB5* Abelian category Kashiwara and Schapira (2006). Thus, in this category, filtered colimits are exact. The underlying C^∞-modules of the colimits in DVS are simply those computed in C^∞-modules, so it remains to check that the C^∞-morphisms arising from the colimit in the category of C^∞-modules

$$0 \to \operatorname*{colim}_I A \to \operatorname*{colim}_I B \to \operatorname*{colim}_I C \to 0$$

respect the flat connections. This, however, is almost immediate.

For example, for any manifold X and any section of $s \in \operatorname{colim}_I B(X)$, there is some $i \in I$ such that s arises from an element $s_i \in B_i(X)$. The map g_i intertwines the flat connections on B_i and C_i, so we know that $\nabla^{C_i}(g_i s_i) = g_i(\nabla^{B_i} s_i)$. Moreover, this equality holds for the images $s_j \in B_j$ for all $j \geq i$. Hence, it will hold for the section s itself, by the defining property of the connection on the colimits $\operatorname{colim} B$ and $\operatorname{colim} C$. $\square$

C.2.2 Homological Algebra in DVS

To summarize the results so far, we have shown the following.

Theorem C.2.3 *The category* DVS *is a Grothendieck Abelian category.*

This result makes it possible to deploy standard arguments from homological algebra. In fact, Grothendieck's original work, aimed at unifying homological algebra for modules over a ring and sheaves of Abelian groups on a topological space, singled out these properties as particularly useful, as they guarantee the existence of injective objects. For those readers familiar with the basics of sheaf cohomology, this theorem ensures that our homological constructions behave as they expect. For those readers who would like deeper treatments, we point to Chapter 14 of Kashiwara and Schapira (2006), which treats the unbounded derived category of a Grothendieck Abelian category, and to section 1.3 of Lurie (2016), which studies the stable ∞-category associated to unbounded chain complexes in a Grothendieck Abelian category.

C.3 Spectral Sequences

We will often use versions of spectral sequence arguments in the category of differentiable complexes.

Definition C.3.1 A *filtered differentiable cochain complex* is a sequence of differentiable cochain complexes

$$\cdots \to F_{n-1}C \to F_nC \to F_{n+1}C \to \cdots$$

where each map $F_{n-1}C \to F_nC$ is a monomorphism (explicitly, a monomorphism in each cohomological degree). A *map of filtered differentiable cochain complexes* is a sequence of cochain maps $\{f_n : F_nC \to F_nD\}_{n\in\mathbb{Z}}$ that commute with the maps of each sequence.

Typically, one views these cochain complexes F_nC as subcomplexes of a given cochain complex C. We will always identify C as $\mathrm{colim}_n F_nC$, the differentiable cochain complex that is $\mathrm{colim}_n(F_nC)^k$ in cohomological degree k.

The following condition is essential for taking advantage of spectral sequences.

Definition C.3.2 A filtered differentiable cochain complex is *complete* if $C \cong \lim_n C/F_nC$ along the canonical map induced by the quotient maps $C \to C/F_nC$.

Recall the following useful theorem. (For a proof and further discussion, see Eilenberg and Moore (1962).)

Theorem C.3.3 (Eilenberg–Moore Comparison Theorem) *Let $f : C \to D$ be a map of complete filtered cochain complexes in an AB4 Abelian category. Suppose for each integer n, there exists an integer p such that $F_pC^n = 0 = F_pD^n$ (i.e., in each cohomological degree n, the filtration is bounded below).*

If there is a natural number r such that the induced map between the rth pages of the spectral sequences

$$f_r^{pq} : E_r^{pq}C \to E_r^{pq}D$$

is an isomorphism for all p, q, then $f : C \to D$ is a quasi-isomorphism.

As DVS is Grothendieck, we can (and do throughout this book) invoke the Eilenberg–Moore comparison theorem.

For ease of later use, we state an immediate corollary.

Lemma C.3.4 *Let C, D be filtered differentiable cochain complexes in which $F_n C = 0 = F_n D$ for $n < 0$. Let $f : C \to D$ be a map of filtered differentiable cochain complexes.*

If there is a natural number r such that the map f_r between the rth pages of the spectral sequences is an isomorphism of differentiable cochain complexes, then f is a quasi-isomorphism of differentiable cochain complexes.

In particular, the following two cases show up several times in the text:

(i) If the maps $\mathrm{Gr} f_n : F_n C / F_{n-1} C \to F_n D / F_{n-1} D$ are quasi-isomorphisms for all n, then f is a quasi-isomorphism.
(ii) If the maps $f_n : F_n C \to F_n D$ are quasi-isomorphisms for all n, then f is a quasi-isomorphism.

These two cases amount to the assertion that the E_2 or E_1 pages are isomorphisms.

Remark: There is another approach to proving these kinds of results that is, in some sense, more concrete. To verify that a map of differentiable cochain complexes is a quasi-isomorphism, it is enough to exhibit the quasi-isomorphism on stalks. Taking stalks, however, sends a differentiable cochain complex to a dg vector space. Applying these observations to the situation of a map of filtered differentiable cochain complexes, we see that we can invoke the Eilenberg–Moore comparison theorem for dg vector spaces at the level of stalks. $\Diamond$

We have a similar statement for inverse systems, but only under a stronger hypothesis. A *tower* is a sequence

$$\cdots \to V_{n+1} \xrightarrow{v_{n+1}} V_n \xrightarrow{v_n} V_{n-1} \to \cdots$$

where each map v_n is an epimorphism (explicitly, an epimorphism in each cohomological degree). There is an associated filtration for the limit $V = \lim_n V_n$ by taking the kernel of the canonical map

$$F_n V = \ker (V \to V_n).$$

We can thus hope to use this filtration to learn something about V.

Lemma C.3.5 *Let $V_\bullet, W_\bullet$ be two towers of differentiable cochain complexes such that*

(i) *The towers are eventually constant: there is some integer n such that epimorphisms v_p and w_p are isomorphisms for all $p \geq n$.*

(ii) *The towers are bounded: there is some integer m such that $V_q = 0 = W_q$ for all $q \leq m$.*

If $f_\bullet : V_\bullet \to W_\bullet$ is a map of sequences that induces a quasi-isomorphism $\mathrm{Ker}\, v_n \to \mathrm{Ker}\, w_n$ for all n, then the map $\lim f : \lim_n V_n \to \lim_n W_n$ is a quasi-isomorphism.

Proof Under the hypothesis that the towers are eventually constant, we know that the associated filtrations $F_\bullet V$ and $F_\bullet W$ are bounded below. Under the hypothesis that they are bounded, the associated filtrations are exhaustive: $V = \mathrm{colim}_n F_n V$ and $W = \mathrm{colim}_n F_n W$.

The map $f_\bullet$ then induces a map of filtered differentiable cochain complexes $\widehat{f} : V \to W$. As $F_{n+1} V / F_n V \cong \mathrm{Ker}\, v_n$ and likewise for the filtration on W, the spectral sequence argument above implies that $\widehat{f}$ is a quasi-isomorphism. But $\lim f = \widehat{f}$, so we are finished. □

C.4 Differentiable Pro-cochain Complexes

Most of the examples of factorization algebras in this book take values not in the category of cochain complexes but in the category of complete filtered cochain complexes. (For free field theories, it is natural to work just with differentiable cochain complexes, but we approach interacting theories via formal geometry, *aka* perturbative methods, and hence work with completed symmetric algebras.) Some straightforward modifications of certain constructions thus appear, such as colimits. This section is devoted to spelling out some important aspects of the category of complete filtered differentiable cochain complexes that appears in this book.

Definition C.4.1 A *differentiable pro-cochain cochain complex* is a differentiable cochain complex V with a filtration

$$\cdots \hookrightarrow F_n V \hookrightarrow F_{n+1} V \hookrightarrow \cdots \hookrightarrow F_0 V = V$$

such that the canonical map $V \to \lim_n V / F_n V$ is a quasi-isomorphism.

A *map of differentiable pro-cochain complexes* is a cochain map $f : V \to W$ that preserves the filtration. Such a map is a *filtered weak equivalence* if the associated graded map $\mathrm{Gr}\, f : \mathrm{Gr}\, V \to \mathrm{Gr}\, W$ is a quasi-isomorphism: for each n, the map $\mathrm{Gr}_n f : F_n V / F_{n-1} V \to F_n W / F_{n-1} W$ is a quasi-isomorphism.

C.4.1 Colimits

Colimits are not just given by colimits as differentiable cochain complexes: one must complete the "naive" colimit. To be explicit, there is an inclusion functor of differentiable pro-cochain complexes into negatively filtered differentiable cochain complexes (i.e., filtered objects such that $F_n V = F_0 V$ for all positive n). This inclusion functor is right adjoint to a "completion functor" that sends a filtration

$$\cdots \hookrightarrow F_n V \hookrightarrow F_{n+1} V \hookrightarrow \cdots \hookrightarrow F_0 V = V$$

to its completion $\widehat{V} = \lim_n V/F_n V$ equipped with the filtration $F_n \widehat{V} = \ker(\widehat{V} \to V/F_n V)$. Note that completion is "idempotent" in the sense that a complete filtered cochain complex is isomorphic to its completion.

We define the *completed* colimit of a diagram of differentiable pro-cochain complexes as the completion of the colimit in the category of filtered differentiable cochain complexes. To emphasize the role of completion, we will sometimes write $\widehat{\mathrm{colim}}$. For example, we have the following.

Definition C.4.2 Let a set A index a collection of differentiable pro-cochain complexes $\{V_\alpha \mid \alpha \in A\}$. The *completed direct sum* is the limit

$$\widehat{\bigoplus}_{\alpha \in A} V_\alpha = \lim_n \left(\bigoplus_{\alpha \in A} V_\alpha / F_n V_\alpha \right),$$

where on the right-hand side we use the direct sum of differentiable cochain complexes.

C.4.2 Stalks

The n-dimensional stalk of a differentiable pro-cochain complex

$$\mathrm{Stalk}_n(V) = \widehat{\mathrm{colim}}_{0 \in U \subset \mathbb{R}^n} V(U)$$

satisfies $\mathrm{Stalk}_n(V)/F^i \mathrm{Stalk}_n(V) = \mathrm{Stalk}_n(V/F^i V)$.

An immediate consequence is thus the following.

Lemma C.4.3 *A map $f : V \to W$ of differentiable pro-cochain complexes is a filtered weak equivalence if and only if the maps $\mathrm{Stalk}_n(f) : \mathrm{Stalk}_n(V) \to \mathrm{Stalk}_n(W)$ are filtered weak equivalences for every n.*

C.4.3 Spectral Sequences

We have analogs of the results from Section C.3 in Appendix C.

Lemma C.4.4 *Let $V_\bullet, W_\bullet$ be two towers of differentiable pro-cochain complexes such that*

(i) *The towers are eventually constant: there is some integer n such that epimorphisms v_p and w_p are isomorphisms for all $p \geq n$.*

(ii) *The towers are bounded: there is some integer m such that $V_q = 0 = W_q$ for all $q \leq m$.*

If $f_\bullet : V_\bullet \to W_\bullet$ is a map of sequences that induces a quasi-isomorphism $\operatorname{Ker} v_n \to \operatorname{Ker} w_n$ for all n, then the map $\lim f : \lim_n V_n \to \lim_n W_n$ is a filtered weak equivalence of differentiable pro-cochain complexes.

Proof This result follows immediately from Lemma C.3.5. □

C.4.4 Multilinear Maps

Differentiable pro-cochain complexes form a multicategory, just like differentiable cochain complexes.

Definition C.4.5 Let $V_1, \ldots, V_k, W$ be differentiable pro-cochain complexes. In the multicategory of differentiable pro-cochain complexes, an element of $\operatorname{Hom}(V_1, \ldots, V_k; W)$ is a multilinear map of differentiable cochain complexes

$$\Phi : V_1 \times \cdots \times V_k \to W = \lim W / F^i W$$

that preserves filtrations: if $v_i \in F^{r_i}(V_i)$, then $\Phi(v_1, \ldots, v_k) \in F^{r_1 + \cdots + r_k} W$.

C.4.5 Working over a Differentiable dg Ring

The category of differentiable cochain complexes is a multicategory. Thus, we can talk about commutative differentiable dg algebras R: it is a differentiable cochain complex R and a bilinear map $m \in \operatorname{Hom}(R, R; R)$, defining the multiplication, that satisfies the axioms of a commutative algebra. Similarly, a commutative differentiable pro-algebra is a commutative algebra in the multicategory of differentiable pro-cochain complexes.

In either context, we can define an R-module M to be a differentiable (pro-)cochain complex equipped with an action of the commutative differentiable (pro-)algebra R, in the obvious way. We say a map $M \to M'$ is a weak equivalence if it is a weak equivalence (as defined earlier) in the category of differentiable (pro-)cochain complex. In either context, we say a sequence of R-modules $0 \to M_1 \to M_2 \to M_3 \to 0$ is exact if it is exact in the category of differentiable (pro-)cochain complexes.

The category of modules over a differentiable (pro-)dg algebra R is, as earlier, a multicategory. In either case, the multimaps

$$\operatorname{Hom}_R(M_1, \ldots, M_n; N)$$

are the multimaps in the category of differentiable (pro-)cochain complexes whose underlying multilinear map $M_1 \times \cdots \times M_n \to N$ are R-multilinear.

C.5 Homotopy Colimits

A factorization algebra is, in particular, a cosheaf with respect to the Weiss topology. When working with precosheaves in cochain complexes, the conceptually correct version of the cosheaf gluing axiom uses the *homotopy colimit* rather than the usual colimit. Explicitly, a precosheaf $\mathcal{F}$ of cochain complexes is a cosheaf if for any open set V and any Weiss cover $\mathfrak{U} = \{U_i\}_{i\in I}$ of V, the canonical map

$$\mathrm{hocolim}_{\check{C}\mathfrak{U}}\, \mathcal{F} \to \mathcal{F}(V)$$

is a quasi-isomorphism, where the left-hand side denotes the homotopy colimit over the Čech cover of $\mathfrak{U}$.

In our situation, the homotopy colimit of a diagram $F : I \to \mathrm{Ch(DVS)}$ can be described by a familiar cochain complex, as we explain in the text that follows. In the case of the cosheaf gluing axiom, we recover precisely the formula that appears in the definition of a factorization algebra (see Section 6.1 in Chapter 6).

We begin by discussing a general situation that includes $\mathrm{Ch(DVS)}$ as a special case. In particular, we explain the homotopical universal property characterizing a homotopy colimit and then state a theorem providing an explicit construction of a homotopy colimit in this situation. In the next subsection, we give the proof of this theorem. Finally, we treat the case of differentiable pro-cochain complexes.

Remark: The Čech complex we use is long-established in homological algebra and easy to motivate. Readers satisfied with it should probably look no further, as the arguments that follow are highly technical. The goal of this section, though, is to ensure compatibility between our work in this book and the perspective emphasized by higher category theory. It bridges an odd gap in the literature between homological and homotopical algebra.   ◇

C.5.1 Reminder on Homotopy Colimits

Let $\mathcal{A}$ denote an Abelian category, and let $\mathbf{A} = \mathrm{Ch}(\mathcal{A})$ denote the category of unbounded cochain complexes in $\mathcal{A}$. Let I be a category and let $F : I \to \mathbf{A}$ be a functor, which we will call a *diagram of cochain complexes* (or I-diagram to emphasize the shape of the diagram).

Let's start by reviewing the definition of a colimit in order to motivate the definition of a homotopy colimit. There are several different definitions that characterize the same notion.

Note that any cochain complex C provides a "constant" diagram $p^*C : I \to$ **A** given by pullback along the constant functor $p : I \to \{*\}$: for every i, $p^*C(i) = C$, and every map in this diagram is the identity. There is thus a "constant diagram" functor

$$p^* : \mathbf{A} \to Fun(I, \mathbf{A})$$

given by pullback along $p : I \to \{*\}$. Given a diagram $F : I \to \mathbf{A}$, a *cocone for F* is a natural transformation $\alpha : F \Rightarrow p^*C$ for some cochain complex C. In other words, it is the data of a map from every $F(i)$ to C such that for every arrow $i \to j$ in I, the associated composition $F(i) \to F(j) \to C$ equals the map $F(i) \to C$. A *colimit for F* is an initial cocone, if it exists. If the colimit exists, it is unique up to isomorphism. Every cocone of F thus factors uniquely through the colimit of F.

In the case of the category **A**, every diagram admits a colimit. (There are categories where only certain diagrams have a colimit.) We can, in fact, construct a functor "take the colimit"

$$\operatorname*{colim}_{I} : Fun(I, \mathbf{A}) \to \mathbf{A}$$

as the left adjoint to p^*. The universal property satisfied by the colimit of a diagram ensures that we have the adjunction. This kind of characterization of the colimit is *global* in the sense that it works with all I-diagrams simultaneously, rather than being *local*, like the cocone definition, which only involves a fixed diagram F.

In the setting of homological algebra, this notion of colimit is not appropriate because we want to consider cochain complexes that are quasi-isomorphic as equivalent, even though not every quasi-isomorphism is invertible. This consideration leads to two different notions of *homotopy colimit*, a global and a local one. The construction we give will be global in nature, because the ∞-category of cochain complexes is nice enough that homotopy colimits exist for all diagrams. In a moment, we will sketch the (local) definition of colimit used in the setting of ∞-categories, and we will provide an explicit construction of it in Theorem C.5.11. There is, however, a more concrete notion of homotopy colimit requiring no higher category theory that we now describe. (This notion is a shadow in the homotopy category of an ∞-categorical notion.)

Since we want to view quasi-isomorphisms as equivalences, we want to work with the localization of **A** with respect to the quasi-isomorphisms, which we call the *homotopy category of* **A** and denote Ho(**A**). Likewise, we want to consider two I-diagrams F and G as equivalent if there is a natural transformation $\eta : F \Rightarrow G$ such that $\eta(i) : F(i) \to G(i)$ is a quasi-isomorphism for all i. Let Ho $Fun(I, \mathbf{A})$ denote the homotopy category of I-diagrams.

Observe that the constant diagram functor p^* preserves weak equivalences and hence induces a functor $Ho(p^*) : Ho(\mathbf{A}) \to Ho\,Fun(I, \mathbf{A})$.

Definition C.5.1 The *homotopy colimit over I* is the left adjoint to Ho(p^*), if it exists. The homotopy colimit of a diagram $F : I \to \mathbf{A}$ is denoted hocolim$_I$ F.

In other words, the homotopy colimit of a diagram satisfies, at the level of homotopy categories, a version of adjunction in the global definition of colimit. It is the initial cochain complex, up to quasi-isomorphism, among all cochain complexes, up to quasi-isomorphism, that receive a map from the diagram, up to quasi-isomorphism of diagrams.

In Section C.5.3, we describe the ∞-category of cochain complexes in several ways, as a model category and as a quasi-category, following Lurie. Given a diagram $F : I \to \mathcal{C}$ in an ∞-category, there is an ∞-category of cocones for F, which Lurie denotes as $\mathcal{C}_{F/}$. A cocone is essentially a homotopy coherent version of the notion of cocone given in ordinary categories. A colimit for F is then an initial object in $\mathcal{C}_{F/}$, which means that any cocone admits a contractible space of maps from the colimit. (Having a contractible space of morphisms is the ∞-categorical replacement of the notion of unique map.) This ∞-categorical colimit is often called the homotopy colimit to distinguish it from the 1-categorical colimit, when there is the possibility of confusion. See section 1.2.13 of Lurie (2009b) for the beginning of a systematic discussion.

C.5.2 A Cochain-Level Construction

We will now explain how to construct the homotopy colimit when $\mathcal{A}$ is a Grothendieck Abelian category. In fact, this construction produces an explicit cochain complex from an actual diagram of cochain complexes. It is a functor from diagrams to **A** that preserves weak equivalences and that induces the homotopy colimit at the level of homotopy categories. In short, we are in the best possible situation.

Let $F : I \to \mathbf{A}$ be a diagram. Consider the double cochain complex $C(F)^{*,*}$ with

$$C_I(F)^{-p,q} = \bigoplus_{\sigma:[p]\to I} F(\sigma(0))^q,$$

where $[p]$ denotes the category of the totally ordered set $\{0 < \cdots < p\}$ and σ varies over all functors from $[p]$ to I, the vertical differential (in the q-grading) is simply the internal differentials of the complexes $F(i)$, and the horizontal differential is

$$\begin{array}{rccc} d_{hor} : & C_I(F)^{-p,q} & \to & C_I(F)^{1-p,q} \\ & (x)_\sigma & \mapsto & \sum_{i=0}^{p}(-1)^i(F(\sigma \circ \delta_i(0)))_{\sigma \circ \delta_i}, \end{array}$$

where $\delta_i : [p-1] \to [p]$ is the order-preserving injection that avoids the element $i \in [p]$. (Note that this double complex is concentrated in the "left" half-space.)

Definition C.5.2 The *cone of F* is $\mathrm{Cone}_I(F) = \mathrm{Tot}^{\oplus}(C_I(F))$, the totalization (using the direct sum) of the double complex defined earlier.

Remark: When $I = \Delta^{op}$, $\mathrm{Cone}_I(F)$ is precisely the direct sum totalization of the double complex that the Dold–Kan correspondence produces out of the simplicial cochain complex F. ◇

Our goal is to prove the following result (indeed, its ∞-categorical strengthening).

Theorem C.5.3 *For a Grothendieck Abelian category $\mathcal{A}$,*

$$\mathrm{hocolim}_I F = \mathrm{Ho}(\mathrm{Cone}_I)(F).$$

In other words, the functor $\mathrm{Cone}_I : Fun(I, \mathbf{A}) \to \mathbf{A}$ induces, at the level of homotopy categories, an adjunction

$$\mathrm{Ho}(\mathrm{Cone}_I) : \mathrm{Ho}\ Fun(I, \mathbf{A}) \leftrightarrows \mathrm{Ho}(\mathbf{A}) : \mathrm{Ho}(p^*),$$

where the right adjoint $\mathrm{Ho}(p^*)$ comes from p^*, the "constant diagram" functor given by pullback along $p : I \to \{*\}$.

We learned the proof given below from Rune Haugseng, who explained how standard arguments with simplicial model categories admit a modest modification to include this case. His approach is stronger than the theorem stated above: it is a statement at the level of ∞-categories. (Any mistakes in the presentation below are ours.)

The proof of this theorem is quite technical, using ideas and results from higher category theory that are not used elsewhere in this book, so we will not give expository background. Our main references will be Lurie (2009b) and Lurie (2016), but the essential ingredients from higher categories are due to Bousfield and Kan (1972) and Dwyer and Kan (1980a, 1980b).

Remark: Lecture notes of Hörmann (2014) describe an alternative, very concrete proof of this theorem, requiring only that $\mathcal{A}$ is an *AB*4 Abelian category. The proof that he outlines in the problem sets relies on the existence of certain spectral sequences and exhibits explicitly the units and counits of the adjunction at the level of homotopy categories.   ◊

C.5.3 Getting Oriented

For $\mathcal{A}$ a Grothendieck Abelian category, let $\mathbf{A}$ denote the ordinary (i.e., discrete) category $\mathrm{Ch}(\mathcal{A})$ of unbounded cochain complexes in $\mathcal{A}$. Let W denote the collection of quasi-isomorphisms in $\mathbf{A}$ as the appropriate notion of weak equivalences. We will describe here an ∞-categorical refinement of the earlier notion of the derived category, which is simply the localization $\mathbf{A}[W^{-1}]$, following Lurie (2016). In fact, Lurie discusses two different but equivalent ways to construct this ∞-category, each of which will be useful in the proof.

Via the dg Nerve

The following theorem gives us a good grip on the situation.

Proposition C.5.4 (Lurie (2016), Proposition 1.3.5.3) *There is a left proper combinatorial model structure on* $\mathbf{A}$ *in which a map* $f : M \to N$ *is*

- *A weak equivalence if it is a quasi-isomorphism.*
- *A cofibration if in every degree* k, *the map* $f^k : M^k \to N^k$ *is a monomorphism in* $\mathcal{A}$.
- *A fibration if it satisfies the right lifting property with respect to every acyclic cofibration.*

Note that every cochain complex is automatically cofibrant.

Following Lurie, we define the *derived ∞-category of* $\mathcal{A}$, denoted $\mathcal{D}(\mathcal{A})$, to be $\mathrm{N}_{dg}(\mathbf{A}^\circ)$, the differential graded nerve of the subcategory $\mathbf{A}^\circ$ of fibrant objects of this model category $\mathbf{A}$. As Lurie shows in Proposition 1.3.5.9, $\mathcal{D}(\mathcal{A})$ is a stable ∞-category. It also satisfies a universal property: $\mathcal{D}(\mathcal{A})$ is the ∞-categorical localization of $\mathbf{A}$ at the quasi-isomorphisms. We now explain this property precisely.

Recall first the ∞-categorical version of the localization of a category, given in Definition 1.3.4.1 of Lurie (2016). Let W be a collection of morphisms in an ∞-category $\mathcal{C}$. A functor $q : \mathcal{C} \to \mathcal{C}'$ between ∞-categories *exhibits* $\mathcal{C}'$

as the ∞-*category obtained by inverting* W if, for any other ∞-category $\mathcal{C}''$, precomposition with q induces a fully faithful embedding $q^* : Fun(\mathcal{C}', \mathcal{C}'') \to Fun(\mathcal{C}, \mathcal{C}'')$ whose essential image consists of the functors from $\mathcal{C}$ to $\mathcal{C}''$ that send a morphism in W to an equivalence in $\mathcal{C}''$. For $\mathcal{C}$ an ordinary category and W a collection of morphisms in $\mathcal{C}$, we use $\mathcal{C}[W^{-1}]$ to denote the ∞-category obtained by inverting W in the ∞-category $N(\mathcal{C})$, the nerve of $\mathcal{C}$.

The inclusion functor $\mathbf{A}^\circ \hookrightarrow \mathbf{A}$ induces a functor $N_{dg}(\mathbf{A}^\circ) \to N_{dg}(\mathbf{A})$. The composite functor

$$\mathcal{D}(\mathcal{A}) \to N_{dg}(\mathbf{A}^\circ) \to N_{dg}(\mathbf{A})$$

admits a left adjoint that we will denote by L.

Proposition C.5.5 (Lurie (2016), Proposition 1.3.5.15) *The composite functor*

$$N(\mathbf{A}) \to N_{dg}(\mathbf{A}) \xrightarrow{L} \mathcal{D}(\mathcal{A})$$

exhibits $\mathcal{D}(\mathcal{A})$ *as the* ∞-*category obtained by inverting the quasi-isomorphisms.*

In particular, the homotopy category of $\mathcal{D}(\mathcal{A})$ is equivalent to the usual derived category.

Via Dold–Kan

There is a different procedure for extracting an ∞-category from $\mathbf{A}$ that is perhaps better known than the dg nerve construction. First, the subcategory of fibrant objects $\mathbf{A}^\circ$ admits a natural enrichment over cochain complexes of Abelian groups, by the usual "hom-complexes." Next, apply Dold–Kan to obtain a simplicially enriched category $\mathbf{A}_\Delta^\circ$. Finally, take the homotopy coherent (or simplicial) nerve $N(\mathbf{A}_\Delta^\circ)$.

We now make more precise those aspects of this construction that we use in the proof.

Let Ab denote the category of Abelian groups, and let $\mathcal{B}$ be an *AB4* Abelian category (i.e., $\mathcal{B}$ possesses infinite coproducts and hence is a cocomplete Ab-enriched category). We say that $\mathcal{B}$ is *tensored over* Ab to mean that the following construction is well defined. For any Abelian group M and object $A \in \mathcal{B}$, define $M \otimes A \in \mathcal{B}$ by the property that there is an isomorphism of Abelian groups

$$\mathcal{B}(M \otimes A, B) \cong Ab(M, \mathcal{B}(A, B)),$$

natural in B. This object $M \otimes A$ exists, as can be seen in a few steps. First, this "tensor product" clearly exists for finitely generated free Abelian groups via finite direct sum in $\mathcal{B}$, so $\mathbb{Z}^n \otimes A = A^{\oplus n}$ (this holds in any Abelian category). As $\mathcal{B}$ possesses infinite coproducts, we also have the tensor product with any infinitely generated free Abelian group. Second, any Abelian group M is the cokernel of a morphism between free Abelian groups $f : F_1 \to F_2$, so define $M \otimes A$ as the cokernel of $f \otimes \mathrm{id} : F_1 \otimes A \to F_2 \otimes A$. (Note that this morphism $f \otimes \mathrm{id}$ makes sense since there is a canonical map $\mathbb{Z} \to \mathcal{B}(A, A)$ that defines what "multiplication by an integer" in A means.)

Let $\mathbf{Ch} = \mathrm{Ch}(\mathrm{Ab})$ denote the category of cochain complexes of Abelian groups where morphisms are cochain maps. The category $\mathbf{A} = \mathrm{Ch}(\mathcal{A})$ has a natural enrichment over $\mathbf{Ch}$ that we will denote $\mathbf{A}_{\mathrm{Ch}}$. Between any two complexes $A, B \in \mathbf{A}$, there is a cochain complex of Abelian groups $\mathbf{A}_{\mathrm{Ch}}(A, B)$ where a degree n element is a levelwise linear map $f = (f^m : A^m \to B^{m+n})_{m \in \mathbb{Z}}$ and where the differential sends f to $d_B \circ f - (-1)^m f \circ d_A$. In parallel with the preceding argument, we see that $\mathbf{A} = \mathrm{Ch}(\mathcal{A})$ is tensored over $\mathbf{Ch}$: for every cochain complex $M \in \mathbf{Ch}$, there is a natural isomorphism

$$\mathbf{A}(M \otimes A, B) \cong \mathbf{Ch}(M, \mathbf{A}_{\mathrm{Ch}}(A, B)).$$

We now use the Dold–Kan correspondence to obtain a natural relationship with simplicial sets.

Recall that every simplicial set X has an associated cochain complex $N_* X = \mathbf{N}_\bullet \mathbb{Z} X$, by applying the Dold–Kan correspondence after taking the level-wise free Abelian group. By Dold–Kan, we can thus produce a natural simplicially enriched category $\mathbf{A}_\Delta$: for K a simplicial set, we have

$$\mathrm{sSet}(K, \mathbf{A}_\Delta(A, B)) \cong \mathbf{Ch}(N_* K, \mathbf{A}_{\mathrm{Ch}}(A, B)).$$

As $\mathbf{A}$ is Grothendieck and hence presentable, we know that $\mathbf{A}_\Delta$ is actually tensored over simplicial sets, so that it is meaningful to write $K \otimes A$, with $K \in \mathrm{sSet}$ and $A \in \mathbf{A}$.

Thus we have a model category $\mathbf{A}$ with a natural simplicial enrichment $\mathbf{A}_\Delta$. (It does *not* form a simplicial model category, though. See Warning 1.3.5.4 of Lurie (2016).) Note that in this enrichment, the hom spaces are always simplicial Abelian groups and hence Kan complexes. Thus, $\mathbf{A}_\Delta$ is a fibrant simplicial category.

The Equivalence

Proposition 1.3.1.17 of Lurie (2016) states that for a dg category $\mathcal{C}$, there is an equivalence of ∞-categories between its dg nerve $N_{dg}(\mathcal{C})$ and $N(\mathcal{C}_\Delta)$, the homotopy coherent nerve of it simplicial enrichment. Hence, we see that $\mathcal{D}(\mathcal{A}) = N_{dg}(\mathbf{A}^\circ)$ is equivalent to $N(\mathbf{A}_\Delta^\circ)$.

Propositions 1.3.5.13 and 1.3.5.15 imply that $\mathcal{D}(\mathcal{A})$ is an ∞-category obtained by inverting the quasi-isomorphisms W. Hence we see that $\mathcal{D}(\mathcal{A}) \simeq N(\mathbf{A}_\Delta)[W^{-1}]$ as well.

C.5.4 The ∞-Categorical Version of the Theorem

Our goal is to find a way of describing colimits in the ∞-category $\mathcal{D}(\mathcal{A})$ in a concrete way. We will see that the construction Cone$_I$ from the preceding text provides this colimit.

As an example for the kind of result we want, we review the following construction of the homotopy limit of a diagram of simplicial sets, since we will use it in our proof.

The *twisted arrow category* Tw($\mathcal{C}$) of a category $\mathcal{C}$ has objects the morphisms in $\mathcal{C}$ but a morphism from $f \in \mathcal{C}(x, y)$ to $f' \in \mathcal{C}(x', y')$ is a commuting diagram

$$
\begin{array}{ccc}
X & \longrightarrow & X' \\
f \downarrow & & \downarrow f' \\
Y & \longleftarrow & Y'
\end{array}
$$

Note that there is a natural functor from Tw($\mathcal{C}$) to $\mathcal{C} \times \mathcal{C}^{op}$ sending an object to its source and target. Given a functor $F : \mathcal{C} \times \mathcal{C}^{op} \to \mathcal{D}$, the *end of F* is the limit of the composition Tw($\mathcal{C}$) $\to \mathcal{C} \times \mathcal{C}^{op} \to \mathcal{D}$. (The coend is the colimit.)

A *cosimplicial simplicial set* is a functor $X^\bullet : \Delta \to$ sSet. A key example is $\Delta^\bullet$ whose n-cosimplex is precisely the simplicial set $\Delta[n]$. It is a kind of fattened model of a point.

Recall that the *totalization* of a cosimplicial object $X^\bullet : \Delta \to$ sSet is the end of sSet$_\Delta(\Delta^\bullet, X^\bullet)$, where sSet$_\Delta$ will denote the internal hom of sSet. This construction is dual to the more familiar geometric realization. (Geometric realization assembles a space by attaching cells, so it is a colimit. Totalization assembles a space as a tower of extensions, so it is a limit.)

Definition C.5.6 Let $F : I \to$ sSet be a diagram of simplicial sets. The *cosimplicial cobar construction* $\check{\mathfrak{B}}^\bullet F$ is the cosimplicial simplicial set whose n-cosimplices are

$$
\check{\mathfrak{B}}^n F = \prod_{\sigma : [n] \to I} F(\sigma(n)),
$$

where σ runs over all functors from the poset $[n]$ into I. The *cobar construction* $\check{B}(F)$ is the totalization of $\check{\mathfrak{B}}^\bullet F$.

Remark: The cosimplicial cobar construction keeps track, in a very explicit sense, of all the ways that a given object $F(i)$ of the diagram is mapped into. It remembers every finite sequence of arrows whose terminus is $F(i)$. The cobar construction then looks at ways of coherently mapping a "fat point" into this enormous arrangement.   ◊

Here is the statement at the level of homotopy categories.

Proposition C.5.7 (Bousfield and Kan (1972), XI.8.1) *Let* $F : I \to$ sSet *be a diagram such that* $F(i)$ *is a Kan complex for every* $i \in I$ *(i.e., F is object-wise fibrant). Then* $\check{B}(F)$ *is a representative for the homotopy limit* holim$_I$ F *in* Ho(sSet).

There is a corresponding ∞-categorical statement. Let $\mathcal{S}$ denote the ∞-category of *spaces*, which is N(sSet$^\circ$), i.e., the homotopy coherent nerve of sSet$^\circ$, the category of Kan complexes, which is the subcategory of fibrant objects in sSet. As an immediate corollary of the proceeding proposition and Theorem 4.2.4.1 of Lurie (2009b), we have the following.

Proposition C.5.8 *Let* $F : I \to$ sSet$^\circ$ *be a diagram. Then* $\check{B}(F)$ *is a limit of the associated diagram in spaces* $\mathcal{S}$.

We will parallel this development to show that our cone construction Cone$_I$ is a kind of bar construction and hence a colimit in ∞-categories.

Above, we used the fact that a diagram of simplicial sets naturally produced a cosimplicial simplicial set, and we could then talk about mapping the "fat point" into this cosimplicial cobar construction. Here, we will start with a diagram of cochain complexes in $\mathcal{A}$ and then produce a simplicial cochain complex in $\mathcal{A}$. To do a bar construction, we need to find a replacement for the "fat point" $\Delta^\bullet$ in the setting of cochain complexes.

We will use Δ^n to denote the cochain complex $N_*\Delta[n] \in \mathbf{Ch}$, the natural cochain complex associated to the simplicial set $\Delta[n]$ by the Dold–Kan correspondence. There is thus a natural cosimplicial cochain complex $\Delta^\bullet$ that provides our "fat point." Given a simplicial cochain complex $X_\bullet : \Delta^{op} \to \mathbf{A} = \mathbf{Ch}(\mathcal{A})$, we thus have a functor

$$X_\bullet \otimes \Delta^\bullet : \Delta^{op} \times \Delta \to \mathbf{A},$$

using the fact that $\mathbf{A}$ is tensored over $\mathbf{Ch}$. The *realization* of $X_\bullet$ is the colimit of the composite functor $\mathrm{Tw}(\Delta^{op}) \to \Delta^{op} \times \Delta \to \mathbf{A}$.

Definition C.5.9 Let $F : I \to \mathbf{A}$ be a diagram in $\mathbf{A}$. The *simplicial bar construction* $\mathfrak{B}_\bullet F$ is the simplicial object in $\mathbf{A}$ whose n-simplices are

$$\mathfrak{B}_n F = \bigoplus_{\sigma:[n]\to I} F(\sigma(0)),$$

where σ runs over all functors from the poset $[n]$ into I. The *bar construction* $B(F)$ is the realization of $\mathfrak{B}^\bullet F$.

It is a combinatorial exercise to verify the following.

Lemma C.5.10 *For a diagram $F : I \to$ **A**, there is an isomorphism $B(F) \cong$ $\mathrm{Cone}_I(F)$.*

Our theorem can then be stated as follows.

Theorem C.5.11 *Let $F : I \to$ **A** be a diagram of unbounded cochain complexes in the Grothendieck Abelian category $\mathcal{A}$. Then the bar construction $B(F)$ is a colimit of the associated diagram in $\mathcal{D}(\mathcal{A})$.*

Taking the underlying homotopy category of $\mathcal{D}(\mathcal{A})$, we obtain Theorem C.5.3.

C.5.5 The Proof of Theorem C.5.11

Let $\mathbf{A}_\Delta$ be the natural simplicial enrichment of **A**, as described in Section C.5.3. We have seen that the derived ∞-category $\mathcal{D}(\mathcal{A})$ is equivalent to $N(\mathbf{A}_\Delta)[W^{-1}]$, with W the quasi-isomorphisms. By Theorem 4.2.4.1 of Lurie (2009b), we can compute the colimit in $\mathcal{D}(\mathcal{A})$ by computing the homotopy colimit in the simplicially enriched category $\mathbf{A}_\Delta$ with quasi-isomorphisms as the weak equivalences.

To show that $B(F)$ is the homotopy colimit in $\mathbf{A}_\Delta$, it suffices to show that for every fibrant object $Z \in$ **A**, the space $\mathbf{A}_\Delta(B(F), Z)$ is a homotopy limit in sSet of the diagram

$$\mathbf{A}_\Delta(F(-), Z) : I \to \mathrm{sSet}.$$

Indeed, this property characterizes the homotopy colimit of F. (See Remark A.3.3.13 of Lurie (2009b).)

Observe that for any simplicial set K, we have natural isomorphisms

$$\mathrm{sSet}(K, \mathbf{A}_\Delta(B(F), Z)) \cong \mathbf{A}(K \otimes B(F), Z)$$
$$\cong \lim_{\mathrm{Tw}(\Delta^{op})} \mathbf{A}(K \otimes \Delta^\bullet \otimes \mathfrak{B}_\bullet F, Z)$$
$$\cong \mathrm{sSet}(K, \lim_{\mathrm{Tw}(\Delta^{op})} \mathbf{A}_\Delta(\Delta^\bullet \otimes \mathfrak{B}_\bullet F, Z)).$$

The Yoneda lemma then implies we have a natural isomorphism

$$\mathbf{A}_\Delta(B(F), Z) \cong \lim_{\mathrm{Tw}(\Delta^{op})} \mathbf{A}_\Delta(\Delta^\bullet \otimes \mathfrak{B}_\bullet F, Z).$$

We now analyze the right-hand side in more detail.

The Alexander–Whitney and Eilenberg–Zilber maps provide a cochain homotopy equivalence

$$\mathbf{A}_{\mathbf{Ch}}(\Delta^n \otimes \mathfrak{B}_m F, Z) \leftrightarrows \mathbf{Ch}(\Delta^n, \mathbf{A}_\Delta(\mathfrak{B}_m F, Z))$$

for every m and n. Applying Dold–Kan thus produces a simplicial homotopy equivalence

$$\mathbf{A}_\Delta(\Delta[n] \otimes \mathfrak{B}_m F, Z) \leftrightarrows \mathrm{sSet}_\Delta(\Delta[n], \mathbf{A}_\Delta(\mathfrak{B}_m F, Z))$$

for every m and n. Taking the limit over the twisted arrow category, we obtain a simplicial homotopy equivalence

$$\mathbf{A}_\Delta(\mathrm{B}(F), Z) \leftrightarrows \lim_{\mathrm{Tw}(\Delta^{op})} \mathrm{sSet}_\Delta(\Delta^\bullet, \mathbf{A}_\Delta(\mathfrak{B}_\bullet F, Z)),$$

and hence a weak equivalence of simplicial sets. The right-hand side is isomorphic to the cobar construction,

$$\lim_{\mathrm{Tw}(\Delta^{op})} \mathrm{sSet}_\Delta(\Delta^\bullet, \mathbf{A}_\Delta(\mathfrak{B}_\bullet F, Z)) \cong \check{\mathrm{B}}(F),$$

which computes the homotopy limit in sSet by Lemma C.5.7.

C.5.6 Homotopy Colimits in Differentiable Pro-cochain Complexes

We often work as well with diagrams of differentiable pro-cochain complexes. The construction Cone_l makes sense in this setting, so long as one replaces the direct sum totalization by its natural completion (i.e., use the completed direct sum rather than the direct sum). We want it to satisfy the homotopical universal property that ensures this cone is a homotopy colimit.

Let $\mathcal{A}$ denote a Grothendieck Abelian category and $\mathbf{A} = \mathrm{Ch}(\mathcal{A})$. We will describe homotopy colimits of complete filtered objects in $\mathbf{A}$. Again, our argument is quite technical. We start with some preliminaries and then deduce the result from our earlier work.

Remark: These arguments arose in conversations with Dmitri Pavlov.　　　◇

Sequences, Filtrations, and Completions

Let $\mathbb{Z}_{\leq 0}$ denote the category whose objects are the nonpositive integers $\{n \leq 0\}$ with a single morphism $n \to m$ if $n \leq m$ and no morphism if $n > m$. Let Seq denote the functor category $Fun(\mathbb{Z}_{\leq 0}, \mathbf{A})$. We call an object

$$\cdots X(n) \to X(n+1) \to \cdots \to X(0)$$

in Seq a *sequence*. Equip Seq with the projective model structure: a map of sequences $f : X \to Y$ is

- A weak equivalence if $f(n) : X(n) \to Y(n)$ is a quasi-isomorphism for every n.
- A fibration if $f(n) : X(n) \to Y(n)$ is a fibration for every n.

Note that we are working with *levelwise* weak equivalences, which we denote W_L.

The following observation gives a nice explanation for the role of filtered cochain complexes: they are the cofibrant sequences.

Lemma C.5.12 *A sequence X is cofibrant if and only if every map* $X(n) \to X(n+1)$ *is a cofibration (i.e., a monomorphism in every cohomological degree).*

Hence, we introduce the following definition.

Definition C.5.13 Let $\mathbf{A}_{fil}$ denote the category Seq^c of cofibrant objects, i.e., the filtered cochain complexes.

We want to work with a different notion of weak equivalence than W_L. We say a map of sequences $f : X \to Y$ is a *filtered weak equivalence* if the induced maps

$$\text{hocofib}(X(n) \to X(n+1)) \to \text{hocofib}(Y(n) \to Y(n+1))$$

are quasi-isomorphisms for every n. Here "hocofib" is the homotopical version of the cokernel, and it can be computed using the standard cone construction of homological algebra. This condition is the homotopical version of saying we have a quasi-isomorphism on the associated graded, since we are taking the homotopy quotient rather than the naive quotient. On the filtered cochain complexes, it agrees with our earlier notion of a filtered weak equivalence, since the homotopy cofiber agrees with the cokernel. (This observation gives an explanation for why one should work with filtered complexes: one can do the naive computations.)

Let W_F denote the collection of filtered weak equivalences. We want to get a handle on the ∞-category $\mathbf{A}_{fil}[W_F^{-1}]$, which can be called the *filtered derived* ∞-*category of* $\mathcal{A}$.

The complete filtered cochain complexes will be the crucial tool for us. Recall that a filtered cochain complex $X \in \mathbf{A}_{fil}$ is *complete* if the canonical map $X(0) \to \lim_n X(0)/X(n)$ is a quasi-isomorphism. (We apologize for the slight change of notation from $F_n X$ to $X(n)$, but this notation makes clearer the comparison with sequences, which is useful in this context.)

Definition C.5.14 Let $\mathbf{A}_{cfil}$ denote the category of completed filtered cochain complexes in $\mathcal{A}$ (i.e., the pro-cochain complexes in $\mathcal{A}$, in analogy with the terminology of Section C.4).

The inclusion functor $\mathbf{A}_{cfil} \hookrightarrow \mathbf{A}_{fil}$ is right adjoint to a "completion functor" that sends a filtration

$$X = (\cdots \hookrightarrow X(n) \hookrightarrow X(n+1) \hookrightarrow \cdots \hookrightarrow X(0))$$

to its completion $\widehat{X}$ where $\widehat{X}(0) = \lim_n X(0)/X(n)$ equipped with the filtration $\widehat{X}(n) = \ker(\widehat{X} \to X(0)/X(n))$. Note that completion is "idempotent" in the sense that a complete filtered cochain complex is isomorphic to its completion.

Lemma C.5.15 *The inclusion* $(\mathbf{A}_{cfil}, W_F) \hookrightarrow (\mathbf{A}_{fil}, W_F)$ *induces an equivalence between their Dwyer–Kan simplicial localizations, and hence we have an equivalence*

$$\mathbf{A}_{cfil}[W_F^{-1}] \simeq \mathbf{A}_{fil}[W_F^{-1}]$$

between the ∞-categories presented by these categories with weak equivalences.

Proof Under this inclusion-completion adjunction, we see that filtered weak equivalences are preserved. Moreover, for every filtered cochain complex X, the canonical map $X \to \widehat{X}$ is a filtered weak equivalence. By Corollary 3.6 of Dwyer and Kan (1980a), these have weakly equivalent hammock localizations, which implies that the standard simplicial localizations are weakly equivalent. $\qquad\square$

The Theorem

Definition C.5.16 The *completed cone construction* $\widehat{\mathrm{Cone}}_I(F)$ of a diagram $F : I \to \mathbf{A}_{cfil}$ is given by $\mathrm{Tot}^{\widehat{\oplus}}(C_I(F))$, the totalization (via the completed direct sum) of the double complex $C_I(F)$ of Section C.5.2.

Theorem C.5.17 *Let $F : I \to \mathbf{A}_{cfil}$ be a diagram. The completed cone construction $\widehat{\mathrm{Cone}}_I(F)$ is a model for the colimit of F in the ∞-category $\mathbf{A}_{cfil}[W_F^{-1}]$.*

Proof We can view this diagram as living in $\mathbf{A}_{fil}$ (and hence also the category Seq), and by Lemma C.5.15, it suffices to compute the colimit in $\mathbf{A}_{fil}[W_F^{-1}]$. We will thus verify the (uncompleted) cone construction provides a model for this colimit. Completing this cone will provide a complete filtered cochain complex that is filtered weak equivalent, and hence also a model for the desired colimit.

We compute this colimit in filtered cochain complexes by using model categories. Let Seq_L denote Seq equipped with the projective model structure, so the weak equivalences are W_L. Observe that $W_L \subset W_F$ as collections of morphisms inside Seq. Then the left Bousfield localization at W_F, which we will denote Seq_F, has the same cofibrations, by definition. Hence, if one wants to compute homotopy colimits in Seq_F, they agree with the homotopy colimits in Seq_L. (More explicitly, the projective model structures on $Fun(I, \text{Seq}_F)$ and $Fun(I, \text{Seq}_L)$ have the same cofibrant replacement functor, so homotopy colimits can be computed in either category.) In Seq_L, our computation becomes simple: a homotopy colimit of diagram in such a functor category can be computed objectwise, so we can apply Theorem C.5.11 objectwise. This construction naturally produces Cone_I with the natural filtration. $\qquad \square$

Appendix D

The Atiyah–Bott Lemma

Atiyah and Bott (1967) shows that for an elliptic complex $(\mathscr{E}, Q)$ on a compact closed manifold M, with $\mathscr{E}$ the smooth sections of a $\mathbb{Z}$-graded vector bundle, there is a homotopy equivalence $(\mathscr{E}, Q) \hookrightarrow (\overline{\mathscr{E}}, Q)$ into the elliptic complex of distributional sections. The argument follows from the existence of parametrices for elliptic operators. This result was generalized to the noncompact case in Tarkhanov (1987).

Let M be a smooth manifold, not necessarily compact. Let $(\mathscr{E}, Q)$ be an elliptic complex on M. Let $\overline{\mathscr{E}}$ denote the complex of distributional sections of $\mathscr{E}$. We will endow both $\mathscr{E}$ and $\overline{\mathscr{E}}$ with their natural topologies.

Lemma D.1 (Lemma 1.7, Tarkhanov (1987)) *The natural inclusion $(\mathscr{E}, Q) \hookrightarrow (\overline{\mathscr{E}}, Q)$ admits a continuous homotopy inverse.*

The continuous homotopy inverse

$$\Phi : \overline{\mathscr{E}} \to \mathscr{E}$$

is given by a kernel $K_\Phi \in \mathscr{E}^! \otimes \mathscr{E}$ with proper support. The homotopy $S : \overline{\mathscr{E}} \to \overline{\mathscr{E}}$ is a continuous linear map with

$$[Q, S] = \Phi - \mathrm{Id}.$$

The kernel K_S for S is a distribution, that is, an element of $\overline{\mathscr{E}}^! \otimes \overline{\mathscr{E}}$ with proper support.

Proof We will reproduce the proof in Tarkhanov (1987). Choose a metric on E and a volume form on M. (If E is a complex vector bundle, use a Hermitian metric.) Let Q^* be the formal adjoint to Q, which is a degree -1 differential operator, and form the graded commutator $D = [Q, Q^*]$. This operator D is elliptic on each space $\mathscr{E}^i$, the cohomological degree i part of $\mathscr{E}$. Thus, by standard results in the theory of pseudodifferential operators, there is a parametrix

P for D. The kernel K_P is an element of $\overline{\mathscr{E}}^{!} \otimes \overline{\mathscr{E}}$, and the corresponding operator $P : \mathscr{E}_c \to \overline{\mathscr{E}}$ is an inverse for D up to smoothing operators.

By multiplying K_P by a smooth function on $M \times M$ that is 1 in a neighborhood of the diagonal, we can assume that K_P has proper support. Hence, we can extend P to a map $P : \mathscr{E} \to \overline{\mathscr{E}}$ that is still a parametrix: thus $P \circ D$ and $D \circ P$ both differ from the identity by smoothing operators.

The homotopy S is now defined by $S = Q^*P$. $\qquad\qquad\square$

Note that the homotopy inverse $\Phi : \overline{\mathscr{E}} \to \mathscr{E}$ and the homotopy $S : \overline{\mathscr{E}} \to \overline{\mathscr{E}}$ produce a cochain homotopy equivalence in DVS, after applying the functor $dif_t : \text{LCTVS} \to \text{DVS}$.

The following corollary is quite useful in the study of free field theories.

Let $\mathscr{E}_c$ denote the cosheaf that assign to an open $U \subset M$ the cochain complex of differentiable vector spaces $(\mathscr{E}_c(U), Q)$, namely the compactly supported smooth sections on U of the graded vector bundle E. Note that, somewhat abusively, we are viewing these as differentiable vector spaces rather than topological vector spaces. Similarly, let $\overline{\mathscr{E}}_c$ denote the cosheaf sending U to $(\overline{\mathscr{E}}_c(U), Q)$, namely the compactly supported distributional sections on U of the graded vector bundle E, viewed as a differentiable cochain complex. (See Section B.7.2 in Appendix B and Lemma 6.5.2 for the proof that we have a cosheaf.)

The canonical map $\mathscr{E}_c \to \overline{\mathscr{E}}_c$ is a quasi-isomorphism of cosheaves of differentiable cochain complexes.

Proof The Atiyah–Bott–Tarkhanov lemma assures us that $\mathscr{E}_c(U) \to \overline{\mathscr{E}}_c(U)$ is a quasi-isomorphism on every open U. $\qquad\qquad\square$

References

Adámek, Jiří, and Rosický, Jiří. 1994. *Locally presentable and accessible categories.* London Mathematical Society Lecture Note Series, Vol. 189. Cambridge: Cambridge University Press.

Atiyah, M., and Bott, R. 1967. A Lefschetz fixed point formula for elliptic complexes: I. *Ann. Math., (2)*, **86**(2), 374–407.

Atiyah, Michael. 1988. Topological quantum field theories. *Inst. Hautes Études Sci. Publ. Math.*, 175–186 (1989).

Ayala, David, and Francis, John. 2015. Factorization homology of topological manifolds. *J. Topol.*, **8**(4), 1045–1084.

Ayala, David, Francis, John, and Rozenblyum, Nick. 2015. Factorization homology from higher categories. Available at http://arxiv.org/abs/1504.04007.

Ayala, David, Francis, John, and Tanaka, Hiro Lee. 2016. Factorization homology for stratified manifolds. Available at http://arxiv.org/abs/1409.0848.

Behnke, Heinrich, and Stein, Karl. 1949. Entwicklung analytischer Funktionen auf Riemannschen Flächen. *Math. Ann.*, **120**, 430–461.

Beilinson, Alexander, and Drinfeld, Vladimir. 2004. *Chiral algebras.* American Mathematical Society Colloquium Publications, Vol. 51. Providence, RI: American Mathematical Society.

Ben Zvi, David, Brochier, Adrien, and Jordan, David. 2016. Integrating quantum groups over surfaces: Quantum character varieties and topological field theory. Available at http://arxiv.org/abs/1501.04652.

Boardman, J. M., and Vogt, R. M. 1973. *Homotopy invariant algebraic structures on topological spaces.* Lecture Notes in Mathematics, Vol. 347. Berlin and New York: Springer-Verlag.

Boavida de Brito, Pedro, and Weiss, Michael. 2013. Manifold calculus and homotopy sheaves. *Homol. Homot. Appl.*, **15**(2), 361–383.

Bödigheimer, C.-F. 1987. Stable splittings of mapping spaces. *Algebraic topology (Seattle, Wash., 1985)*, pp. 174–187. Lecture Notes in Mathematics, Vol. 1286. Berlin: Springer-Verlag.

Borceux, Francis. 1994a. *Handbook of categorical algebra. 1.* Encyclopedia of Mathematics and Its Applications, Vol. 50. Cambridge: Cambridge University Press. Basic category theory.

Borceux, Francis. 1994b. *Handbook of categorical algebra. 2.* Encyclopedia of Mathematics and Its Applications, Vol. 51. Cambridge: Cambridge University Press. Categories and structures.

Bott, Raoul, and Tu, Loring W. 1982. *Differential forms in algebraic topology.* Graduate Texts in Mathematics, Vol. 82. Berlin and New York: Springer-Verlag.

Bousfield, A. K., and Kan, D. M. 1972. *Homotopy limits, completions and localizations.* Lecture Notes in Mathematics, Vol. 304. Springer-Verlag, Berlin-New York.

Calaque, Damien, and Willwacher, Thomas. 2015. Triviality of the higher formality theorem. *Proc. Am. Math. Soc.*, **143**(12), 5181–5193.

Costello, Kevin. 2004. The A_∞ operad and the moduli space of curves. Available at http://arxiv.org/abs/math/0402015.

Costello, Kevin. 2007a. Renormalisation and the Batalin-Vilkovisky formalism. Available at http://arxiv.org/abs/0706.1533.

Costello, Kevin. 2007b. Topological conformal field theories and Calabi-Yau categories. *Adv. Math.*, **210**(1), 165–214.

Costello, Kevin. 2010. A geometric construction of the Witten genus, I. In: *Proceedings of the International Congress of Mathematicians (Hyderabad, 2010).*

Costello, Kevin. 2011a. A geometric construction of the Witten genus, II. Available at http://arxiv.org/abs/1112.0816.

Costello, Kevin. 2011b. *Renormalization and effective field theory.* Mathematical Surveys and Monographs, Vol. 170. Providence, RI: American Mathematical Society.

Costello, Kevin. 2013a. Notes on supersymmetric and holomorphic field theories in dimensions 2 and 4. *Pure Appl. Math. Q.*, **9**(1), 73–165.

Costello, Kevin. 2013b. Supersymmetric gauge theory and the Yangian. Available at http://arxiv.org/abs/1303.2632.

Costello, Kevin. 2014. Integrable lattice models from four-dimensional field theories. *String-Math 2013*, pp. 3–23. Proceeding of Symposia in Pure Mathematics, Vol. 88. Providence, RI: American Mathematical Society.

Costello, Kevin, and Li, Si. 2011. Quantum BCOV theory on Calabi-Yau manifolds and the higher genus B-model. Available at http://arxiv.org/abs/1201.4501.

Costello, Kevin, and Scheimbauer, Claudia. 2015. Lectures on mathematical aspects of (twisted) supersymmetric gauge theories. *Mathematical aspects of quantum field theories*, pp. 57–87. Mathematical Physics Studies. Berlin and New York: Springer.

Curry, Justin. 2014. Sheaves, Cosheaves and applications. Available at http://arxiv.org/abs/1303.3255.

Deligne, Pierre, Etingof, Pavel, Freed, Daniel S., Jeffrey, Lisa C., Kazhdan, David, Morgan, John W., Morrison, David R., and Witten, Edward (eds). 1999. *Quantum fields and strings: A course for mathematicians.* Vols. 1, 2. American Mathematical Society, Providence, RI; Institute for Advanced Study (IAS), Princeton, NJ. Material from the Special Year on Quantum Field Theory held at the Institute for Advanced Study, Princeton, NJ, 1996–1997.

Dugger, Daniel. 2008. A primer on homotopy colimits. Available at http://pages.uoregon.edu/ddugger/.

Dwyer, W. G., and Kan, D. M. 1980a. Calculating simplicial localizations. *J. Pure Appl. Algebra*, **18**(1), 17–35.

Dwyer, W. G., and Kan, D. M. 1980b. Simplicial localizations of categories. *J. Pure Appl. Algebra*, **17**(3), 267–284.

Eilenberg, Samuel, and Moore, John C. 1962. Limits and spectral sequences. *Topology*, **1**, 1–23.

Etingof, Pavel, Gelaki, Shlomo, Nikshych, Dmitri, and Ostrik, Victor. 2015. *Tensor categories*. Mathematical Surveys and Monographs, Vol. 205. Providence, RI: American Mathematical Society.

Félix, Yves, Halperin, Stephen, and Thomas, Jean-Claude. 2001. *Rational homotopy theory*. Graduate Texts in Mathematics, Vol. 205. New York: Springer-Verlag.

Forster, Otto. 1991. *Lectures on Riemann surfaces*. Graduate Texts in Mathematics, Vol. 81. New York: Springer-Verlag. Translated from the 1977 German original by Bruce Gilligan, Reprint of the 1981 English translation.

Francis, John. 2013. The tangent complex and Hochschild cohomology of E_n-rings. *Compos. Math.*, **149**(3), 430–480.

Freed, Daniel S. 1994. Higher algebraic structures and quantization. *Comm. Math. Phys.*, **159**(2), 343–398.

Frenkel, Edward, and Ben-Zvi, David. 2004. *Vertex algebras and algebraic curves*, 2nd ed. Mathematical Surveys and Monographs, Vol. 88. Providence, RI: American Mathematical Society.

Fresse, Benoit. 2016. *Homotopy of operads and the Grothendieck-Teichmüller group*. Available at `http://math.univ-lille1.fr/ fresse/OperadHomotopyBook/`.

Friedman, Greg. 2012. Survey article: An elementary illustrated introduction to simplicial sets. *Rocky Mountain J. Math.*, **42**(2), 353–423.

Gelfand, Sergei I., and Manin, Yuri I. 2003. *Methods of homological algebra*, 2nd ed. Springer Monographs in Mathematics. Berlin: Springer-Verlag.

Getzler, E. 1994. Batalin-Vilkovisky algebras and two-dimensional topological field theories. *Comm. Math. Phys.*, **159**(2), 265–285.

Ginot, Grégory. 2015. Notes on factorization algebras, factorization homology and applications. *Mathematical aspects of quantum field theories*, pp. 429–552. Mathematical Physics Studies. Berlin and New York: Springer.

Ginot, Grégory, Tradler, Thomas, and Zeinalian, Mahmoud. 2012. Derived higher Hochschild homology, Topological chiral homology and factorization algebras. Available at `http://arxiv.org/abs/1011.6483`.

Ginot, Grégory, Tradler, Thomas, and Zeinalian, Mahmoud. 2014. Derived higher Hochschild cohomology, Brane topology and centralizers of E_n-algebra maps. Available at `http://arxiv.org/abs/1205.7056`.

Glimm, James, and Jaffe, Arthur. 1987. *Quantum physics*, 2nd ed. New York: Springer-Verlag. A functional integral point of view.

Goerss, Paul G., and Jardine, John F. 2009. *Simplicial homotopy theory*. Modern Birkhäuser Classics. Basel: Birkhäuser Verlag. Reprint of the 1999 edition.

Goerss, Paul, and Schemmerhorn, Kristen. 2007. Model categories and simplicial methods. Pages 3–49 of: *Interactions between homotopy theory and algebra*, pp. 3–49. Contemporary Mathematics, Vol. 436. Providence, RI: American Mathematical Society.

Grady, Ryan, Li, Qin, and Li, Si. 2015. BV quantization and the algebraic index. Available at http://arxiv.org/abs/1507.01812.

Griffiths, Phillip, and Harris, Joseph. 1994. *Principles of algebraic geometry*. Wiley Classics Library. New York: John Wiley & Sons Inc. Reprint of the 1978 original.

Grothendieck, A. 1952. Résumé des résultats essentiels dans la théorie des produits tensoriels topologiques et des espaces nucléaires. *Ann. Inst. Fourier Grenoble*, **4**, 73–112 (1954).

Grothendieck, Alexander. 1957. Sur quelques points d'algèbre homologique. *Tôhoku Math. J.*, **9**(2), 119–221.

Grothendieck, Alexandre. 1955. Produits tensoriels topologiques et espaces nucléaires. *Mem. Am. Math. Soc.*, **1955**(16), 140.

Gunning, Robert C., and Rossi, Hugo. 1965. *Analytic functions of several complex variables*. Englewood Cliffs, NJ: Prentice-Hall.

Gwilliam, Owen. 2012. *Factorization algebras and free field theories*. ProQuest LLC, Ann Arbor, MI. PhD thesis, Northwestern University.

Gwilliam, Owen, and Grady, Ryan. 2014. One-dimensional Chern-Simons theory and the $\hat{A}$ genus. *Algebr. Geom. Topol.*, **14**(4), 2299–2377.

Gwilliam, Owen, and Johnson-Freyd, Theo. 2012. How to derive Feynman diagrams for finite-dimensional integrals directly from the BV formalism. Available at http://arxiv.org/abs/1202.1554.

Hatcher, Allen. 2007. Notes on basic 3-manifold topology. Available at http://www.math.cornell.edu/~hatcher/.

Horel, Geoffroy. 2015. Factorization homology and calculus à la Kontsevich Soibelman. Available at http://arxiv.org/abs/1307.0322.

Hörmander, Lars. 2003. *The analysis of linear partial differential operators. I*. Classics in Mathematics. Berlin: Springer-Verlag. Distribution theory and Fourier analysis, Reprint of the second (1990) edition [Springer, Berlin].

Hörmann, Fritz. 2014. Homotopy limits and colimits in nature: A motivation for derivators. Available at http://home.mathematik.uni-freiburg.de/hoermann/holimnature.pdf.

Huang, Yi-Zhi. 1997. *Two-dimensional conformal geometry and vertex operator algebras*. Progress in Mathematics, Vol. 148. Boston: Birkhäuser Boston.

Johansen, A. 1995. Twisting of $N = 1$ SUSY gauge theories and heterotic topological theories. *Int. J. Mod. Phys. A*, **10**(30), 4325–4357.

Kapustin, Anton. 2010. Topological field theory, higher categories, and their applications. *Proceedings of the International Congress of Mathematicians*, Vol. III, pp. 2021–2043. Hindustan Book Agency, New Delhi: Hindustan Book Agency.

Kapustin, Anton, and Saulina, Natalia. 2011. Topological boundary conditions in abelian Chern-Simons theory. *Nucl. Phys. B*, **845**(3), 393–435.

Kashiwara, Masaki, and Schapira, Pierre. 2006. *Categories and sheaves*. Grundlehren der Mathematischen Wissenschaften [Fundamental Principles of Mathematical Sciences], Vol. 332. Berlin: Springer-Verlag.

Knudsen, Ben. 2014. Higher enveloping algebras. Available at http://www.math.northwestern.edu/~knudsen/.

Kontsevich, Maxim. 1994. Feynman diagrams and low-dimensional topology. *First European Congress of Mathematics*, Vol. II (Paris, 1992), pp. 97–121. Progress in Mathematics, Vol. 120. Basel: Birkhäuser.

Kriegl, Andreas, and Michor, Peter W. 1997. *The convenient setting of global analysis*. Mathematical Surveys and Monographs, Vol. 53. Providence, RI: American Mathematical Society.

Leinster, Tom. 2004. *Higher operads, higher categories*. London Mathematical Society Lecture Note Series, Vol. 298. Cambridge: Cambridge University Press.

Li, Qin, and Li, Si. 2016. On the B-twisted topological sigma model and Calabi–Yau geometry. *J. Differ. Geom.*, **102**(3), 409–484.

Loday, Jean-Louis. 1998. *Cyclic homology*, 2nd ed. Grundlehren der Mathematischen Wissenschaften [Fundamental Principles of Mathematical Sciences], Vol. 301. Berlin: Springer-Verlag. Appendix E by María O. Ronco, Chapter 13 by the author in collaboration with Teimuraz Pirashvili.

Loday, Jean-Louis, and Vallette, Bruno. 2012. *Algebraic operads*. Grundlehren der Mathematischen Wissenschaften [Fundamental Principles of Mathematical Sciences], Vol. 346. Heidelberg: Springer.

Lurie, Jacob. 2009a. Derived Algebraic Geometry X: Formal Moduli Problems. Available at `http://math.mit.edu/~lurie/papers`.

Lurie, Jacob. 2009b. *Higher topos theory*. Annals of Mathematics Studies, Vol. 170. Princeton, NJ: Princeton University Press.

Lurie, Jacob. 2009c. On the classification of topological field theories. *Current developments in mathematics, 2008*, pp. 129–280. Somerville, MA: Internation Press.

Lurie, Jacob. 2016. Higher Algebra. Available at `http://math.mit.edu/~lurie/papers`.

MacLane, Saunders. 1998. *Categories for the working mathematician*, 2nd ed. Graduate Texts in Mathematics, Vol. 5. New York: Springer-Verlag.

Matsuoka, Takuo. 2014. Descent and the Koszul duality for factorization algebras. Available at `http://arxiv.org/abs/1312.2562`.

McDuff, Dusa. 1975. Configuration spaces of positive and negative particles. *Topology*, **14**, 91–107.

Morrison, Scott, and Walker, Kevin. 2012. Blob homology. *Geom. Topol.*, **16**(3), 1481–1607.

Ramanan, S. 2005. *Global calculus*. Graduate Studies in Mathematics, Vol. 65. Providence, RI: American Mathematical Society.

Salvatore, Paolo. 2001. Configuration spaces with summable labels. *Cohomological methods in homotopy theory (Bellaterra, 1998)*, pp. 375–395. Progress in Mathematics, Vol. 196. Basel: Birkhäuser.

Scheimbauer, Claudia. 2014. On fully extended topological field theories. PhD thesis, ETH Zurich.

Segal, Graeme. 1973. Configuration-spaces and iterated loop-spaces. *Invent. Math.*, **21**, 213–221.

Segal, Graeme. 1991. Geometric aspects of quantum field theory. Pages 1387–1396 of: *Proceedings of the International Congress of Mathematicians, Vol. I, II (Kyoto, 1990)*. Math. Soc. Japan, Tokyo.

Segal, Graeme. 2004. The definition of conformal field theory. *Topology, geometry and quantum field theory*, pp. 421–577. London Mathematical Society Lecture Note Ser., Vol. 308. Cambridge: Cambridge Univ. Press. First circulated in 1988.

Segal, Graeme. 2010. Locality of holomorphic bundles, and locality in quantum field theory. *The many facets of geometry*, pp. 164–176. Oxford: Oxford University Press.

Serre, Jean-Pierre. 1953. Quelques problèmes globaux relatifs aux variétés de Stein. *Colloque sur les fonctions de plusieurs variables, tenu à Bruxelles, 1953*, pp. 57–68. Liège: Georges Thone.

Stacks Project Authors, The. 2016. *Stacks Project*. http://stacks.math.columbia.edu.

Stasheff, Jim. 2004. What is . . . an operad? *Notices Am. Math. Soc.*, **51**(6), 630–631.

Tarkhanov, N.N. 1987. On Alexander duality for elliptic complexes. *Math. USSR Sbornik*, **58**(1): 62–85.

Trèves, François. 1967. *Topological vector spaces, distributions and kernels*. New York: Academic Press.

Vallette, Bruno. 2014. Algebra + homotopy = operad. *Symplectic, Poisson, and noncommutative geometry*, pp. 229–290. Mathematical Sciences Research Institute Publications, Vol. 62. New York: Cambridge University Press.

Wallbridge, James. 2015. Homotopy theory in a quasi-abelian category. Available at http://arxiv.org/abs/1510.04055.

Weibel, Charles A. 1994. *An introduction to homological algebra*. Cambridge Studies in Advanced Mathematics, Vol. 38. Cambridge: Cambridge University Press.

Weiss, Michael. 1999. Embeddings from the point of view of immersion theory. I. *Geom. Topol.*, **3**, 67–101 (electronic).

Wells, Jr., Raymond O. 2008. *Differential analysis on complex manifolds*, 3rd ed. Graduate Texts in Mathematics, Vol. 65. New York: Springer. With a new appendix by Oscar Garcia-Prada.

Williams, Brian. 2016. The Virasoro vertex algebra and factorization algebras on Riemann surfaces. Available at http://www.math.northwestern.edu/~bwill/ now at http://arXiv.org/abs/1603.02349.

Witten, Edward. 1988a. Topological quantum field theory. *Comm. Math. Phys.*, **117**(3), 353–386.

Witten, Edward. 1988b. Topological quantum field theory. *Comm. Math. Phys.*, **117**(2), 014, 21 pp. (electronic).

Witten, Edward. 1992. Mirror manifolds and topological field theory. *Essays on mirror manifolds*, pp. 120–158. Hong Kong: International Press.

Index

Printed in the USA
CPSIA information can be obtained
at www.ICGtesting.com
LVHW040019120124
768759LV00003B/54